Anaerobiosis and Stemness

Anaerobiosis and Stemness

An Evolutionary Paradigm

Zoran Ivanovic

Marija Vlaski-Lafarge

Aquitaine-Limousin Branch of French Blood Institute (EFS-AQLI)/UMR 5164 CNRS/Bordeaux University, France

AMSTERDAM • BOSTON • HEIDELBERG • LONDON
NEW YORK • OXFORD • PARIS • SAN DIEGO
SAN FRANCISCO • SINGAPORE • SYDNEY • TOKYO
Academic Press is an imprint of Elsevier

Academic Press is an imprint of Elsevier
125 London Wall, London EC2Y 5AS, UK
525 B Street, Suite 1800, San Diego, CA 92101-4495, USA
225 Wyman Street, Waltham, MA 02451, USA
The Boulevard, Langford Lane, Kidlington, Oxford OX5 1GB, UK

ISBN: 978-0-12-800540-8

British Library Cataloguing-in-Publication Data
A catalog record for this book is available from the British Library

Library of Congress Cataloging-in-Publication Data
A catalog record for this book is available from the Library of Congress

For information on all Academic Press publications
visit our website at http://store.elsevier.com/

Publisher: Mica Haley
Editorial Project Manager: Lisa Eppich
Production Project Manager: Lucía Pérez
Designer: Maria Inês Cruz

Typeset by TNQ Books and Journals
www.tnq.co.in

Contents

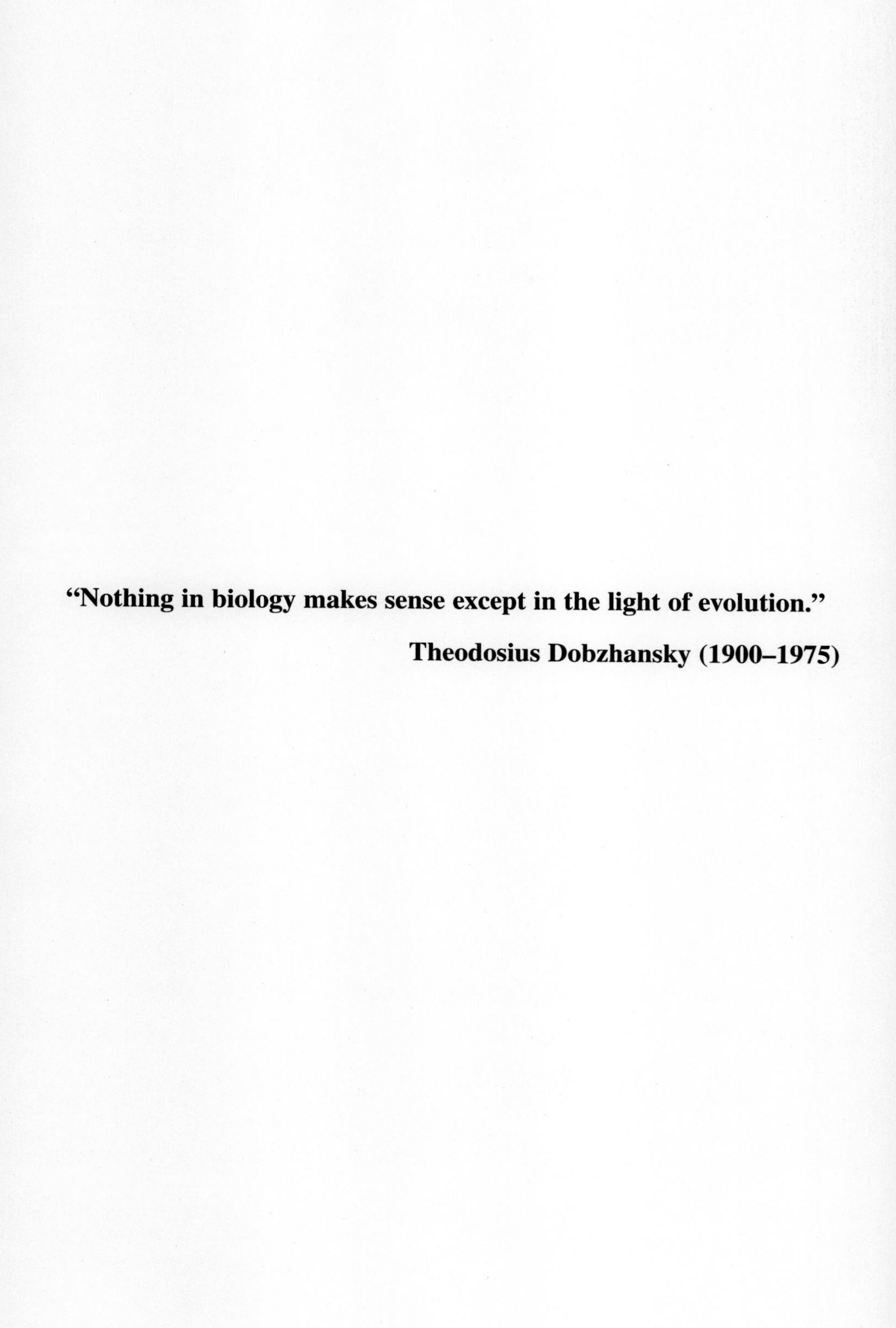

"Nothing in biology makes sense except in the light of evolution."

Theodosius Dobzhansky (1900–1975)

Preface

During a 20-year period of studying the "hypoxic" nature of hematopoietic stem cells (HSC), an observation came to light: These cells behave as facultative anaerobic single-celled organisms. When we published for the first time, back in 2000, the notion that low O_2 concentration favors HSC self-renewal, we became aware of some analogies in HSC behavior with one of the facultatively anaerobic protist. This still largely speculative link between stem cell physiology and evolution of the first eukaryotes was articulated in our "Oxygen Stem Cell Paradigm" published in 2009. In the meantime, the same features (facultative anaerobiosis related to self-renewal, differentiation in function of oxygen availability, etc.) were evidenced literally for all categories of stem cells. Furthermore, the evolution of stem cell entity in metazoa was better explored and documented, and some key features of stemness were recognized in the life cycle of single-cell eukaryotes.

In this book, we review the available data and the cues, trying to put each piece in its place, to build our "evolutionary stem cell paradigm" based on the relationship between anaerobiosis and stemness. Important to mention is that our work presented here has, as any hypothesis does, in some parts, a still-speculative character.

In our regular laboratory work in cell engineering, we have been applying this concept for several years, and it has proved to have an excellent predictive value. Based on these data, we present its potential application in cell engineering and cell therapy, as well as its compatibility with the cancer stem cell concept and cancer evolution.

Zoran Ivanovic

Acknowledgments

We have greatly benefited from critical readings and final English styling of all chapters by Ivana Gadjanski, PhD, as well as critical readings by Jean-Charles Massabuau, PhD (Chapter 10), Vladimir Niculescu, PhD (Chapters 10 and 12), and Philippe Brunet de la Grange, PhD (Chapter 10). We also wish to acknowledge the "first line" English reading done by Elisabeth Doutreloux-Volkmann, as well as the great contribution of Laura Rodriguez, MSc, who realized the artwork presented in this book.

Introduction

Special Remarks

What Entity Could Be Called a Stem Cell?

1

Chapter Outline

1.1 First Notions of Morphologically Nonrecognizable Cells Exhibiting a High Proliferative and Differentiation Potential

Discovery of the first functional cellular entity capable of differentiating into the mature (morphologically recognizable) cells of blood lineages, named CFU-S (Colony Forming Unit-Spleen) [1,2], indicated the morphologically nonrecognizable lymphoid-like cell as a "stem cell candidate." These cells, contained in murine bone marrow, are capable of giving rise to colonies comprising cells of all hematopoietic lineages in the spleen of lethally irradiated recipient mice. This discovery initiated a completely new approach to the identification of different classes of functionally heterogeneous cells that exhibit a high proliferative capacity and a high differentiation potential. As evidenced later on the basis of the functional assays, the high proliferative capacity results from a property typical of stem cells: the self-renewal. This property, enabling the maintenance of the primitive nondifferentiated (stem) cells in parallel with production of descendant still-primitive and morphologically nonrecognizable cell populations ("committed progenitors") relies mainly on asymmetric cell divisions. This means that, after the division of a stem cell, one of the two daughter cells undergoes the program of commitment and serves to amplify committed progenitor cell populations, while the other one maintains the primitive character of the mother cell, thus preserving the regenerative potential (i.e., renewing the stem cell population). The other model is the so-called "symmetric cell division." In this case, the self-renewal and differentiation could be maintained only at the population level since a single stem cell could have the capacity either to divide and give two cell daughters that are both primitive stem cells (i.e., identical to the mother cell), or to give two cells capable of undergoing the commitment program, thus, both different from the mother cell [3,4].

Anaerobiosis and Stemness. http://dx.doi.org/10.1016/B978-0-12-800540-8.00001-6

1.2 Hematopoiesis as a Paradigmatic Case

As the first and most investigated tissue with respect to the stem cells, the hematopoietic tissue is a paradigmatic case. Intensive investigations revealed a great heterogeneity in population of cells considered as "stem cells" and defined on the basis of their capacities to reconstitute the hematopoiesis in vivo. Schematically they were classified as short- and long-term engrafting stem cells, and the first "wave" of hematopoietic reconstitution depends on committed progenitors initially called "Erythroid Repopulation Ability" [5] and "Granulocyte Repopulation Ability" [6]. These committed progenitors are more precisely defined by their capacity to give rise to colonies composed of mature cells or morphologically recognizable precursors in semisolid cultures (Colony Forming Units (CFUs) or Colony Forming Cells (CFCs)). Depending on their differentiation capacity these progenitors can be distinguished as follows: CFU-GEMM (CFU-Mix), CFU-Mk, CFU-GM, CFU-G, CFU-M, BFU-E, and CFU-E.

However, in specific conditions, the colonies in primary cultures could be issued from the cells more primitive than committed progenitors. Typical examples are delayed big-diameter macrophage colonies issued form High Proliferative Potential–Colony Forming Cells (HPP-CFCs) [7–9] or colonies composed of blasts, obtained in semisolid medium (CFU-Blast) [10–13]. The cells at the origin of these colonies are associated mainly with short-term repopulating capacity. The short-term engraftment capacity could be also estimated by functional tests for detection of heterogeneous stem cell population either by in vitro cultures on stromal layer (Cobblestone areas–forming cells (CAFCs) [14–16], Long-term Culture–Initiating Cells (LTC–ICs) [17,18]) or by appropriately stimulated primary and secondary liquid cultures ("Pre-CFC") [19,20]. The two last listed techniques are exploiting the commitment and differentiation of hematopoietic stem cells (HSCs) in primary culture and consequent detection of committed progenitors (CFC) by methylcellulose cultures. However, the best way to detect the short-term HSC remains to test their in vivo repopulating potential. The most developed experimental approach to perform this with human cells is the immunodeficient mice model (stem cells detected this way are usually called "Scid Repopulating Cells" (SRC)) [21–23].

In spite of numerous technical variations, it can be said, for simplicity purposes, that the in vivo approach allows estimation of at least two subpopulations of HSC: (1) short-term repopulating HSC and (2) long-term repopulating HSC. The latter HSC population could be assayed either by analyzing the presence of human cells in hematopoietic tissue of recipient mice, several months after injection, or by serial transplantation. In fact, the serial transplantation is widely accepted as a "gold standard" approach to assay for the long-term repopulating stem cells [24–26].

Early works demonstrated that the treatment of mice with cytotoxic agents HU, 5FU, and Cytarabine selects more primitive subpopulations of CFU-S exhibiting a higher self-renewal capacity and proliferative activity, sparing the quiescent cells (concept of G_0 phase of cycle), thus enriching these more primitive stem cells. Furthermore, experiments with HU and 5FU demonstrated not only that CFU-S compartment is heterogeneous with respect to primitiveness but that a stem cell more primitive than CFU-S exists—a pre-CFU-S. Pre-CFU-S are the cells that do not form the colonies in spleen of irradiated mice but can proliferate and differentiate into hematopoietic tissue of lethally irradiated recipient mice to the stage of CFU-S that is then detectable by "CFU-S Assay" [27,28]. Pre-CFU-S are concentrated in bone marrow of mice treated by 5FU (Ref. [29]; confirmation, Ref. [30]). The same functional entity is also known as "Marrow Repopulating Ability (MRA)" [31]. Of note is that the CFU-S compartment is also heterogeneous: within the CFU-S population capable of giving rise to the colonies found after 12 days (CFU-Sd12) more primitive cells are contained with respect to the CFU-S giving rise to the colonies after 8 days (CFU-Sd8), although there is over 50% overlapping between these two categories [32,33]. Also CFU-Sd12 are, at least in part, overlapping with the pre-CFU-S [34].

The functional heterogeneity of the hematopoietic stem cell compartment is explained by the "generation-age" hypothesis [35], implying a hierarchical model that fits very well with the asymmetric cell division model.

1.3 Functional Definition of Stem Cell Entity

Based on the above, we can conclude that the main problem in studying HSC is related to the functional definition of a stem cell entity; that is, to the inevitable functional assays that are always indirect, providing *a posteriori* information on the HSC. For example, the long-term engraftment of a recipient mouse means that, in the injected population, the long-term reconstituting stem cells were present among the donor cells. Repopulation of long-term stroma-based cultures by committed progenitors means that in the added cell population some more primitive cells preceding the stage of committed progenitor were present, and so on. The "gold standard" to reveal them is to test HSC capacity; that is, to establish the short- or long-term hematopoietic reconstitution in vivo (or even the capacity of reconstitution of secondary recipients), applying the limiting dilution or single-cell-injection approach. Using this approach for murine cells (syngeneic recipients allowing to trace donor cells either by an isoenzyme mismatch or by tracing a transgene imported in the injected cells) or human cells (xenograft of immunodeficient mice recipients), it is possible to estimate a frequency of stem cells in the cell populations studied.

It is generally considered that these models regularly underestimate the stem cell frequencies in the cell populations studied due to the fact that not every stem cell potentially capable of reconstituting hematopoiesis is allowed to implant and proliferate in the recipient tissue. This depends on the intrinsic and extrinsic factors to the stem cell itself. For example, a stem cell could exhibit all stem cell capacities but lack the membrane molecules mandatory for seeding in the tissue that will prevent the engraftment and consequent reconstitution or, conversely, can exhibit all the properties needed but still get nonspecifically removed from the circulation or get mechanically sequestered in a nonsupportive tissue. These extrinsic factors may be especially pronounced in a xenogeneic set-up [36]. However, one-cell syngeneic engraftment experiments with almost absolutely enriched mouse bone marrow HSC population ("TIP"-SPCD34-KSL) [37] show that the marrow seeding efficiency of HSC can be nearly 100% [38].

1.4 Quest for the Phenotype Definition of HSC: Quest for the Holy Grail

In parallel with gathering data related to the functional heterogeneity of hematopoietic stem and progenitor cells, important efforts were invested in identifying their phenotypic characteristics in order to realize a more direct approach in studying the stem cell subpopulations.

The first attempts, based on the physical properties of cells in bone marrow (and other sources) mononuclear cell fraction (murine, rat, and human cells), were completed by affinity to bind lektines [39–41], fluorescent labeling techniques, and monoclonal antibodies technology and techniques of supravital staining [42], allowing for the analysis and isolation of cells by flow cytometry (Flow-Activated Cell Sorter). These approaches resulted in enrichment of multicriteria-defined cell populations in stem cells to various extents. These important advancements enabled exclusion of mature cells as well as some differentiated cell populations by a negative selection and further enrichment of stem cells by positive selection of antigens associated with the stem cell subpopulations.

These criteria for enrichment of stem cells in a population are based on:

1. Expression of some (CD34, CD133, CD90 (Thy1), CD49, Sca-1, CD117 (c-kit), etc. membrane markers and nonexpression or low expression of others (Lineage antigens, CD38, CD45RA, etc.).
2. High expression of some enzymes as lactate dehydrogenase (LDH^{+} cells).
3. Low retention of some supravital stains such as Rhodamine and HOECHST and related low mitochondria content and expression of the ABC proteins.
4. Low ROS Content [43].
5. Other factors.

Using the "phenotypic" or "combined" approaches, great advancements were achieved in enrichment of HSC [44] both in murine and human tissues. The most efficient protocol published as of 2015 claims to enable a real "isolation" of murine HSC (capable of the in vivo hematopoietic reconstitution)—96% in SP CD34-KSL population [37]. One of the most sophisticated approaches is based on the "SLAM family" markers enabling to physically separate and highly enrich (engraftment experiments essays performed with five cells only) in different fractions the long- and short- term reconstituting HSC, multipotent progenitors, and committed progenitors [45]. On the basis of further analysis of the "SLAM" membrane markers (CD150, CD48, CD229, CD244) on the LSK ($Lin^{-}Sca\text{-}1^{+}c\text{-}Kit^{+}$) population, the authors isolated two HSC fractions termed HSC-1 ($CD150^{+}CD48^{-/low}CD229^{-/low}$-$CD224^{-}LSK$) and HSC-2 ($CD150^{+}CD48^{-/low}CD229^{+}CD224^{-}LSK$) highly enriched in long- and short-term reconstitution capacity, respectively. HSC-1 fraction contains a relatively higher proportion of quiescent cells and HSC exhibiting the myeloid differentiation potential, whereas HSC-2 fraction was composed of mostly proliferating cells enriched in HSC exhibiting the lymphoid differentiation potential. In addition, using this approach it was shown that the LSK population can be fractionated in an additional three fractions containing highly enriched functionally distinct sets of multipotent progenitors (MPP): MPP-1 ($CD150^-CD48^{-/low}CD229^{-/low}CD224^-LSK$), MPP-2 ($CD150^-CD48^{-/low}CD229^+CD224^-LSK$), and MPP-3 ($CD150^-CD48^{-/low-}CD229^+CD224^+LSK$) enriched in hematopoietic cells exhibiting a transient reconstituting capacity up to 8, 6, and 4 weeks, respectively [45].

Concerning human cells, the most efficient protocol allows for enrichment of the HSC population (robust in vivo engraftment of NSG mice) up to 1 per 5 $Lin^-CD34^+CD38^-CD45RA^-Thy^+Rho^{lo}CD49^+$ cells [46]. Of note is that the xenogeneic human/mouse model, even if NSG mice represent an extremely low threshold of engraftment [47], probably does not allow the absolute seeding efficiency of HSC and, hence, underestimates the real HSC frequency. Therefore, it can be assumed that the protocol in question approaches the absolute human HSC selection.

These advancements, however, do not make room for the generalization of the use of the term "HSC" to other phenotypically defined populations in which the real frequencies of HSC (this term should be attributed only to the cells fulfilling the functional criteria) are much lower and in which a significant proportion is represented by the committed hematopoietic progenitors. In addition, as stated by Valent et al. [48] "It should be noted that many of phenotypic markers found to be empirically useful for isolating particular subsets of cells in unperturbed normal tissue have proved to be functionally dispensable or differently regulated when they are activated in vivo and in vitro."

Tendency of noncritical use of the term "HSC" is widely present in literature with both human and murine cells. Even the populations in which the real HSC are represented in infinitesimal frequencies ($CD34^+$ or $CD133^+$ cells) have HSC as an habitual "adjective." The same is true for the $CD34^+CD38^-$ population, as well as for many other phenotypically defined cell populations. The similar situation is with murine cells. In order to illustrate the problem, as an instructive example we are discussing here the "LSK" cell case, since it is prevalent in literature to draw conclusions related to HSC activity on the basis of this cell population.

1.5 LSK (KLS) Case

$Lin^{-/low}Sca\text{-}1^+c\text{-}kit^{hi(+)}$ (LSK or KLS depending on the author) cells represent only $0.08 \pm 0.05\%$ of bone marrow cells [49]. It has been shown that the expression of Sca-1 is determinant for HSC since in the fraction $Lin^-c\text{-}kit^+Sca\text{-}1^-$ is highly enriched in committed progenitors and CFU-Sd8, but does not contain CFUSd12 and pre-CFU-S [49] (note, however that >50% CFU-Sd8 and CFU-Sd12 grow from the same cells [32,33]). LSK population exhibits a high pre-CFU-S activity and contains an important proportion of CFU-Sd12 estimated to be about 80% on the basis of a supposed seeding efficiency factor. One-third of LSK cells are capable of giving long-lasting colonies of undifferentiated cells ("cobblestone areas") on the stromal cell layer, another way to detect a subset of HSC [50] highly overlapping with the CFU-Sd12 and the pre-CFU-S [28]. Surprisingly, the authors detected, in classical multicytokine-cocktail-stimulated methylcellulose cultures, that 46% of LSK cells exhibit direct colony-forming ability [49]. They did not give the details concerning the cell content in the colonies (presence of "blast" colonies?), but given that the cultures were analyzed after 8 days and that the "colony types" were confirmed by "lifting the cells from colonies" and a cytospin analysis, it is obvious that the standard colonies grown from committed progenitors are in question. Within the c-Kit$^+$ Sca-1$^+$ Lin$^-$ population, the frequency of interleukin-3 (IL-3)-dependent colony-forming cells (in the presence of either IL-3 alone or IL-3$^+$SCF in methylcellulose cultures)—the typical committed progenitors—is 20.0 ± 3.9% [51]. Also, the others consistently found a proportion of committed progenitors in LSK cell population (e.g., Ref. [52]). All these data corroborate the presence of a substantial number of committed progenitors in the LSK population. Even in the populations considered to be extremely enriched in LT-HSC (e.g., CD34^{low}LSK) only one out of five cells can be considered a "true" HSC [51], so it is obvious that their frequency in LSK population is lower. Indeed, Bryder, Rossi, and Weissman [53] stated that only 1/30 LSK cells are actually stem cells capable of long-term multilineage repopulation while the vast majority of LSK cells are multipotent progenitors. The functional heterogeneity of LSK cell population is schematically presented in Figure 1.1.

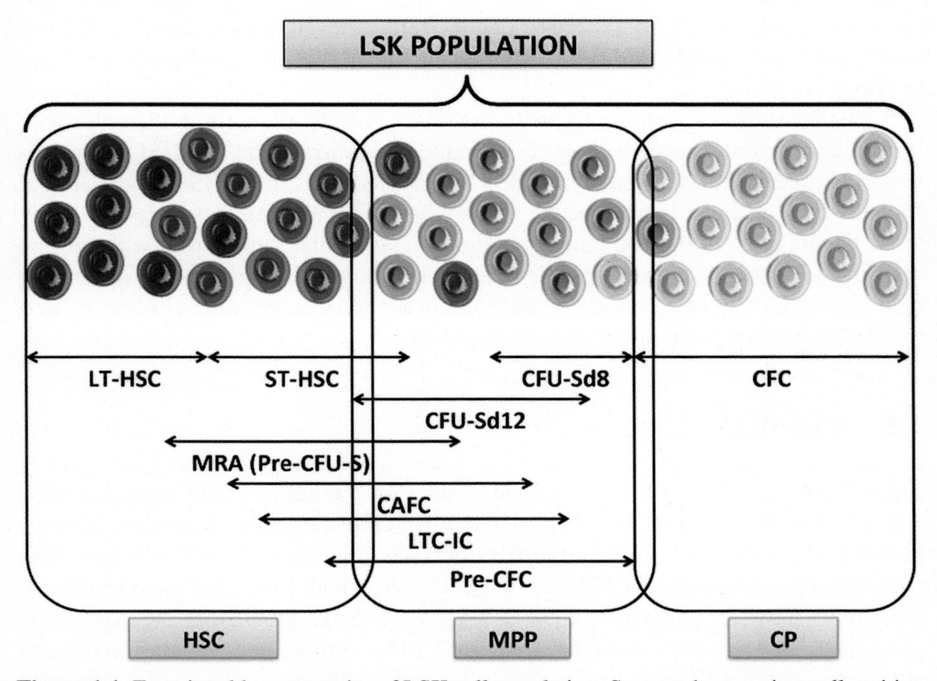

Figure 1.1 Functional heterogeneity of LSK cell population. Stem and progenitor cell entities as detected in in vivo and ex vivo functional assays. In vivo assays: LT-HSC, Long-Term Repopulating Hematopoietic Stem Cells; ST-HSC, Short-Term Repopulating Hematopoietic Stem Cells; MRA, Marrow Repopulating Ability; CFU-Sd12, Colony Forming Units–Spleen day 12; CFU-Sd8, Colony Forming Units–Spleen day eight; pre-CFU-S, Pre-Colony Forming Units–Spleen; Ex vivo essays: CFC, Colony Forming Cells; pre-CFC, Pre-Colony Forming Cells; LTC–IC, Long-Term Culture–Initiating Cells; CAFC, Cobblestone Area–Forming Cells.

Thus, in view of all these data it is obvious that, even in steady state, analyzing the numerical changes of the LSK population may not be representative for HSC due to the functional heterogeneity of the LSK population since real HSC do not still represent the majority of this population, and due to the presence of a nonnegligible proportion of committed progenitors. This approach is even more problematic in situations out of steady state. Frequencies and function of different subsets in the LSK compartment can be significantly altered in different physiological (aging) [54] or ex vivo (expansion culture) settings. In fact, the phenomenon called "dissociation phenotype/function" was pointed out a long time ago [55,56]. Due to this problem, even the strategies approaching the absolute purity of HSC (SP CD34-KSL population [37]) (if confirmed by other groups) for steady-state cells may not be relevant in the situations out of steady state, especially after ex vivo manipulation.

In addition to all of these problems, the "prospective purification of HSC" results in the loss of a substantial number of functional HSC not endowed by the "target" phenotype. A long time ago it was shown that the HSC in murine bone marrow can be Lin^+ [51,57,58]. Nilsson et al. [59] showed that the majority of HSC potential is, in fact, discarded during the purification. Recently, this problem was actualized [60], even seriously questioning the pertinence of prospective phenotype-based purification of HSC. The important issue related to this consideration is this: can the HSC contained in these "hyperpurified" phenotypic cell populations be representative of a standard HSC?

For this reason, extreme caution should be applied in the interpretation of the results obtained with the phenotypically "purified" HSC population before attributing a result to the "hematopoietic stem cell."

1.6 Stem Cells from Other Tissues: Mesenchymal Stem Cell Example

A similar situation exists with the other stem cell populations due to their functional heterogeneity, elusive phenotypic character, and "changing of face" (dissociation phenotype-function), depending on the microenvironment conditions. Typical examples are "Mesenchymal Stem Cells (MSCs)", which were discovered by the Friedenstein group [61,62] on the basis of their capacity to form the colonies of adherent fibroblast-like cells. This initial definition, although revealing only one of the stem cell capacities (high proliferative potential) remains as the only functional one-cell–based essay (i.e., "clonogenic," since a colony grows from one cell), allowing the MSC detection. Later, the term "Mesenchymal Stem Cells" was noncritically extended to all fibroblast-like cells obtained after one or more culture passages starting from primary bone marrow (and later fat tissue, cord blood, umbilical cord, etc.) mononuclear cells (the population that we are going to term "Mesenchymal Stromal Cells" (MStroCs) in this book).

From the pragmatic viewpoint, a quality control–oriented definition of human MSC is based on minimal classification criteria that were established by the International

Society for Cell Therapy (ISCT): plastic adherence; osteo-, chondro-, and adipogenic differentiation; and cell-surface expression of CD73, CD90, and CD105 concurrent with absent expression of CD11b or CD14, CD45, CD34, CD79a or CD19, and human leukocyte antigen (HLA)-DR [63]. However, the problem is that even if in a bulk population corresponding to these criteria some cells are able to undergo osteo-, chondro-, and adipogenic differentiation this does not automatically imply that all the cells of the population in question are able to do the same.

The practice of proving the capacity of the bulk ISCR criteria-defined "MSC" population and attribution of these capacities to all the cells in the population in question generated a real confusion in the perception of the MSC, and many conclusions are attributed to MSC noncritically and without a real scientific basis. In general, in total cell content after "detaching" fibroblast-like cells (i.e., MStroC, wrongly called "MSC") from plastic, less than 10% exhibit a colony-forming capacity (i.e., represent the CFU-F, entity initially defined as "MSC"). Different approaches were used to enable concentrating the CFU-F in some phenotypically defined populations (both positive and negative selection) [64]. As a "canonical marker" of MSC the STRO-1 is pointed out [65] since the total bone marrow CFU-F activity is concentrated in the fraction of STRO-1 positive cells. At the same time, it is not expressed by hematopoietic committed progenitors [65]. These important findings, however, were followed by numerous articles in which an STRO-1 positive cell was simply considered as "MSC." For example, if multipotent MSCs were functionally detected in Stro-1$^+$ cell population then the title of the article "The STRO-1$^+$ marrow cell population is multipotential" [66], although not wrong, is still misleading since it suggests that all STRO-1$^+$ cells are multipotent, which is not the case. To what extent it is misleading can testify to the frequency of CFU-F in STRO-1$^+$ bone marrow fraction published in the same Simmons and Torok-Storb paper: 2 per 1000 cells. The CFU-F (MSC) frequency can be increased if in STRO-1$^+$ population were selected CD140b$^+$ cells [64], but even in STRO-1$^+$CD140b$^+$ population the functional MSC remains the absolute cell minority.

Here, it is interesting to evoke some well-established facts concerning MStroC heterogeneity: minor subpopulations of quiescent small/agranular cells and proliferating small/granular cells were identified (yielding 12 to 42 CFU-F/100 MStroC, depending on the condition). The cells belonging to these two sub-populations of MStroC exhibit 82% CFU-F cloning efficiency and were considered by authors as "the earliest progenitors" in culture [67]. The same group revealed, one year later, that between these "early progenitors" a third distinct subpopulation exists [68]. These are very small, round, highly refractal in phase-contrast [69] cells that rapidly self-renew and exhibit a very high proliferative potential. Also, within ex vivo-expanded bone marrow-derived MStroC exists a discrete subpopulation (5–20%) of MSC with properties of uncommitted and undifferentiated cells that are quiescent [70]. These cells can either self-renew or generate the committed progenitors, which can, upon appropriate stimulation, differentiate at least in adipogenic and osteogenic lineages [70].

It is interesting to mention that 35 years ago Mets and Verdonk [71] noticed that early passage cultures contained two morphologically distinct cell types: (1) large, slowly dividing cells and (2) rapidly dividing spindle-shaped cells. This morphologic heterogeneity almost disappeared in the late passage cultures in which the cells were large and slowly proliferating. An approach based on the light scattering only enabled isolation of the population composed of up to 90% clonogenic cells (CFU-F) evidenced in a single-cell assay [72]. The up-to-date data obtained with the one-cell approach confirm the heterogeneity in proliferative potential of the MSC and persistence of some high proliferative capacity clones isolated in 20–21% O_2 [73]. Furthermore, this article claims that "the combination of three cell-surface markers (LNGFR, THY-1, and VCAM-1) allows for the selection of highly enriched clonogenic cells (one out of three isolated cells)" [73]. Even in this hyperselected MSC (CFU-F) population the authors found the clones exhibiting different proliferative and differentiation capacities. We should remember that, 15 years ago, Muraglia et al. [74] provided, by analyzing human bone marrow individual clones (splitting the cell content originating from a single CFU-F in three separate dishes to challenge the osteogenic, chondrogenic, and adipose differentiation), the data demonstrated that only ~34% of CFU-F exhibit trilineage potential, ~60% osteogenic and chondrogenic, and 6% can differentiate into only one line. It should be stressed that these data were obtained for the cultures supplemented with FGF. In cultures without FGF the proportion of trilineage CFU-F is only 18% versus 84% of bilineage ones (osteogenic and chondrogenic differentiation potential) [74].

Hence, the situation with MSC seems to be similar to the one concerning HSC elaborated above: (1) the results obtained on cell populations where the cell entities that are considered "stem cells" on the basis of at least one functional parameter represent the absolute minority while the non–stem cells make the absolute majority, are being frequently wrongly attributed to MSC and (2) the prospective enrichment of functional MSC is possible [75] and is loaded with the similar problems and limitations revealed with the HSC enrichment.

Particular attention to this problem will be paid in the following text in order to avoid the misinterpretation of the data and attribution of the properties exhibited by non–stem-cell populations to the "stem cells."

References

[1] Till JE, McCulloch EA. A direct measurement of the radiation sensitivity of normal mouse bone marrow cells. Radiat Res 1961;14:213–22.

[2] Siminovitch L, McCuloch EA, Till JE. The distribution of colony-forming cells among spleen colonies. J Cell Physiol 1963;62:327–36.

[3] Huttner WB, Kosodo Y. Symmetric versus asymmetric cell division during neurogenesis in the developing vertebrate central nervous system. Curr Opin Cell Biol 2005;17:648–57.

[4] Morrison SJ, Kimble J. Asymmetric and symmetric stem-cell divisions in development and cancer. Nature 2006;29(441):1068–74.

[5] Milenković P, Pavlović-Kentera V. Erythroid repopulating ability of bone marrow cells in polycythaemic mice. Acta Haematol 1979;61:258–63.

[6] Marsh JC, Blackett NM. A direct assay of granulocytic repopulating ability. Exp Hematol 1978;6:135–40.

[7] Bradley TR, Hodgson GS. Detection of primitive macrophage progenitor cells in mouse bone marrow. Blood 1979;54:1446–50.

[8] McNiece IK, Stewart FM, Deacon DM, Temeles DS, Zsebo KM, Clark SC, et al. Detection of a human CFC with a high proliferative potential. Blood 1989;74:609–12.
[9] Ivanović Z, Bartolozzi B, Bernabei PA, Cipolleschi MG, Milenkovic P, Praloran V, et al. A simple, one-step clonal assay allows the sequential detection of committed (CFU-GM-like) progenitors and several subsets of primitive (HPP-CFC) murine progenitors. Stem Cells 1999;17:219–25.
[10] Nakahata T, Ogawa M. Identification in culture of a class of hemopoietic colony-forming units with extensive capability to self-renew and generate multipotential hemopoietic colonies. Proc Natl Acad Sci USA 1982;79:3843–7.
[11] Nakahata T, Ogawa M. Hemopoietic colony-forming cells in umbilical cord blood with extensive capability to generate mono- and multipotential hemopoietic progenitors. J Clin Invest 1982;70:1324–8.
[12] Nakahata T, Akabane T. Multipotential hemopoietic colonies in single-cell cultures. Nihon Ketsueki Gakkai Zasshi 1984;47:1764–71.
[13] Keller G, Holmes W, Phillips RA. Clonal generation of multipotent and unipotent hemopoietic blast cell colonies in vitro. J Cell Physiol 1984;120:29–35.
[14] Fruehauf S, Breems DA, Knaän-Shanzer S, Brouwer KB, Haas R, Löwenberg B, et al. Frequency analysis of multidrug resistance-1 gene transfer into human primitive hematopoietic progenitor cells using the cobblestone area-forming cell assay and detection of vector-mediated P-glycoprotein expression by rhodamine-123. Hum Gene Ther 1996;7:1219–31.
[15] Schrezenmeier H, Jenal M, Herrmann F, Heimpel H, Raghavachar A. Quantitative analysis of cobblestone area-forming cells in bone marrow of patients with aplastic anemia by limiting dilution assay. Blood 1996;88:4474–80.
[16] van Os RP, Dethmers-Ausema B, de Haan G. In vitro assays for cobblestone area-forming cells, LTC-IC, and CFU-C. Methods Mol Biol 2008;430:143–57.
[17] Weaver A, Ryder WD, Testa NG. Measurement of long-term culture initiating cells (LTC-ICs) using limiting dilution: comparison of endpoints and stromal support. Exp Hematol 1997;25:1333–8.
[18] Pettengell R, Luft T, Henschler R, Hows JM, Dexter TM, Ryder D, et al. Direct comparison by limiting dilution analysis of long-term culture-initiating cells in human bone marrow, umbilical cord blood, and blood stem cells. Blood 1994;84:3653–9.
[19] Ivanović Z, Dello Sbarba P, Trimoreau F, Faucher JL, Praloran V. Primitive human HPCs are better maintained and expanded in vitro at 1 percent oxygen than at 20 percent. Transfusion 2000;40:1482–8.
[20] Hammoud M, Vlaski M, Duchez P, Chevaleyre J, Lafarge X, Boiron JM, et al. Combination of low O(2) concentration and mesenchymal stromal cells during culture of cord blood CD34(+) cells improves the maintenance and proliferative capacity of hematopoietic stem cells. J Cell Physiol 2012;227:2750–8.
[21] Larochelle A, Vormoor J, Hanenberg H, Wang JC, Bhatia M, Lapidot T, et al. Identification of primitive human hematopoietic cells capable of repopulating NOD/SCID mouse bone marrow: implications for gene therapy. Nat Med 1996;2:1329–37.
[22] Dick JE, Bhatia M, Gan O, Kapp U, Wang JC. Assay of human stem cells by repopulation of NOD/SCID mice. Stem Cells 1997;15(Suppl. 1):199–203. Discussion 204–7.
[23] Wang JC, Doedens M, Dick JE. Primitive human hematopoietic cells are enriched in cord blood compared with adult bone marrow or mobilized peripheral blood as measured by the quantitative in vivo SCID-repopulating cell assay. Blood 1997;89:3919–24.
[24] Cheung AM, Nguyen LV, Carles A, Beer P, Miller PH, Knapp DJ, et al. Analysis of the clonal growth and differentiation dynamics of primitive barcoded human cord blood cells in NSG mice. Blood 2013;122:3129–37.

[25] Park CY, Majeti R, Weissman IL. In vivo evaluation of human hematopoiesis through xenotransplantation of purified hematopoietic stem cells from umbilical cord blood. Nat Protoc 2008;3:1932–40.
[26] Ivanovic Z, Duchez P, Chevaleyre J, Vlaski M, et al. Clinical-scale cultures of cord blood CD34(+) cells to amplify committed progenitors and maintain stem cell activity. Cell Transplant 2011;20:1453–63.
[27] Spangrude GJ, Johnson GR. Resting and activated subsets of mouse multipotent hematopoietic stem cells. Proc Natl Acad Sci USA 1990;87:7433–7.
[28] Zotin AA, Domaratskaja EI, Prianishnikova OB. Existence of pre-CFU-S in the liver of mouse embryos. Ontogenez 1988;19:21–9.
[29] Hodgson GS, Bradley TR. Properties of haematopoietic stem cells surviving 5-fluorouracil treatment: evidence for a pre-CFU-S cell? Nature 1979;281:381–2.
[30] Van Zant G. Studies of hematopoietic stem cells spared by 5-fluorouracil. J Exp Med 1984;159:679–90.
[31] Ploemacher RE, Brons NH. Cells with marrow and spleen repopulating ability and forming spleen colonies on day 16, 12, and 8 are sequentially ordered on the basis of increasing rhodamine 123 retention. J Cell Physiol 1988;136:531–6.
[32] Priestley GV, Wolf NS. A technique for the daily examination of spleen colonies in mice. Exp Hematol 1985;13:733–5.
[33] Wolf NS, Priestley GV. Kinetics of early and late spleen colony development. Exp Hematol 1986;14:676–82.
[34] Ploemacher RE, Brons RH. Separation of CFU-S from primitive cells responsible for reconstitution of the bone marrow hemopoietic stem cell compartment following irradiation: evidence for a pre-CFU-S cell. Exp Hematol 1989;17:263–6.
[35] Rosendaal M, Hodgson GS, Bradley TR. Organization of haemopoietic stem cells: the generation-age hypothesis. Cell Tissue Kinet 1979;12:17–29.
[36] van Hennik PB, de Koning AE, Ploemacher RE. Seeding efficiency of primitive human hematopoietic cells in nonobese diabetic/severe combined immune deficiency mice: implications for stem cell frequency assessment. Blood 1999;94:3055–61.
[37] Matsuzaki Y, Kinjo K, Mulligan RC, Okano H. Unexpectedly efficient homing capacity of purified murine hematopoietic stem cells. Immunity 2004;20:87–93.
[38] Ema H, Nakauchi H. "Homing to Niche," a new criterion for hematopoietic stem cells? Immunity 2004;20:1–2.
[39] Johnson GR, Nicola NA. Characterization of two populations of CFU-S fractionated from mouse fetal liver by fluorescence-activated cell sorting. J Cell Physiol 1984;118:45–52.
[40] Ploemacher RE, Brons RH, Leenen PJ. Bulk enrichment of transplantable hemopoietic stem cell subsets from lipopolysaccharide-stimulated murine spleen. Exp Hematol 1987;15:154–62.
[41] Ploemacher RE, Brons NH. Isolation of hemopoietic stem cell subsets from murine bone marrow: I. Radioprotective ability of purified cell suspensions differing in the proportion of day-7 and day-12 CFU-S. Exp Hematol 1988;16:21–6.
[42] Baines P, Visser JW. Analysis and separation of murine bone marrow stem cells by H33342 fluorescence-activated cell sorting. Exp Hematol 1983;11:701–8.
[43] Jang YY, Sharkis SJ. A low level of reactive oxygen species selects for primitive hematopoietic stem cells that may reside in the low-oxygenic niche. Blood 2007;110: 3056–63.
[44] Sudo T, Yokota T, Ishibashi T, Ichii M, Doi Y, Oritani K, et al. Canonical HSC markers and recent achievements. In: Alimoghaddam K, editor. Stem cell biology in normal life and diseases. InTech; 2013. p. 51–64. Chapter 4.

[45] Oguro H, Ding L, Morrison SJ. SLAM family markers resolve functionally distinct subpopulations of hematopoietic stem cells and multipotent progenitors. Cell Stem Cell 2013;13:102–16.
[46] Notta F, Doulatov S, Laurenti E, Poeppl A, Jurisica I, Dick JE. Isolation of single human hematopoietic stem cells capable of long-term multilineage engraftment. Science 2011;8(333):218–21.
[47] Miller PH, Cheung AM, Beer PA, Knapp DJ, Dhillon K, Rabu G, et al. Enhanced normal short-term human myelopoiesis in mice engineered to express human-specific myeloid growth factors. Blood 2013;121:e1–4.
[48] Valent P, Bonnet D, De Maria R, Lapidot T, Copland M, Melo JV, et al. Cancer stem cell definitions and terminology: the devil is in the details. Nat Rev Cancer 2012;12:767–75.
[49] Okada S, Nakauchi H, Nagayoshi K, Nishikawa S, Miura Y, Suda T. In vivo and in vitro stem cell function of c-kit- and Sca-1-positive murine hematopoietic cells. Blood 1992;80:3044–50.
[50] Weilbaecher K, Weissman I, Blume K, Heimfeld S. Culture of phenotypically defined hematopoietic stem cells and other progenitors at limiting dilution on Dexter monolayers. Blood 1991;78:945–52.
[51] Osawa M, Hanada K, Hamada H, Nakauchi H. Long-term lymphohematopoietic reconstitution by a single CD34-low/negative hematopoietic stem cell. Science 1996;273:242–5.
[52] Bäumer N, Bäumer S, Berkenfeld F, Stehling M, Köhler G, Berdel WE, et al. Maintenance of leukemia-initiating cells is regulated by the CDK inhibitor Inca1. PLoS One 2014;9:e115578.
[53] Bryder D, Rossi DJ, Weissman IL. Hematopoietic stem cells: the paradigmatic tissue-specific stem cell. Am J Pathol 2006;169:338–46.
[54] Rossi DJ, Bryder D, Zahn JM, Ahienius H, Sonu R, Wagers AJ, et al. Cell intrinsic alterations underlie hematopoietic stem cell aging. Proc Natl Acad Sci USA 2005; 102:9194–9.
[55] Dorrell C, Gan OI, Pereira DS, Hawley RG, Dick JE. Expansion of human cord blood CD34(+)CD38(−) cells in ex vivo culture during retroviral transduction without a corresponding increase in SCID repopulating cell (SRC) frequency: dissociation of SRC phenotype and function. Blood 2000;95:102–10.
[56] Donaldson C, Denning-Kendall P, Bradley B, Hows J. The CD34(+)CD38(neg) population is significantly increased in haemopoietic cell expansion cultures in serum-free compared to serum-replete conditions: dissociation of phenotype and function. Bone Marrow Transplant 2001;27:365–71.
[57] Danet GH, Lee HW, Luongo JL, Simon MC, Bonnet DA. Dissociation between stem cell phenotype and NOD/SCID repopulating activity in human peripheral blood CD34(+) cells after ex vivo expansion. Exp Hematol 2001;29:1465–73.
[58] Onishi M, Nagayoshi K, Kitamura K, Hirai H, Takaku F, Nakauchi H. $CD4^{dull+}$ hematopoietic progenitor cells in murine bone marrow. Blood 1993;81:3217–25.
[59] Nilsson SK, Donner MS, Tiarks CY, Weiner HU, Quessenberry PJ. Potential and distribution of transplanted hematopoietic stem cells in a nonablated mouse model. Blood 1997;89:4013–20.
[60] Quessenberry PJ, Goldberg L, Aliotta J, Dooner M. Marrow hematopoietic stem cells revisited: they exist in a continuum and are not defined by standard purification approaches; then there are the microvesicles. Front Oncol 2014;56:1–11.
[61] Friedenstein AJ, Chailakhjan RK, Lalykina KS. The development of fibroblast colonies in monolayer cultures of guinea-pig bone marrow and spleen cells. Cell Tissue Kinet 1970;3:393–403.

[62] Luria EA, Panasyuk AF, Friedenstein AY. Fibroblast colony formation from monolayer cultures of blood cells. Transfusion 1971;11:345–9.

[63] Dominici M, Le Blanc K, Mueller I, Slaper-Cortenbach I, Marini F, Krause D, et al. Minimal criteria for defining multipotent mesenchymal stromal cells. The International Society for Cellular Therapy position statement. Cytotherapy 2006;8:315–7.

[64] Sivasubramaniyan K, Lehnen D, Ghazanfari R, Sobiesiak M, Harichandan A, Mortha E, et al. Phenotypic and functional heterogeneity of human bone marrow- and amnion-derived MSC subsets. Ann NY Acad Sci 2012;1266:94–106.

[65] Simmons PJ, Torok-Storb B. Identification of stromal cell precursors in human bone marrow by a novel monoclonal antibody, STRO-1. Blood 1991;78:55–62.

[66] Dennis JE, Carbillet JP, Caplan AI, Charbord P. The STRO-1$^+$ marrow cell population is multipotential. Cells Tissues Organs 2002;170:73–82.

[67] Colter DC, Class R, DiGirolamo CM, Prockop DJ. Rapid expansion of recycling stem cells in cultures of plastic-adherent cells from human bone marrow. Proc Natl Acad Sci USA 2000;97:3213–8.

[68] Colter DC, Sekiya I, Prockop DJ. Identification of a subpopulation of rapidly self-renewing and multipotential adult stem cells in colonies of human marrow stromal cells. Proc Natl Acad Sci USA 2001;98:7841–5.

[69] Prockop DJ, Sekiya I, Colter DC. Isolation and characterization of rapidly self-renewing stem cells from cultures of human marrow stromal cells. Cytotherapy 2001;3:393–6.

[70] Conget PA, Allers C, Minguell JJ. Identification of a discrete population of human bone marrow-derived mesenchymal cells exhibiting properties of uncommitted progenitors. J Hematother Stem Cell Res 2001;10:749–58.

[71] Mets T, Verdonk G. In vitro aging of human bone marrow derived stromal cells. Mech Ageing Dev 1981;16:81–9.

[72] Smith JR, Pochampally R, Perry A, Hsu SC, Prockop DJ. Isolation of a highly clonogenic and multipotential subfraction of adult stem cells from bone marrow stroma. Stem Cells 2004;22:823–31.

[73] Mabuchi Y, Morikawa S, Harada S, Niibe K, Suzuki S, Renault-Mihara F, et al. LNGFR(+) THY-1(+)VCAM-1(hi+) cells reveal functionally distinct subpopulations in mesenchymal stem cells. Stem Cell Reports 2013;11(1):152–65.

[74] Muraglia A, Cancedda R, Quarto R. Clonal mesenchymal progenitors from human bone marrow differentiate in vitro according to a hierarchical model. J Cell Sci 2000;113:1161–6.

[75] Mabuchi Y, Houlihan DD, Akazawa C, Okano H, Matsuzaki Y. Prospective isolation of murine and human bone marrow mesenchymal stem cells based on surface markers. Stem Cells Int 2013;2013:507301.

In Situ Normoxia versus "Hypoxia"

2

Chapter Outline

2.1 Physioxia (Normoxia In Situ)

The oxygen concentration in human tissues varies between 1% and 11% [1]. While a classical gas analysis of healthy humans' bone marrow aspirates revealed a mean value of 7.2% (59.9 mm Hg) [2], the most recent direct measurements of O_2 concentration using a very sophisticated approach (two-photon phosphorescence lifetime microscopy (2PLM)) in mouse bone marrow revealed values from 1.3% to 4.2% [3]. The modeling studies, taking into account the consumption of O_2, provide a reasonable suggestion that its concentration could approach 0% in some micro-environmental niches as for example the bone marrow ones in which hematopoietic stem cells (HSC) reside [4]. Another example is the skin of human fingernail folds where the O_2 concentrations of 4–5.2% and 3.2% have been measured in dermal papillae and subpapillary plexus, respectively [5]. In some tissues, the O_2 concentration is relatively high, as in kidney tissue (9.5% ± 2.6%) [6], while most other tissues are mainly less oxygenated. In fact the highest O_2 concentration in the body is in arterial blood (13.2%). This arterial O_2 concentration (when breathing the atmospheric air at "zero" altitude) is considered by classical physiology as *normoxemia* (*normoxia*). The venous O_2 concentration, which is much lower (5.3%), in fact, represents the integral value of the tissue concentrations after O_2 consumption. The O_2 concentration of "in situ *normoxia*" or "*physioxia*" can be much lower in some niches. The physiologic level of oxygen provides at the cellular level an appropriate oxygenation (1.3–2.5%) [7] or even <1% (5–25 μM) [8–10]. It is considered that, depending on cell type, anaerobic metabolism (oxidative phosphorylation) starts in the range of 2–6 μM; that is, 1.6–4.8 mm Hg (0.21–0.63%) [10], thus in some tissues, even in steady state, the cells are orientated toward anaerobic glycolysis and/or anaerobic mitochondrial respiration. The schematic presentation of O_2 gradient from atmosphere to "in situ" tissue niche is given in Figure 2.1.

Anaerobiosis and Stemness. http://dx.doi.org/10.1016/B978-0-12-800540-8.00002-8

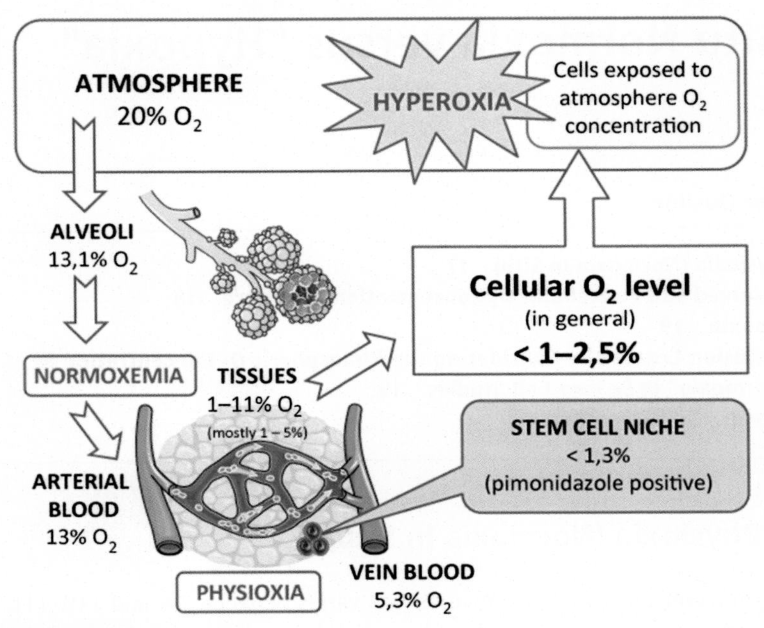

Figure 2.1 Oxygen gradient in vivo.

2.2 Dissolved and Pericellular O_2 Concentration in Culture

At an atmospheric pressure of 760 mm Hg, a gas mixture containing 5% CO_2, 20% O_2, and 75% N_2, equilibrated in a flask with medium, will result in a water vapor partial pressure (pH_2O) of 47, partial pressure of CO_2 (pCO_2) of 35.6 mm Hg, partial pressure of pO_2 of 142.6 mm Hg (18.7%), and that of N_2 534.8 mm Hg [11]. The measured medium pO_2 in mammalian fibroblast cultures exposed to atmosphere O_2 concentration was 125–135 mm Hg (16.4–17.7%) and in those exposed to 1% O_2 was 40–60 mm Hg (5.25–7.87%) [12]. In general, if expressed in molar values, the dissolved O_2 concentration in liquid cultures is always >200 µM. During the mid-logarithmic phase of cell growth, the O_2 is being rapidly consumed in static cultures equilibrated by atmosphere O_2 concentration at the beginning of the culture period, leading to an O_2 decrease (67–70 mm Hg; i.e., 8.58–9.2%) [13] due to the O_2 consumption. If the medium was equilibrated with 0 or 1% O_2 gas mixture, the medium-dissolved O_2 concentrations were 20–40 mm Hg (2.6–5.25%).

Of course, with a permanent maintenance of a stable O_2 concentration in the gas phase, the relative decrease in dissolved O_2 may not be so pronounced. The real O_2 pericellular concentration in culture is the function of the medium depth, cell density, cell type, cell activation state, and so on [14,15]. This especially concerns the adherent cell layer cultures and cultures in which the formation of the tissue-like structures occurs. For example, the pericellular O_2 concentration in confluent bovine endothelial cell cultures is measured at 10.23% (78 mm Hg) and that in rat renal mesangial cell culture at 14.44% (110 mm Hg) [16]. However, in the most cases, especially if the medium depth is low (1–2 mm) and in the case of low cell density suspension cultures, the dissolved O_2 concentration in cultures exposed to atmosphere O_2 concentrations is more than sufficient to provide a highly aerobic environment. It should be emphasized that the tissue O_2 concentrations ≥5% O_2, much lower than arterial blood one (~13%), are fully sufficient to support the aerobic processes. Thus, the O_2 concentrations that result from exposing the cultures to atmospheric O_2 concentrations cannot be considered as physiologic and, therefore, cannot and should not be, by any means, termed "normoxia."

2.3 Hypoxia

The term "*hypoxemia*" in classical physiology is reserved for the values of arterial blood O_2 concentration inferior to physiologic one (usually <13%) and consequent decrease in its tissue- and cell-level concentrations ("*hypoxia*"), specific for each tissue and cell niche. Thus, by its basic definition, "hypoxia" represents a decrease of oxygen availability in tissues (oxygen starvation).

Hypoxemia/hypoxia induced by a decrease in atmospheric O_2 concentration (high altitude, hypobaric conditions) is usually a temporary state since it is, if the physiologic control is operational, compensated by erythron reaction through a well-known feedback mechanism based on the erythropoietin secretion (see [17] for a comprehensive and didactic monograph). However, hypoxia without hypoxemia can also exist. The typical case is the hypoxia developed in tumor tissues, where the O_2 concentration is always lower compared to the normal tissue. For example, in the renal tumors the mean O_2 concentration was measured to be 1.05–1.3% [18], almost 10 times lower than in the normal human kidney tissue. The other example is hypoxia developed in the course of inflammation.

To study the normal cell physiology, it is necessary to take into consideration the oxygenation issue. The molecular, metabolic, and secretory profile of a cell is impacted by actual O_2 concentration of its environment and it is not the same in physioxia and in hypoxia [1]. Furthermore, the cellular response to cytokines and growth factors can be completely different. The same occurs for higher O_2 concentrations with respect to physioxia where the secretory pattern and the response to cytokines are being highly influenced by hyperoxia [19,20,21].

Based on the above, we can conclude that in order to avoid assimilating artifacts to the normal cell properties, the cells should be studied in physioxia.

2.4 Confusion Created by Considering the Atmospheric O_2 Concentration as "Normoxia" in Ex Vivo Cell Studies

Although the classical physiology clearly defines the terms "normoxia" and "hypoxia" as the tissular/cellular phenomena, in cell physiology this described "shortcut" introduced a severe confusion: the cells were usually studied ex vivo in atmospheric O_2 concentrations (20–21%), a condition usually named "normoxia." With that respect, the typically physiologic O_2 concentration values (1–11%) are considered "hypoxia," while the hyperoxic values, not existing in any tissue in vivo, atmospheric 20–21% O_2 (yielding some lower real values of dissolved O_2 in culture medium which are, in most cultures, still higher than physiological ones), are considered a basal "normoxic" condition. This way, not only the distinction between "hypoxia" and "physioxia" disappeared, but also the notion of physiologic versus pathological state was overshadowed. The consequence of this logical mistake is that, for example, the degradation of HIFs transcripts in nonphysiologic hyperoxic situation (20–21% O_2) was adopted as a basal state, the relative stabilization of these transcripts in physiologic O_2 concentrations (1–5%) is considered "hypoxic cellular adaptation" occulting the real hypoxic adaptation occurring at lower O_2 concentrations (the level depending on the tissue). Conversely, a culture in atmospheric O_2 concentration induces in the cells an adaptation to hyperoxygenation. As demonstrated by Gnaiger [8], the pronounced oxygen uptake above mitochondrial saturation at air-level oxygen pressure cannot be inhibited by rotenone and antimycin A, amounting to >20% of routine respiration in fibroblasts. Furthermore, it represents a hyperoxic environment increasing the ROS concentration with consequent lipid, proteins, and nucleic acid damages [22,23], senescence activation, and death [24]. It is clear that the cells with such metabolic activity cannot be considered as a "physiological standard."

This misplaced terminology resulted in an incorrect perception of basal cellular physiology features, which is particularly evident for stem cells since their primitive state ("stemness") is intimately related to their anaerobic/facultative anaerobic metabolic features [25,26].

In this book, we discuss the relationship between the anaerobiosis/facultative anaerobiosis and stemness, and therefore the notion of physiologic order is of paramount importance. To ensure a coherent reasoning, the abovementioned facts will be taken into consideration.

References

[1] Carreau A, El Hafny-Rahbi B, Matejuk A, Grillon C, Kieda C. Why is the partial oxygen pressure of human tissues a crucial parameter? Small molecules and hypoxia. J Cell Mol Med 2011;15:1239–53.

[2] Harrison JS, Rameshwar P, Chang V, Bandari P. Oxygen saturation in the bone marrow of healthy volunteers. Blood 2002;99:394.

[3] Spencer JA, Ferraro F, Roussakis E, Klein A, Wu J, Runnels JM, et al. Direct measurement of local oxygen concentration in the bone marrow of live animals. Nature 2014;508:269–73.

[4] Chow DC, Wenning LA, Miller WM, Papoutsakis ET. Modeling pO(2) distributions in the bone marrow hematopoietic compartment. II. Modified Kroghian models. Biophys J 2001;81:685–96.

[5] Wang W, Winlove CP, Michel CC. Oxygen partial pressure in outer layers of skin of human finger nail folds. J Physiol 2003;549:855–63.

[6] Müller M, Padberg W, Schindler E, Sticher J, Osmer C, Friemann S, et al. Renocortical tissue oxygen pressure measurements in patients undergoing living donor kidney transplantation. Anesth Anal 1998;87:474–6.
[7] Gleadle J, Ratcliffe PJ. Hypoxia. In: Encyclopedia of life sciences. Chichester: John Wiley and Sons, Ltd; 2001.
[8] Gnaiger E. Oxygen conformance of cellular respiration. A perspective of mitochondrial physiology. Adv Exp Med Biol 2003;543:39–55.
[9] Costa LE, Méndez G, Boveris A. Oxygen dependence of mitochondrial function measured by high-resolution respirometry in long-term hypoxic rats. Am J Physiol 1997;273:C852–8.
[10] Boveris A, Costa LE, Cadenas E, Poderoso JJ. Regulation of mitochondrial respiration by adenosine diphosphate, oxygen, and nitric oxide. Methods Enzymol 1999;301:188–98.
[11] Balin AK, Fisher AJ, Carter DM. Oxygen modulates growth of human cells at physiologic partial pressures. J Exp Med 1984;160:152–66.
[12] Taylor WG, Camalier RF, Sanford KK. Density-dependent effects of oxygen on the growth of mammalian fibroblasts in culture. J Cell Physiol 1978;95:33–40.
[13] Taylor WG, Camalier RF. Modulation of epithelial cell proliferation in culture by dissolved oxygen. J Cell Physiol 1982;111:21–7.
[14] Yarmush ML, Toner M, Dunn JC, Rotem A, Hubel A, Tompkins RG. Hepatic tissue engineering. Development of critical technologies. Ann N Y Acad Sci 1992;665:238–52.
[15] Tokuda Y, Crane S, Yamaguchi Y, Zhou L, Falanga V. The levels and kinetics of oxygen tension detectable at the surface of human dermal fibroblast cultures. J Cell Physiol 2000;182:414–20.
[16] Metzen E, Wolff M, Fandrey J, Jelkmann W. Pericellular PO_2 and O_2 consumption in monolayer cell cultures. Respir Physiol 1995;100:101–6.
[17] Wintrobe MM. Blood pure and eloquent. New York: McGraw-Hill; 1980.
[18] Lawrentschuk N, Poon AMT, Foo SS, Putra LG, Murone C, Davis ID, et al. Assessing regional hypoxia in human renal tumours using 18F-fluoromisonidazole positron emission tomography. BJU Int 2005;96:540–6.
[19] Kovacević-Filipović M, Petakov M, Hermitte F, Debeissat C, Krstić A, Jovcić G, et al. Interleukin-6 (IL-6) and low O(2) concentration (1%) synergize to improve the maintenance of hematopoietic stem cells (pre-CFC). J Cell Physiol 2007;212:68–75.
[20] Krstić A, Vlaski M, Hammoud M, Chevaleyre J, Duchez P, Jovcić G, et al. Low O_2 concentrations enhance the positive effect of IL-17 on the maintenance of erythroid progenitors during co-culture of $CD34^+$ and mesenchymal stem cells. Eur Cytokine Netw 2009;20:10–6.
[21] Ivanovic Z. Physiological, ex vivo cell oxygenation is necessary for a true insight into cytokine biology. Eur Cytokine Netw 2009;20:7–9.
[22] Lee HC, Wei YH. Mitochondrial biogenesis and mitochondrial DNA maintenance of mammalian cells under oxidative stress. Int J Biochem Cell Biol 2005;37:822–34.
[23] Saretzki G, Armstrong L, Leake A, Lako M, von Zglinicki T. Stress defense in murine embryonic stem cells is superior to that of various differentiated murine cells. Stem Cells 2004;22:962–71.
[24] Parrinello S, Samper E, Krtolica A, Goldstein J, Melov S, Campisi J. Oxygen sensitivity severely limits the replicative lifespan of murine fibroblasts. Nat Cell Biol 2003;5:741–7.
[25] Ivanovic Z. Hypoxia or in situ normoxia: the stem cell paradigm. J Cell Physiol 2009;219:271–5.
[26] Ivanovic Z. Respect the anaerobic nature of stem cells to exploit their potential in regenerative medicine. Regen Med 2013;8:677–80.

Part One

Anaerobiosis and Stem Cell Entity

The "Hypoxic" Stem Cell Niche

3

Chapter Outline

3.1 Embryonic and Fetal Development

The oxygen concentrations, much lower compared to those in atmospheric air, were measured in the lumen and tissue of female reproductive organs. Oviductal oxygen concentrations range from 1.3% to 8% (10–60mmHg) in the rabbit, rodent, and rhesus monkey [1–3]. In the uterine environment, the oxygen concentrations are even lower: 1.5–2% (11–14mmHg) [3]. It is important to stress that the O_2 concentrations in the uterus are cycle-dependent and decrease at the time of implantation [3]. In guinea pigs, for example, the intrauterine O_2 concentrations vary, depending on cycle phase, from 2.7%±0.7% (20.5±5.5mmHg) to 5.8%±1.4% (44±10.5mmHg) [4]. The values between 0.7% and 6.5% (5–50mmHg) and 3.3–6.5% (25–50mmHg) were measured in the uterus of hamsters and rats [5]. In sheep, after implantation of embryos the amniotic fluid oxygen concentration is very low (1.4%±0.1% in the early gestation and 1.5%±0.1% in mid-gestation [6]). Even during the late gestation (onset of placental gas exchange) the O_2 concentrations are still below maternal venous levels (3.0, 3.9, and 1.6%, for umbilical artery, vein, and amniotic fluid, respectively [6–9]).

The embryo development in humans also occurs in an "oxygen poor" environment: at 8–10weeks in the placental tissue the O_2 concentration is 2.3%±0.9% (17.9±6.9mmHg) and in the endometrial tissue, 5.2%±1.6% (39.6±12.3mmHg), while it increases but remains still relatively low at 12–13weeks (7.9%±1.1% and 6.1%±2.3% in placental and endometrial tissue, respectively) [10]. At 16weeks, a mean pO_2 in intervillous blood was measured to be 66mmHg (8.7%) [11]. In second-trimester human fetus, the pO_2 in the umbilical artery is 34±4mmHg (4.5%±0.5%) [12]. Similar pO_2 values were found long ago in umbilical arteries of mare and ewe [13], while in the right guinea pig fetus atrium the values ranged from 7.9 to 22.2 mmHg (1% and 2.9%) [14]. The umbilical arterial cord blood pO_2 varies from 2.26 kPa±08 kPa to 2.53 kPa±1.05kPa (2.24%±0.79% to 2.49%±0.13%) and those of vein cord blood, 3.79 kPa±1.02 kPa to 3.88 kPa±1.29kPa (3.73%±1.00% to

Anaerobiosis and Stemness. http://dx.doi.org/10.1016/B978-0-12-800540-8.00003-X

3.82% ± 1.26%) (reviewed in [15]). Although the left shift in the fetal O_2-hemoglobin dissociation curve partly compensates for very low pO_2 values, genesis of each system is related to the areas of embryonary/fetal tissue with very low O_2 concentrations ("hypoxic niche").

The example of neurogenesis is instructive: Chen et al. [16] provided the data (using a hypoxic marker EF5, a nitroimidazole derivative that binds covalently to protein, RNA, and DNA in cells exposed to a hypoxic environment [17]) that the neural tube in both the hindbrain and midbrain regions O_2 tensions are substantially below 7.6 mmHg (1%), which is corroborated by the data of Lee et al. [18] showing that the regions of very low O_2 tensions represent the foci of neurogenesis through the whole embryonic development of mice. An important point from their study is that the very low O_2 tension regions correspond to the areas in which the neural stem and progenitor cells are actively proliferating. A similar situation occurs with vascular and kidney development (reviewed in [19,20]). In this context it is not surprising that HIF-1α is a regulator of HSC generation and function in hypoxic sites of the mouse embryo beginning at the earliest embryonic stages [21]. It should be stressed that the "stem cell niche" has been described in worms [22] and flies (*Drosophila* ovary) [23]. It seems that in the last one "hypoxia"-related signaling (Notch) maintains a balance between self-renewal and differentiation [24].

3.2 "Hypoxic" Stem Cell Niche during the Postnatal Life

As already discussed in Chapter 2, the physiological oxygen concentration in "adult" (the term used here for all stages after birth, i.e., to distinguish from the terms "embryonic" and "fetal") tissues varies between 1% and 11% [25], while the O_2 concentration at the cellular level is estimated to be 1.3–2.5% or even <1% [26,27]; the physioxia in some particular physiologically segregated tissue areas ("niches") can refer to much lower O_2 concentrations.

The best-known example of "hypoxic" stem cell niche (more appropriately would be "physioxic" instead of "hypoxic" to distinguish from the tumoral one) is a bone marrow hematopoietic stem cell niche. The "hypoxic" aspect was "amended" to the classical concept of hematopoietic niche, a microenvironmental unit composed of cells and soluble factors [28] first by Dello Sbarba's group [29,30] after demonstrating that more primitive HSC subsets (MRA) are more "resistant" to the low O_2 concentrations in culture with respect to descendant, relatively less primitive subpopulations (CFU-S) and especially the committed progenitors (CFC). This concept was strengthened by a pertinent mathematical modeling with respect to the blood perfusion and tissue O_2 consumption suggesting that some bone marrow areas (supposed to be the HSC niche) are exposed to extremely low O_2 concentrations [31]. During more than three decades, this concept evolved and includes now specific cell types, anatomical locations, soluble molecules, signaling cascades and gradients, as well as physical factors such as shear stress, oxygen tension, and temperature (reviewed in [32]). Nevertheless, definition of the niche as a microenvironment that provides spatially and temporally coordinated signals to support stem cell function has still persisted. The niche can be permissive of, or conducive to, self-renewal and differentiation of HSC in homeostasis. In certain pathologies (myelodysplasia, hematologic malignancy) and during aging, the niche can also limit the extent of normal hematopoiesis.

3.3 Location and "Hypoxic" Character of HSC and Their Niche

Nilson et al. [33] found, after transplantation of purified murine bone marrow $Lin^-$and $Lin^-/WGA^{dim\text{-}med}/Rh^{dull}$ cells that they are predominantly (>70%) retained (after 15h) in subendosteal bone marrow areas unlike their Lin^+ and $Lin^-/WGA^{dim\text{-}med}/Rh^{brigh}$ counterparts. Even if these data, as well as that of Askenasy and Farkas [34] (used Lin^- bone marrow cells), are not specific enough to claim that HSC niches are in endosteal areas (see Chapter 1), the other findings strongly imply that this is the case: all the cells of donor origin 6 weeks following a transplantation of $Rh/Hoechst^{dull}$ cells were located in the endosteal region [35]. In line with these data are the observations that 5-FU-resistant $Tie2^+MPL^+$ (Tie2, a tyrosine kinase receptor, is considered as a marker for long-term repopulating HSC [36]; MPL is a thrombopoietin receptor) cells were localized in the osteoblastic niche [37]. Furthermore after 5-FU treatment, the BRdU long-term label-retaining (BrdU-LTR) cells expressing MPL were detected in the same position, adhered to the bone-lining cells [37]. Altogether implies that the cells in the endosteal niche are quiescent and/or slow cycling [38], thus corresponding to the primitive stem cells. The pertinence of the endosteal niche is strengthened by the data suggesting that multiple subsets of osteoblasts and mesenchymal progenitor cells constitute the endosteal niche and regulate HSCs in adult bone marrow [39]. All these cell types enhance HSC repopulating potential, in particular osteoblast-enriched $ALCAM^+Sca\text{-}1^-$ cells, which also express cytokines Angiopoetin-1 (Ang-1) and Tpo and "stem cell marker genes" [39]. It was demonstrated both by a functional in vivo repopulating approach and gene expression (LSK SP cells, highly enriched in HSC) that Ang-1 maintains ex vivo HSC activity [40]. Also, it has been shown, through the manipulation of osteoblast-specific activated PTH/PTHrP receptors, that osteoblastic cells represent a regulatory component of the hematopoietic stem cell niche in vivo that influences stem cell function through Notch activation [41]. Finally, the simple ionic mineral content of the niche may dictate the preferential localization of adult mammalian hematopoiesis in bone [42]. Seven transmembrane-spanning calcium-sensing receptors (CaR) were identified on SLK cells. In CaR-deficient mice hematopoietic cells (LTC-IC) were found in the circulation and spleen, whereas few were found in bone marrow [42]. The authors claim that "CaR has a function in retaining HSCs in close physical proximity to the endosteal surface and the regulatory niche components associated with it" [42].

Altogether, these data suggest that multiple subsets of osteoblasts and mesenchymal progenitor cells constitute the endosteal niche and regulate HSCs in adult bone marrow.

Other microenvironmental structure, vascular sinusoidal niche, has been implicated in HSC maintenance and regeneration [43,44]. Many $CD150^+CD48^-CD41^-Lin^-$ cells (generally considered as HSC) appeared to be in contact with sinusoidal endothelium in bone marrow ("vascular niche"), while other $CD150^+CD48^-CD41^-Lin^-$ cells appeared to be associated with endosteum ("endosteal niche") [43]. However, which of these two niches is preferred by HSC cannot be evaluated since two-thirds of single $CD150^+CD48^-CD41^-$ cells fail to give long-term multilineage reconstitution in irradiated mice and, hence, do not meet the main functional criterion to be considered as "HSC" (see Chapter 1). Later it was shown that hematopoietic stem and progenitor cells also lie adjacent to diverse nonsinusoidal vascular structures including arteries and arterioles so the "vascular niche" may not be limited to sinusoids [45]. This concept is further developed by Kunisaki et al. [46], whose results suggest that the association of $CD150^+CD48^-CD41^-Lin^-$ (considered by authors as "HSCs") with sinusoids is random; on the contrary their association with arterioles is specific and statistically nonrandom, therefore implying an "arterioral niche" in control of HSC quiescence [46]. The authors claim the distinctive phenotype of the stromal cells of these niches: the arteriolar ones are defined as Nestin-GFP bright, $NG2^+$, $LepR^-$ pericytes; whereas the sinusoidal ones are described as Nestin-dim, $NG2^-$, $LepR^+$ cells with reticular morphology. It seems that the $NG2^+$ pericytes exhibit a distinguished role in quiescence maintaining since their depletion led to a decrease in $CD150^+CD48^-CD41^-Lin^-$ cell quiescence [46]. But, this model based on two distinct niches appears to be too simplified. As noticed by Lévesque and Winkler [47] osteoblastic/mesenchymal niches and perivascular niches overlap anatomically. Furthermore, local blood perfusion in a niche alone can functionally separate HSC populations. This viewpoint is supported by the most recent data showing that the O_2 partial pressure is lower in perisinusoidal regions compared to the endosteal regions [48].

In fact, until these most recent data [48] based on direct measurements of pO_2 in bone marrow (<32 mm Hg; i.e., <4.2%) with the lowest values measured in deeper perisinusoidal regions (<9.9 mm Hg; i.e., <1.3%), the "hypoxic" character of HSC niche was deduced indirectly, on the basis of "hypoxic" properties of cell populations related to the functional stem and progenitor cells. In that respect, Parmar et al. [49] demonstrated that cells belonging to the "Side Population" (concentrating cobblestone area-forming cells and in vivo long-term repopulating cells) are placed in the lowest O_2 gradient in bone marrow since these cells bind Pimonidazole, an agent proved to be a useful marker for cells that have experienced hypoxia with $pO_2 < 10$ mm Hg (<1.3%). In addition [50] the most Pimonidazole-positive areas within the BM were the endosteal ones. Furthermore, hypoxia is involved in regulation of the mobilization response since, after G-CSF treatment, Pimonidazole-positive areas extended toward the central BM [51,52] showed that "BrdU label-retaining cells" (LRCs), considered very primitive HSC (slow cycling ones), reside in the sinusoidal "hypoxic" zone distant from the "vascular niche." Another indirect approach suggests that bone marrow HSC niches are exposed to low O_2 concentrations since serially reconstituting hematopoietic stem cells (the most primitive subpopulation of HSC detectable by functional assays) reside in distinct nonperfused niches [52]. This approach, based on staining with the Hoechst 333142 (Ho) after i.v. perfusion simultaneously with a blockade of Ca-dependent pumps (to avoid the "Side Population

Cells" phenomenon; see Chapter 12), is particularly interesting since it shows that the cells exhibiting LSK CD41$^-$CD48$^-$CD150$^-$ phenotype, usually assimilated to "HSC," are, with respect to the perfusion intensity, segregated in two subpopulations: Ho^{neg} and Ho^{med}; the most primitive HSC exhibiting the serial reconstituting ability are found only in Ho^{neg} fraction. These data stress at which point it is important to distinguish at the functional basis the "real" HSC from the multipotent, oligopotent, and unipotent progenitors (see Chapter 1).

As expected, the normal LSK cells (including their subpopulation enriching better the primitive HSC Tie2$^+$ LSK or CD150$^+$CD48$^-$CD41$^-$LSK cells) maintain "intracellular hypoxia" and stabilize hypoxia-inducible factor-1α (HIF-1α) protein confirming their "hypoxic" character [53]. In the same line of arguments belong the findings of Jang and Sharkis [54], showing that cell populations of murine bone marrow containing lower levels of ROS had higher in vivo reconstitution potential. Furthermore, the HIF-1α is essential to mobilization of LSK and LSKCD48-150$^+$ cells, highly enriched in HSC [55], explaining why the exit of these cells from the microenvironmental niche is associated with reduced oxygenation in situ, induced by the "mobilizing" factors [50]. Together these data suggest that a "conservation" of anaerobic/microaerophilic (i.e., "hypoxic") phenotype is ensured before HSC leaves a niche—a prerequisite for maintenance of its stemness. The most direct evidence of a "hypoxic" profile of in situ HSC are certainly the results showing that LSK, CD34$^-$Flk2$^-$, a fraction enriched in long-term repopulating hematopoietic cells (LT-HSC) utilize glycolysis instead of mitochondrial oxidative phosphorylation to meet their energy demands, which is related to Meis1 (HSC-associated transcription factor) activation of HIF-1α [56]. This metabolic profile certainly enables the HSCs to reside in extremely low O_2 areas, but it does not mean necessarily that it is the product of a "hypoxic" environment since the transcriptional regulation of HIF-1α by Meis1 implies that the glycolytic phenotype is rather an intrinsic property of HSCs. This would be in line with previous results showing that the "hypoxic-phenotype" hematopoietic cells can be found in human bone marrow not only relatively far from the blood vessels but also near to them [57].

The recent study of Silberstein et al. [45] appears to be conclusive: they found that in situ position of hematopoietic cell population enriched in the HSC with long-term repopulating capacity (LSK CD34$^-$) and exhibiting a hypoxic profile (strong retention of Pimonidazole and expression of HIF-1α) is not related to the distance from vascular structures or to the cell cycle status. The same seems to be true for LSK cells as well as LKS$^-$ cells (containing mostly committed progenitors; see Chapter 1), which are also dispersed in bone marrow, exhibited the strong "hypoxic" profile (incorporated Pimonidazole). The stabilized HIF-1α transcripts are also found to be characteristic of CD34$^+$ cells (overwhelming majority of committed progenitors) even in circulation (arterial blood O_2 concentration is much higher than in bone marrow) [58]. Thus the "hypoxic" profile seems not to be induced by the low O_2 niche but rather to be cell-specific. This scenario is further supported by the fact that inducible Meis1 (a protein implicated in regulation of primitive hematopoiesis) deletion in the adult mouse, which results in down-regulation of both HIF-1α and HIF-2α and a shift to mitochondrial metabolism leading to the loss of in vivo repopulating capacity of HSC in bone marrow [59] accompanied with increased reactive oxygen species production, and apoptosis of LSK

$Flk2^{-}CD34^{-}$ cells. On the other hand, some factors typically associated with the HSC niche, such as TPO [60,61] and stem cell factor (SCF) [62] also stabilize HIF-1α protein in hematopoietic cells even under atmospheric O_2 conditions; thus, in hyperoxic environment.

In line with these implications is the finding that a compromised function (deficiency) of HIF prolyl hydroxylase 2 (PHD2) (O_2 sensor implied in degradation of HIF-1 transcripts) result in enhanced self-renewal of primitive HSC [63]. The pharmacological stabilization of HIF-1α in vivo by two inhibitors of PHD2, resulting in an enhancement of HSC repopulative potential (total BM cell injected) [64], as well as the similar results (LSK population injected) obtained with HIF-1α-defficient mice [53], support the argument that low O_2 environment can influence the stem cell potential. These results, also pointing to an appropriate HIF-1α level, are corroborated by the results of Du et al. [65] showing that the deletion of Cited2, a negative HIF-1 regulator, severely impairs the HSC reconstituting capacity, which is partially recovered after additional HIF-1 deletion. Also, the functional control of the "niche size" (bone morphogenetic protein signaling) in bone marrow simultaneously controls the HSC pool size [66]. The bone marrow niche environment obviously influences the metabolic state and cell cycle of hematopoietic cells since $Lin^{-}CD34^{+}CD38^{-}$ human cord blood cells acquire a "hypoxic" phenotype and their significant proportion enter in G_0 phase in the recipient (immunodeficient mice) bone marrow between 4 and 8 weeks, and between 8 and 12 weeks after transplantation, respectively [67].

Interestingly, exposure to an O_2 concentration physiologically relevant for the stem cell niche (1%) or HIF overexpression together with TGFβ stimulation resulted in enhanced expression of cyclin-dependent kinase inhibitor (CDKN1C) and enhanced cycle arrest of $CD34^{+}CD38^{-}$ cells. Since $CD34^{+}$ cells cultured in 1% O_2 secreted high levels of latent TGFβ, an auto- or paracrine role of TGFβ in the regulation of quiescence could operate [68]. It would be interesting to assay if in situ niche of stromal cells (especially mesenchymal stem cells) is able to secrete the TGFβ as is the case of human amniotic fluid-derived MSC [69]. It was also shown that one important function of the hypoxic niche is to provide HSCs intrinsically with the survival factor VEGFA, especially during recovery after HSC transplantation [70].

The protective effect of the "hypoxic" niche on committed progenitors (CFU-GM) has been revealed long ago [71]: the CFU-GM isolated from the compact bone areas were resistant to ionizing radiation and sensitive to Misonidazole, an agent that becomes toxic in "hypoxic" conditions (activation of nitro group, ROS generation). Conversely, the CFU-GM isolated from the medullary cavity were sensitive to ionizing radiation and resistant to Misonidazole. These data support the scenario where the committed progenitors can exhibit anaerobic ("hypoxic") phenotype induced by an environmental niche, which also provides the prerequisites for their survival in that particular condition.

Altogether, it seems that the HSC "hypoxic" phenotype is regulated by the cell-specific mechanisms but could be also influenced by the environment; that is, by "hypoxic" niche. This "protection" by "double insurance" implies at which point microaerophilic/anaerobic features are important for the maintenance of stem cells (Figure 3.1). In that respect, and also to avoid the use of the term "hypoxic" for denoting a physiological

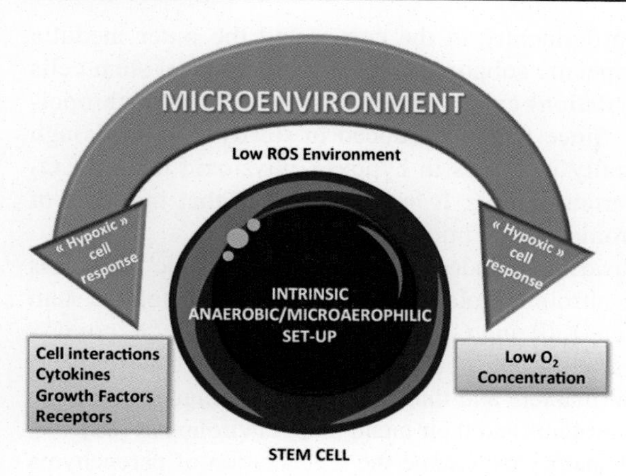

Figure 3.1 Maintaining the anaerobic/microaerophilic metabolic character of a stem cell in niche is operated both by a low O_2 concentration in niche and by influencing the O_2-related cell response (for example, via SCF/c-Kit and TPO/c-MPL, both acting in favor of stabilization of HIF-1α transcripts).

state, the suggestion of Nombella-Arrieta et al. [45] to consider these cells as "low oxidative phosphorylation cells" seems to be very pertinent.

3.4 Low Oxygen Stem Cell Niche in Other Tissues

It was established that in physiologically relevant O_2 concentrations (2.5–5.0%), typical of neural tissues, neural stem cell (NSC) self-renewal is induced; this range of O_2 concentrations applied during ex vivo amplification also favors the success of NSC engraftment upon transplantation into brain of experimental animals (reviewed in [72]). In mice, subependymal zone (SEZ) neural stem cell niche is necessary for regulating adult neurogenesis. It constitutes a specialized microvascular domain defined by unique vessel architecture and reduced rates of blood flow [73]. The authors showed using Pimonidazole that the "hypoxic" conditions are detectable in the ependymal layer that lines the ventricle, and in a subpopulation of neurons throughout the SEZ and striatum.

Inside the cell population restrained to a complex phenotypic fraction c-kit$^+$ Sca-1$^+$ Isl-1$^+$ [74–77] or SP [78] (and exhibiting a similar functional complexity as HSC), the cardiac stem and progenitor cells are contained (for review, [79]). In this population, glycolytic cardiac progenitors (GCPs) that mainly use glycolysis, express HIF-1α, and display a multilineage differentiation potential were lodged in the epicardium and subepicardium as the cardiac hypoxic niche (for review, [80,81]).

Kidney papilla, in which the BrdU LRC were identified, was pointed out as a niche for kidney stem cells [82]. Expression of HIF-1α in the kidney during hypoxia is most pronounced in the papilla, in the same areas where the BrdU LRC reside [83]. It should be stressed that the papilla BrdU LRC do not undergo apoptosis after transient ischemia but start to proliferate instead [82], which reveals their metabolic flexibility characteristic for the stem cells. A 2014 study suggests that this issue is more complex since loading with a DNA label in neonates identified LRC more often in the papilla, while administering the DNA label in adult mice identified LRC prominently in the cortex and the outer medulla [84]. It seems that in kidney a specific subpopulation of mesenchymal stem cells (inside $CD133^+CD73^+$ cell population) exists; these cells are in fact, erythropoietin(Epo)-producing cells [85]. Since Epo is produced in steady state (although a low amount), which dramatically increases in hypoxia, physioxic range of O_2 in their microenvironment is certainly >5%. It has been shown that injection of Epo-positive cells improves chronic kidney injury [86].

The human hair follicle is considered "moderately to severely hypoxic" by Evans et al. [87] established by using 2-nitroimidazole probe. The human hair follicle stem cell candidates expressed markers K19 and CD271, both identified as "hypoxia"-induced genes [88]. These follicles harbor at least two epithelial stem cell compartments that share a number of biological markers and that can also be distinguished by the differential expression of certain proteins and their niche/microenvironment [89].

Even in the liver, where the hepatocytes assume the regeneration of parenchyma tissue, the role of hepatic progenitors/stem cells (term for human liver) and oval cells (term for rodent liver) emerge once the hepatocyte proliferation is overwhelmed by the extent of injury. These cells are at least bipotent (although the functional heterogeneity a priori cannot be ruled out; reviewed in [90]). The liver stem cell issue is very complex. Kuwahara et al. [91] proposed four possible hepatic stem cell niches. Interestingly, in mouse liver the Pimonidazole-fixing cells were not detected in normoxia but only in hypoxia. After only 30 s of hypoxia, the pericentral zones were stained with Pimonidazole [92]. It remains to be elucidated if hypoxia-inducible low O_2 concentration zones are related to the hepatic stem cell niches and implicated in hepatic stem/progenitor cell regulation.

It can be concluded that the maintenance of an undifferentiated state of a broad spectrum of stem cells (embryonic, hematopoietic, mesenchymal, neural, etc.), their proliferation, and cell-fate commitment are related to the microenvironmental niches either directly via low O_2 concentrations or by complex cellular and molecular interactions interfering with oxygen-cell adaptation [93,94].

References

[1] Mastroianni LJ, Jones R. Oxygen tension within the rabbit fallopian tube. J Reprod Fertil 1965;9:99–102.

[2] Maas DHA, Storey BT, Mastroianni L. Oxygen tension in the oviduct of the rhesus monkey. Fertil Steril 1976;27:1312–7.

[3] Fischer B, Bavister BD. Oxygen tension in the oviduct and uterus of rhesus monkeys, hamsters and rabbits. J Reprod Fertil 1993;99:673–9.

[4] Garris DR, Mitchell JA. Intrauterine oxygen tension during the estrous cycle in the guinea pig: its relation to uterine blood volume and plasma estrogen and progesterone levels. Biol Reprod 1979;21:149–59.

[5] Kaufman DL, Mitchell JA. Alterations in intrauterine oxygen tension during the estrous cycle in the rat and hamster and its regulation by ovarian steroid hormones: a comparative study. In: Piiper J, Goldstick TK, Meye M, editors. Oxygen transport to tissue. New York: Plenum Press; 1990. p. 745–50.

[6] Jauniaux E, Kiserud T, Ozturk O, West D, Hanson MA. Amniotic gas exchange and acid—base status during acute maternal hyperoxemia and hypoxemia in the early fetal sheep. Am J Obstet Gynecol 2000;182:661–5.
[7] Eskes T, Jongsma HW, Houx PCW. Percentiles for gas values in human umbilical cord blood. Eur J Obstet Gynecol Reprod Biol 1983;14:341–6.
[8] Yeomans ER, Hauth JC, Gilstrap 3rd LC, Strickland DM. Umbilical cord pH, PCO_2, and bicarbonate following uncomplicated term vaginal deliveries. Am J Obstet Gynecol 1985;151:798–800.
[9] Rurak D, Selke P, Fisher M, Taylor AHT, Whitman B. Fetal oxygen extraction: comparison of the human and sheep. Am J Obstet Gynecol 1987;156:360–6.
[10] Rodesch F, Simon P, Donner C, Jauniaux E. Oxygen measurements in endometrial and trophoblastic tissues during early pregnancy. Obstet Gynecol 1992;80:283–5.
[11] Soothill PW, Nicolaides KH, Rodeck CH, Campbell S. Effect of gestational age on fetal and intervillous blood gas and acid-base values in human pregnancy. Fetal Ther 1986;1:168–75.
[12] Soothill PW, Nicolaides KH, Rodeck CH, Gamsu H. Blood gases and acid-base status of the human second-trimester fetus. Obstet Gynecol 1986;68:173–6.
[13] Comline RS, Silver M. PO_2, PCO_2 and pH levels in the umbilical and uterine blood of the mare and ewe. J Physiol 1970;209:587–608.
[14] Grønlund J, Carter AM. Continuous measurement of blood gas tensions in the fetal guinea pig by mass spectrometry. J Perinat Med 1982;10:226–32.
[15] Armstrong L, Stenson BJ. Use of umbilical cord blood gas analysis in the assessment of the newborn. Arch Dis Child Fetal Neonatal Ed 2007;92:F430–4.
[16] Chen EY, Fujinaga M, Giaccia AJ. Hypoxic microenvironment within an embryo induces apoptosis and is essential for proper morphological development. Teratology 1999;60:215–25.
[17] Lord EM, Harwell L, Koch CJ. Detection of hypoxic cells by monoclonal antibody recognizing 2-nitroimidazole adducts. Cancer Res 1993;53:5721–6.
[18] Lee YM, Jeong CH, Koo SY, Son MJ, Song HS, Bae SK, et al. Determination of hypoxic region by hypoxia marker in developing mouse embryos in vivo: a possible signal for vessel development. Dev Dyn 2001;220:175–86.
[19] Maltepe E, Simon MC. Oxygen, genes, and development: an analysis of the role of hypoxic gene regulation during murine vascular development. J Mol Med (Berl) 1998;76:391–401.
[20] Simon MC, Keith B. The role of oxygen availability in embryonic development and stem cell function. Nat Rev Mol Cell Biol 2008;9:285–96.
[21] Imanirad P, Solaimani Kartalaei P, Crisan M, Vink C, Yamada-Inagawa T, de Pater E, et al. HIF-1α is a regulator of hematopoietic progenitor and stem cell development in hypoxic sites of the mouse embryo. Stem Cell Res 2014;12:24–35.
[22] Kimble JE, White JG. On the control of germ cell development in *Caenorhabditis elegans*. Dev Biol 1981;81:208–19.
[23] Xie T, Spradling AC. A niche maintaining germ line stem cells in the drosophila ovary. Science 2000;290:328–30.
[24] Ward EJ, Shcherbata HR, Reynolds SH, Fischer KA, Hatfield SD, Ruohola-Baker H. Stem cells signal to the niche through the notch pathway in the drosophila ovary. Curr Biol 2006;16:2352–8.
[25] Carreau A, El Hafny-Rahbi B, Matejuk A, Grillon C, Kieda C. Why is the partial oxygen pressure of human tissues a crucial parameter? Small molecules and hypoxia. J Cell Mol Med 2011;15:1239–53.
[26] Gleadle J, Ratcliffe PJ. Hypoxia. In: Encyclopedia of life sciences. Chichester: John Wiley and Sons, Ltd; 2001.

[27] Gnaiger E. Oxygen conformance of cellular respiration. A perspective of mitochondrial physiology. Adv Exp Med Biol 2003;543:39–55.
[28] Schofield R. The relationship between the spleen colony-forming cell and the haemopoietic stem cell. Blood Cells 1978;4:7–25.
[29] Dello Sbarba P, Cipolleschi MG, Olivotto M. Hemopoietic progenitor cells are sensitive to the cytostatic effect of pyruvate. Exp Hematol 1987;15:137–42.
[30] Cipolleschi MG, Dello Sbarba P, Olivotto M. The role of hypoxia in the maintenance of hematopoietic stem cells. Blood 1993;82:2031–7.
[31] Chow DC, Wenning LA, Miller WM, Papoutsakis ET. Modeling pO(2) distributions in the bone marrow hematopoietic compartment. II. Modified Kroghian models. Biophys J 2001;81:685–96.
[32] Wang LD, Wagers AJ. Dynamic niches in the origination and differentiation of haematopoietic stem cells. Nat Rev Mol Cell Biol 2011;12:643–55.
[33] Nilsson SK, Johnston HM, Coverdale JA. Spatial localization of transplanted hemopoietic stem cells: inferences for the localization of stem cell niches. Blood 2001;97:2293–9.
[34] Askenasy N, Farkas DL. Optical imaging of PKH-labeled hematopoietic cells in recipient bone marrow in vivo. Stem Cells 2002;20:501–13.
[35] Nilsson SK, Dooner MS, Tiarks CY, Weier HU, Quesenberry PJ. Potential and distribution of transplanted hematopoietic stem cells in a nonablated mouse model. Blood 1997;89:4013–20.
[36] Arai F, Hirao A, Ohmura M, Sato H, Matsuoka S, Takubo K, et al. Tie2/angiopoietin-1 signaling regulates hematopoietic stem cell quiescence in the bone marrow niche. Cell 2004;23(118):149–61.
[37] Yoshihara H, Arai F, Hosokawa K, Hagiwara T, Takubo K, Nakamura Y, et al. Thrombopoietin/MPL signaling regulates hematopoietic stem cell quiescence and interaction with the osteoblastic niche. Cell Stem Cell 2007;13(1):685–97.
[38] van der Wath RC, Wilson A, Laurenti E, Trumpp A, Liò P. Estimating dormant and active hematopoietic stem cell kinetics through extensive modeling of bromodeoxyuridine label-retaining cell dynamics. PLoS One 2009;4:e6972.
[39] Nakamura Y, Arai F, Iwasaki H, Hosokawa K, Kobayashi I, Gomei Y, et al. Isolation and characterization of endosteal niche cell populations that regulate hematopoietic stem cells. Blood 2010;116:1422–32.
[40] Gomei Y, Nakamura Y, Yoshihara H, Hosokawa K, Iwasaki H, Suda T, et al. Functional differences between two Tie2 ligands, angiopoietin-1 and -2, in regulation of adult bone marrow hematopoietic stem cells. Exp Hematol 2010;38:82–9.
[41] Calvi LM, Adams GB, Weibrecht KW, Weber JM, Olson DP, Knight MC. Osteoblastic cells regulate the haematopoietic stem cell niche. Nature 2003;425:841–6.
[42] Adams GB, Chabner KT, Alley IR, Olson DP, Szczepiorkowski ZM, Poznansky MC, et al. Stem cell engraftment at the endosteal niche is specified by the calcium-sensing receptor. Nature 2006;439:599–603.
[43] Kiel MJ, Yilmaz OH, Iwashita T, Yilmaz OH, Terhorst C, Morrison SJ. SLAM family receptors distinguish hematopoietic stem and progenitor cells and reveal endothelial niches for stem cells. Cell 2005;121:1109–21.
[44] Adams GB, Scadden DT. The hematopoietic stem cell in its place. Nat Immunol 2006;7:333–7.
[45] Nombela-Arrieta C, Pivarnik G, Winkel B, Canty KJ, Harley B, Mahoney JE. Quantitative imaging of haematopoietic stem and progenitor cell localization and hypoxic status in the bone marrow microenvironment. Nat Cell Biol 2013;15:533–43.

[46] Kunisaki Y, Bruns I, Scheiermann C, Ahmed J, Pinho S, Zhang D, et al. Arteriolar niches maintain haematopoietic stem cell quiescence. Nature 2013;502:637–43.
[47] Lévesque JP, Winkler IG. Hierarchy of immature hematopoietic cells related to blood flow and niche. Curr Opin Hematol 2011;18:220–5.
[48] Spencer JA, Ferraro F, Roussakis E, Klein A, Wu J, Runnels JM, et al. Direct measurement of local oxygen concentration in the bone marrow of live animals. Nature 2014;508:269–73.
[49] Parmar K, Mauch P, Vergilio JA, Sackstein R, Down JD. Distribution of hematopoietic stem cells in the bone marrow according to regional hypoxia. Proc Natl Acad Sci USA 2007;104:5431–6.
[50] Lévesque JP1, Winkler IG, Hendy J, Williams B, Helwani F, Barbier V, et al. Hematopoietic progenitor cell mobilization results in hypoxia with increased hypoxia-inducible transcription factor-1 alpha and vascular endothelial growth factor A in bone marrow. Stem Cells 2007;25:1954–65.
[51] Kubota Y, Takubo K, Suda T. Bone marrow long label-retaining cells reside in the sinusoidal hypoxic niche. Biochem Biophys Res Commun 2008;366:335–9.
[52] Winkler IG, Barbier V, Wadley R, Zannettino AC, Williams S, Lévesque JP. Positioning of bone marrow hematopoietic and stromal cells relative to blood flow in vivo: serially reconstituting hematopoietic stem cells reside in distinct nonperfused niches. Blood 2010;116:375–85.
[53] Takubo K, Goda N, Yamada W, Iriuchishima H, Ikeda E, Kubota Y, et al. Regulation of the HIF-1alpha level is essential for hematopoietic stem cells. Cell Stem Cell 2010;7:391–402.
[54] Jang YY, Sharkis SJ. A low level of reactive oxygen species selects for primitive hematopoietic stem cells that may reside in the low-oxygenic niche. Blood 2007;110:3056–63.
[55] Forristal CE, Nowlan B, Jacobsen RN, Barbier V, Walkinshaw G, Walkley CR, et al. HIF-1α is required for hematopoietic stem cell mobilization and 4-prolyl hydroxylase inhibitors enhance mobilization by stabilizing HIF-1α. Leukemia January 12, 2015. http://dx.doi.org/10.1038/leu.2015.8.
[56] Simsek T, Kocabas F, Zheng J, Deberardinis RJ, Mahmoud AI, Olson EN, et al. The distinct metabolic profile of hematopoietic stem cells reflects their location in a hypoxic niche. Cell Stem Cell 2010;7:380–90.
[57] Bourke VA, Watchman CJ, Reith JD, Jorgensen ML, Dieudonnè A, Bolch WE. Spatial gradients of blood vessels and hematopoietic stem and progenitor cells within the marrow cavities of the human skeleton. Blood 2009;114:4077–80.
[58] Piccoli C, D'Aprile A, Ripoli M, Scrima R, Boffoli D, Tabilio A, et al. The hypoxia-inducible factor is stabilized in circulating hematopoietic stem cells under normoxic conditions. FEBS Lett 2007;581:3111–9.
[59] Kocabas F, Zheng J, Thet S, Copeland NG, Jenkins NA, DeBerardinis RJ, et al. Meis1 regulates the metabolic phenotype and oxidant defense of hematopoietic stem cells. Blood 2012;13(120):4963–72.
[60] Kirito K, Fox N, Komatsu N, Kaushansky K. Thrombopoietin enhances expression of vascular endothelial growth factor (VEGF) in primitive hematopoietic cells through induction of HIF-1alpha. Blood 2005;105:4258–63.
[61] Yoshida K, Kirito K, Yongzhen H, Ozawa K, Kaushansky K, Komatsu N. Thrombopoietin (TPO) regulates HIF-1alpha levels through generation of mitochondrial reactive oxygen species. Int J Hematol 2008;88:43–51.
[62] Pedersen M, Löfstedt T, Sun J, Holmquist-Mengelbier L, Påhlman S, Rönnstrand L. Stem cell factor induces HIF-1alpha at normoxia in hematopoietic cells. Biochem Biophys Res Commun 2008;377:98–103.

[63] Singh RP, Franke K, Kalucka J, Mamlouk S, Muschter A, Gembarska A, et al. HIF prolyl hydroxylase 2 (PHD2) is a critical regulator of hematopoietic stem cell maintenance during steady-state and stress. Blood 2013;121:5158–66.
[64] Forristal CE, Winkler IG, Nowlan B, Barbier V, Walkinshaw G, Levesque JP. Pharmacologic stabilization of HIF-1α increases hematopoietic stem cell quiescence in vivo and accelerates blood recovery after severe irradiation. Blood 2013;121:759–69.
[65] Du J, Chen Y, Li Q, Han X, Cheng C, Wang Z, et al. HIF-1α deletion partially rescues defects of hematopoietic stem cell quiescence caused by Cited2 deficiency. Blood 2012;119:2789–98.
[66] Zhang J, Niu C, Ye L, Huang H, He X, Tong WG, et al. Identification of the haematopoietic stem cell niche and control of the niche size. Nature 2003;425:836–41.
[67] Shima H, Takubo K, Tago N, Iwasaki H, Arai F, Takahashi T, et al. Acquisition of G_0 state by CD34-positive cord blood cells after bone marrow transplantation. Exp Hematol 2010;38:1231–40.
[68] Wierenga AT, Vellenga E, Schuringa JJ. Convergence of hypoxia and TGFβ pathways on cell cycle regulation in human hematopoietic stem/progenitor cells. PLoS One 2014;9:e93494.
[69] Jun EK, Zhang Q, Yoon BS, Moon JH, Lee G, Park G, et al. Hypoxic conditioned medium from human amniotic fluid-derived mesenchymal stem cells accelerates skin wound healing through TGF-β/SMAD2 and PI3K/Akt pathways. Int J Mol Sci 2014;15:605–28.
[70] Rehn M, Olsson A, Reckzeh K, Diffner E, Carmeliet P, Landberg G, et al. Hypoxic induction of vascular endothelial growth factor regulates murine hematopoietic stem cell function in the low-oxygenic niche. Blood 2011;118:1534–43.
[71] Allalunis MJ, Chapman JD, Turner AR. Identification of a hypoxic population of bone marrow cells. Int J Radiat Oncol Biol Phys 1983;9:227–32.
[72] De Filippis L, Delia D. Hypoxia in the regulation of neural stem cells. Cell Mol Life Sci 2011;68:2831–44.
[73] Culver JC, Vadakkan TJ, Dickinson ME. A specialized microvascular domain in the mouse neural stem cell niche. PLoS One 2013;8:e53546.
[74] Urbanek K, Torella D, Sheikh F, De Angelis A, Nurzynska D, Silvestri F, et al. Myocardial regeneration by activation of multipotent cardiac stem cells in ischemic heart failure. Proc Natl Acad Sci USA 2005;102:8692–7.
[75] van Vliet P, Roccio M, Smits AM, van Oorschot AA, Metz CH, van Veen TA, et al. Progenitor cells isolated from the human heart: a potential cell source for regenerative therapy. Neth Heart J 2008;16:163–9.
[76] Chimenti I1, Smith RR, Li TS, Gerstenblith G, Messina E, Giacomello A, et al. Relative roles of direct regeneration versus paracrine effects of human cardiosphere-derived cells transplanted into infarcted mice. Circ Res 2010;106:971–80.
[77] Bearzi C, Rota M, Hosoda T, Tillmanns J, Nascimbene A, De Angelis A, et al. Human cardiac stem cells. Proc Natl Acad Sci USA 2007;104:14068–73.
[78] Emmert MY, Emmert LS, Martens A, Ismail I, Schmidt-Richter I, Gawol A, et al. Higher frequencies of $BCRP^+$ cardiac resident cells in ischaemic human myocardium. Eur Heart J 2013;34:2830–8.
[79] Martin-Puig S, Wang Z, Chien KR. Lives of a heart cell: tracing the origins of cardiac progenitors. Cell Stem Cell 2008;2:320–31.
[80] Kimura W, Sadek HA. The cardiac hypoxic niche: emerging role of hypoxic microenvironment in cardiac progenitors. Cardiovasc Diagn Ther 2012;2:278–89.
[81] Kocabas F, Mahmoud AI, Sosic D, Porrello ER, Chen R, Garcia JA, et al. The hypoxic epicardial and subepicardial microenvironment. J Cardiovasc Transl Res 2012;5:654–65.

[82] Oliver JA, Maarouf O, Cheema FH, Martens TP, Al-Awqati Q. The renal papilla is a niche for adult kidney stem cells. J Clin Invest 2004;114:795–804.
[83] Rosenberger C, Mandriota S, Jürgensen JS, Wiesener MS, Hörstrup JH, Frei U, et al. Expression of hypoxia-inducible factor-1alpha and -2alpha in hypoxic and ischemic rat kidneys. J Am Soc Nephrol 2002;13:1721–32.
[84] Rangarajan S, Sunil B, Fan C, Wang PX, Cutter G, Sanders PW, et al. Distinct populations of label-retaining cells in the adult kidney are defined temporally and exhibit divergent regional distributions. Am J Physiol Renal Physiol 2014;307:F1274–82.
[85] Bussolati B, Lauritano C, Moggio A, Collino F, Mazzone M, Camussi G. Renal CD133(+)/CD73(+) progenitors produce erythropoietin under hypoxia and prolyl hydroxylase inhibition. J Am Soc Nephrol 2013;24:1234–41.
[86] Yamaleyeva LM, Guimaraes-Souza NK, Krane LS, Agcaoili S, Gyabaah K, Atala A, et al. Cell therapy with human renal cell cultures containing erythropoietin-positive cells improves chronic kidney injury. Stem Cells Transl Med 2012;1:373–83.
[87] Evans SM, Schrlau AE, Chalian AA, Zhang P, Koch CJ. Oxygen levels in normal and previously irradiated human skin as assessed by EF5 binding. J Invest Dermatol 2006;126:2596–606.
[88] Dayan F, Roux D, Brahimi-Horn MC, Pouyssegur J, Mazure NM. The oxygen sensor factor-inhibiting hypoxia-inducible factor-1 controls expression of distinct genes through the bifunctional transcriptional character of hypoxia-inducible factor-1alpha. Cancer Res 2006;66:3688–98.
[89] Rathman-Josserand M, Genty G, Lecardonnel J, Chabane S, Cousson A, François Michelet J, et al. Human hair follicle stem/progenitor cells express hypoxia markers. J Invest Dermatol 2013;133:2094–7.
[90] Kordes C, Häussinger D. Hepatic stem cell niches. J Clin Invest 2013;123:1874–80.
[91] Kuwahara R, Kofman AV, Landis CS, Swenson ES, Barendswaard E, Theise ND. The hepatic stem cell niche: identification by label-retaining cell assay. Hepatology 2008;47:1994–2002.
[92] Terada N, Ohno N, Saitoh S, Ohno S. Immunohistochemical detection of hypoxia in mouse liver tissues treated with pimonidazole using "in vivo cryotechnique". Histochem Cell Biol 2007;128:253–61.
[93] Mohyeldin A, Garzón-Muvdi T, Quiñones-Hinojosa A. Oxygen in stem cell biology: a critical component of the stem cell niche. Cell Stem Cell 2010;7:150–61.
[94] Perry JM, Li L. To be or not to be a stem cell: dissection of cellular and molecular components of haematopoietic stem cell niches. EMBO J 2012;31:1060–1.

Low O_2 Concentrations and the Maintenance of Stem Cells Ex Vivo

4

Chapter Outline

4.1 First Notions of Oxygenation Ex Vivo

More than one century ago, Alexis Carrel reported the results of experiments aimed to maintain ex vivo tissue of chicken embryo and fetal connective tissue, heart tissue, and portal vein tissue as well as the Rous sarcoma cells [1]. Although controversial, the results of this publication concerning cultures in Ringer's solution supplemented by plasma, serum, or media conditioned by the cells in previous cultures, established the basic principles for further development of ex vivo cultures, which were optimized in order to reproduce, as much as possible, the physiological conditions for each cell type. These experiments introduced the notion of physiological temperature in the ex vivo cultures, but the oxygenation issue ex vivo was not seriously considered during the following half decade until Wright published results with the cells of fetal chick heart, Jensen rat sarcoma, and a mouse carcinoma, carried out in cultures exposed to various low O_2 concentrations [2].

4.2 Oxygenation Level and Ex Vivo Cultures of the Embryonic, Fetal, and Adult Cells

Since it was demonstrated for the first time that successful development of the neural fold by ex utero mouse embryos was dependant on low O_2 levels [3], a lot of

Anaerobiosis and Stemness. http://dx.doi.org/10.1016/B978-0-12-800540-8.00004-1

studies suggested that "reduced oxygen availability" (i.e., the O_2 concentrations near the physiologically relevant ones) were beneficial for germinal cell maturation and embryo development. For example, 5% O_2 atmosphere during ex vivo maturation of bovine oocytes yielded more blastocyst-stage embryos compared to one with 20% O_2, which was correlated with a reduction in the intra-oocyte level of reactive oxygen species (ROS) [4]. Culture of mouse zygotes in 20% O_2, but not in 5% O_2, compromises the developmental potential of resultant blastocysts [5]. Also, under lowered oxygen concentrations (2–12%), development of sheep and cattle embryos can occur through the 8- to 16-cell block in a simple defined medium without somatic cell support [6]. Not in 20% O_2, but under 7% O_2 the goat embryos were morphologically comparable to those developed in vivo, although the mean cell numbers in vitro were only approximately half the numbers obtained in vivo [7]. Indeed, in vitro preimplantation embryo development, gene expression, and offspring production following embryo transfer at 5% O_2, unlike at 20% O_2, approaches those occurring in vivo [8]. If the first 2 to 3 days of human embryo ex vivo development were performed under 5% O_2, the mean number of embryos classified as blastocysts was significantly higher as compared to the 20% O_2. This difference was due to the fact that significantly more blastocysts of the 20% O_2 group had lower cell number compared to the 5% O_2 group [9]. On the basis of several studies, the toxic effects of excessive oxygen on early human embryo have been evidenced as well, implying that prevention of excessive oxygen tension ought to be of special importance for ensuring good embryo vitality [10]. Furthermore, the embryonic culture in 5% O_2 improves the results in the in vitro fertilization/intracytoplasmic injection procedures [11]. These are some examples from the abundant literature showing that the O_2 concentrations allowing the maturation of germinative cells and optimal embryo development ex vivo are much lower in comparison to the atmospheric ones. Some of these data show that the development of embryo (although suboptimal) occurs even in anaerobiosis [6].

A special case represents the cell entity called "embryonic stem cells" (ESCs). This term is used for the cell lines obtained after ex vivo propagation of the cell clones isolated from the internal mass of blastocyst (reviewed in [12]). Using both ES culture at 3% O_2 and genetic interventions (targeted disruption of the Arnt locus—coding for HIF1b) it was shown much earlier that the intact "HIF system," the main part of the "hypoxic cell response machinery," is necessary for multilineage embryonic hematopoiesis [13]. Further experiments with murine ES showed a great metabolic flexibility of these cells, but, although providing some advantage in terms of growth, the physiologically relevant O_2 concentration (dissolved O_2=36 mmHg; i.e., 4.7%) did not significantly decrease the amount of DNA damage nor enhance the primitiveness of these cells (as judged on the basis of morphology and molecular markers) [14]. On the contrary, the O_2 concentration is one of the major factors in murine ES differentiation induced by appropriate culture conditions [14].

Human embryonic cells also can self-renew in cultures exposed to atmospheric O_2 concentrations [15] but the physiologically relevant O_2 concentration cultures ($<5\%$ O_2) prevent spontaneous differentiation, enhance self-renewal, reduce chromosomal aberrations, maintain active X chromosome [16–20], enhance the molecular markers of stemness (SSA and MYC) [21], and NANOG [22]. A very interesting study showed that exposure of already committed ES to 2% O_2 can "push" them back to the pluripotent state [15]. This was not only evident on the basis of cells' morphology, self-renewal capacity, genomewide mRNA and MRNA profiles, Oct4 promotor methylation state, and cell surface markers (TRAI-60 and SSEA) expression, but also, more importantly, by capacity of teratoma formation (which is a specific functional in vivo test) [15]. However, in some cases it was found that a low O_2 concentration (1% in gas phase) down-regulates expression of pluripotency markers in hESC but increases significantly the expression of genes associated with angio- and vasculogenesis including vascular endothelial growth factor and angiopoietin-like proteins [23]. Further differentiation of these cells in a 5% O_2 environment results in generation of functional epithelial progenitor cells (EPC) capable, when transplanted into a rat model of myocardial infarction, of facilitating cardiac function recovery. It was published before that the exposure to 2% O_2 of chondrogenic-differentiation hESC cultures enhanced the ability of the cells to produce functional cartilage constructs [24].

So, both murine and human ESC can self-renew in atmosphere O_2 concentrations. However, while for the murine ESC the low O_2 concentration seems not to be a critical parameter [14], most studies showed the opposite situation for human ESC, although some contradictions were revealed concerning maintenance of undifferentiated state and induction of commitment and differentiation in response to low O_2 concentrations for the latter cells. The differences with respect to the biological properties of these two cell populations [12] at least in part may explain their different responses to the low O_2 concentrations. However, in our opinion, the major problem of ESC in general is the fact that these cells were cultured by many passages in nonphysiologic (high) atmosphere O_2 concentrations, leading to their aerobic adaptation (enhancing their stress-defense capacity [25]). From this point of view, maintenance and differentiation of these cells, as usually studied, may not be completely physiologically relevant. Furthermore, their responses to physiologically relevant O_2 concentrations could be maintained in same lines and not in the other ones. In addition, the fact that only a fraction of cells considered as "ES" can give the in vitro teratoma formations implies a functional heterogeneity of "ESC" population. This is the reason why the interpretation of the data obtained on "bulk" ESC populations should be critically evaluated since a functional heterogeneity of ESC might explain data difference concerning self-renewal versus commitment and differentiation, which may be simply overlooked if the explored cell population is considered homogeneous.

A novel category of pluripotent stem cells are induced pluripotent stem cells (iPSCs), obtained from mature somatic cells after a genetic manipulation [26]. Not surprisingly, cultivation under 5% O_2 favors more efficient iPS generation from mouse and human somatic cells compared to 20–21% O_2 cultures [27]. A substantial acceleration of the reprogramming process was achieved combining low O_2 concentration (5%) with the mesenchymal stromal cells (MStroC) adipose-tissue cell population, substitution of c-Myc with L-Myc, and using a cocktail of chemical inhibitors [28]. Applying the same O_2 concentration, Muchkaeva et al. [29] successfully generated iPS cells from Human Hair Follicle Dermal Papilla Cells. The physiologically relevant O_2 concentrations (2.5% and 5% tested) improve the stemness and quality of iPS cells in 2-month-long cultures [30]. It is possible to dedifferentiate adult human somatic cells (expression of REX1, potentiation of expression of LIN28, translation of OCT4, SOX2, and NANOG, and translocation of these transcription factors to the cell nucleus) into adult human fibroblasts without transgenes, only by modifying the in vitro culture conditions including the O_2 concentration (5% tested by authors) [31]. This phenotype was followed by an enhancement of the proliferative capacity, nearly doubling the number of population doublings before the cells reach replicative senescence. Furthermore, in the absence of viral infection and oncogenic factors, applying the repeated transfection of two expression plasmids (Oct4 and Sox2) into mouse embryonic fibroblasts (MEFs) and in low O_2 concentration (3%) it was possible to generate a novel iPS cell category exhibiting primitive cell morphology, stem cell marker staining, gene expression profiles, embryonic body formation capacity, teratoma formation capacity, and chimeric mouse formation (i.e., pluripotent capacities [32]). It should be stressed that the low O_2 concentration (2%) can revert iPS-derived differentiated cells back to a stem cell state [15]. The maintenance of stemness of iPS in "hypoxia" seems to be primarily related to HIF-2a [33].

It is interesting that a long time ago Wright showed that the lowest O_2 tension allowing proliferation of chicken fetal heart myoblasts was 12 mm Hg (1.57%) [2]. Also, the growth kinetics of the cells of a transformed fetal rabbit kidney cell line is negatively impacted by the O_2 concentrations ≥20% [34]. Balin et al. [35] found that the growth of cells of human fetal lung fibroblasts line WI-38 continues under a partial pressure of oxygen (PO_2) of 7.8 ± 3.5 mmHg (1.02% ± 0.45%) although with a slower growth rate, comparing to PO_2 of 44 ± 7 mmHg (5.77% + 0.91%). With increase in O_2 concentrations up to 134 mmHg (17.6%) population doubling time becomes prolonged, but the higher O_2 concentrations significantly inhibited their growth revealing the toxic effect of oxygen [35]. The same authors later published similar results with optimized cell cultures: at 50 cells/cm^2, growth was maximal at PO_2 9 and 16 mmHg (1.18% and 2.1%) for WI-38 and IMR90 (also a diploid human embryonic lung fibroblast cell line), respectively. Growth was progressively inhibited as the oxygen tension was increased. The population doubling increase at 14 days was 8.6 for pO_2 9 and 16 mmHg (1.18% and 2.1%), 5.8 for pO_2 42 mmHg (5.51%), 3.8 for PO_2 78 mmHg (10.24%), 3.8 for pO_2 104 mmHg (13.65%), and 3 for pO_2 138 mmHg (18.11%) [36]. Even if these cell lines do not represent the "primary" cells, it is obvious that the O_2 concentrations higher than physiologically relevant ones do not allow the optimal growth of the cells.

In experiments with mouse fibroblasts and some cell lines (human liver HLi1, monkey kidney MK2, and human malignant epithelium HeLa) it turned out that the optimal O_2 concentration ranged from 5% to 10% [37]. Using tight bottles filled with the gas mixture for cultures, Clark [38] showed in 1964, on two human adult fibroblast cells lines, that the cells do not grow below oxygen tensions of 1%, and that their survival rate decreased in reduced oxygen tensions, being poorest when oxygen is completely absent. Between 1% and 3% O_2 growth occurs but at a reduced rate compared to growth in air, while above about 3%, growth rate is similar to one in air. Another half-century-old study [39] showed that the growth rate of mouse fibroblast LS cells was limited at extremely low dissolved O_2 partial pressure (1.6 mmHg = 0.21%) comparing with the optimal growth rate in 40–100 mmHg (5.25–13.12%) O_2 ranges.

It should be stressed that the culture systems at the time when these results were published were less perfect compared to those nowadays and that all these experiments were not performed on primary cells but on cell lines. Nevertheless, these old articles (as cited in the introduction of Kilburn et al. [39]) agree that anaerobic conditions generally severely depress (or even halt) the cell growth, but the high oxygen concentration also inhibits cell proliferation (i.e., exhibits a toxic effect). Finally, the intermediary O_2 concentrations, similar to the physioxic ones, allow optimal cell growth.

In spite of numerous limitations related to the first culture approaches, the forthcoming research confirmed their principal points [40,41]. For example, it was shown that the cardiac fibroblasts isolated from adult murine heart ventricle proliferate with lower differentiation rate in cultures exposed to 3% O_2 compared to those in 10% and 21% O_2 where a much higher proportion of cells experienced a remarkable G2/M phase arrest [42]. The higher O_2 concentrations with respect to the physiologic ones enhance routine cell respiration and influence multiple metabolic parameters [43], implying that these conditions are not relevant for a steady-state standard. These studies raise the vital broad-based issue of controlling ambient O_2 during the culture of primary cells isolated from organs and tissues but do not consider another important issue: functional heterogeneity and a hierarchy with respect to the proliferative and differentiation potential of the cells composing the tissues.

4.3 Oxygenation Level and Culture of Stem and Progenitor Cells

The revolutionary results of Till and McCulloch in 1961 [44] and consequent studies, which established a concept of hematopoietic stem and progenitor cells representing a continuum of cellular stages starting with very primitive HSC endowed by a very high proliferative and differentiation potentials, which gradually decline in favor of commitment (reduction of these two potentials) and differentiation toward morphologically recognizable cells of blood lineages, provided a paradigm that was used for other tissues. Also, the principle (hierarchical model) known as "generation-age hypothesis" [45] was adopted as a general concept for all tissues. With these notions, the stem cell biology reveals a new dimension, a complex functional heterogeneity: in a tissue we have to deal with the cells that are substantially different among themselves with respect to their proliferative kinetics, commitment/differentiation rate, cell cycle duration, and so on, but also with respect to their energetic demands, metabolic properties, and ex vivo behavior. In that respect, for example, if 0 or 1% O_2 culture atmosphere (resulting in dissolved 20–40 mmHg O_2) inhibits the proliferation of mature epithelial cells [46], we can conclude that, for these cells, a better O_2 availability is necessary for their physiological purpose. On the other hand, if a stem cell does not proliferate in reduced O_2 availability, this does not necessarily reflect a negative effect since the role of these cells can be to conserve regenerative potential either in quiescent state or by self-renewal (slow intermittent proliferation). For example, the proliferation of limbal epithelial cells is decreased in cultures exposed to 2% O_2, but, simultaneously, the clonogenic efficiency of these cells (i.e., incidence of limbal stem/progenitor cells in limbal epithelial cell population) increased, as well as expression of the phenotypic and molecular markers of stemness [47].

4.3.1 Hematopoietic Stem and Progenitor Cells' Oxygen Issue

Although the early studies on ex vivo production of erythroid cells in 1949 and 1964 [48,49] in and platelets in 1985 [50] revealed the importance of the appropriately adjusted O_2 concentration for optimal proliferation of hematopoietic cells, more specific data concerning the progenitor cells endowed by the capacity of colony formation were not published until the late 1970s. One of the first studies in 1978 [51] found an increase in efficiency of colony formation of granulocyte-macrophage progenitor cells in cultures exposed to 6.8% O_2 comparing to the atmosphere O_2 concentration. Furthermore, the authors detected a progenitor that is more primitive than committed progenitors. It was endowed by a high proliferative capacity producing very large colonies whose maximum size was reached several days later compared to committed progenitors. Later, these primitive progenitors were related to CFU-S and pre-CFC subpopulations and named "high proliferative potential colony forming cells" (HPP-CFC) [52,53]. Similarly, a culture in 5% O_2 enhances the colony growth from erythroid progenitors (BFU-E and CFU-E) [54] compared to the air O_2 concentration. These features were confirmed, and extended to multilineage progenitors (CFU-Mix) [55–57] and megakaryocyte progenitors (CFU-Meg) [56]. Concerning the human cord blood-derived committed progenitors in population of $CD133^+$ cells, a short exposure to 1.5% O_2 atmosphere (24 h) before clonogenic analyses demonstrated a moderate increase in total colony numbers [58], but there are no data on the effect of continuous exposure during several days.

All of these studies, in fact, confirmed the negative effects of hyperoxygenation (atmospheric O_2 concentrations: 20–21% in the gas phase) for different classes of hematopoietic progenitors, as evidenced by ex vivo assays. The O_2 concentrations, ranging from 5% to 7% (these values should be probably lowered to some extent to get the real dissolved O_2 concentrations), approximate the mean tissue O_2 concentrations and allow nonrestricted oxygen supply to the progenitors. At the same time, this range of O_2 concentration prevents oxidative stress induced by atmospheric O_2 concentration and consequently induced senescence [59].

As already stated, in all studies cited so far, the main criterion to judge the effect of a culture condition was the intensive colony growth or intensive cell proliferation. However, understanding of functional complexity of the heterogeneous hematopoietic stem and progenitor cell compartments provided a new insight in physiological behavior of very primitive stem cells. Since these cells represent a functional reserve for regeneration of hematopoietic tissue, their physiological state is not related to an extensive proliferation but to the maintenance of their full proliferative and differentiation potential either in quiescent state or during a restrained number of divisions (self-renewal). These considerations turn out to be necessary for understanding the forthcoming development in the issue of low O_2-stem cell relationship. The crucial points of this development are certainly the results of Cipolleschi et al. in 1993 [60], who demonstrated in the culture of murine bone marrow cells that the O_2 concentration of 1% was not enough to support the growth of committed progenitors, which virtually disappeared from the culture (of note, that they are enhanced at 5% O_2 as previously cited). This low O_2 concentration (1%) was also restrictive to CFU-Sd8 and, at a lower extent to CFU-S d12. In contrast, the more primitive HSC subpopulation MRA (cells endowed by capacity of marrow repopulating ability) was very well maintained. This was the first explicit ex vivo experimental proof obtained by direct O_2 lowering that the HSCs do not have the same demands for O_2 as committed progenitors; that is, that the stem cells have some "hypoxic" (microaerophilic) features. The authors hypothesized that this HSC characteristic comes from their quiescent state conditioning their low energetic demand [60].

It should be stressed that these cultures, stimulated by condition media, were not optimized in terms of survival and proliferation of all cell categories. Optimization of ex vivo cultures of murine bone marrow cells exposed to air oxygen concentration (20–21%) by adding recombinant cytokines allowing an important amplification of committed progenitors, resulted in almost total exhaustion of HSC potential (as measured by MRA and pre-CFC assays) [61,62]. In these conditions, 1% O_2 concentration still allowed an important amplification of committed progenitors (although at a lower extent compared to 20% O_2) in parallel with the full maintenance of HSCs-MRA [61]. Also, in the same cultures, the HPP-CFC progenitors, which were amplified after 8 days of cultures with respect to Day 0, maintained their full Day-0 potential to generate other colonies in secondary cultures, the phenomenon that was completely lost after the cultures were exposed to air. On the basis of these results in 1% O_2 culture, it was obvious that the stem cell maintenance was not a static phenomenon (coming from quiescent cells) but rather a dynamic one (occurring in parallel with proliferation). This publication [61] was the first that related HSC self-renewal to low O_2 concentration (1%). The same phenomenon was confirmed on human $CD34^+$ cells mobilized to peripheral blood, cultured in 1% O_2. In this condition, amplification of committed progenitors was evident although to a lesser extent than in 20% O_2, in parallel with a much better maintenance of pre-CFC activity (reflecting the cells more primitive than committed progenitors; see Chapter 1). It should be mentioned that the total content of pre-CFC and CFC belongs to $CD34^+$ population.

After one week of culture, all $CD34^+$ cells divided one or more times, implying that the maintenance of pre-CFC activity was paralleled with proliferation and not related to the maintenance of quiescent cells in culture [63]. This conclusion was supported with the murine bone marrow cells by two different approaches: (1) the enhancement of pre-CFC activity in liquid cultures exposed to 0.9–1.0% O_2 was completely or almost completely abrogated by adding 5-fluorouracil (5FU) (killing the cells in active cell cycle and sparing only those in G0) in cultures [62,64] and (2) sorting of cells that were not divided, divided once, or divided more than once after 8 days of culture exposed to 0.9% O_2 showed that pre-CFC activity was maintained for all three fractions, formally proving that the high proliferative capacity of pre-CFC was maintained in parallel with cell divisions, unlike in the cultures exposed to 20–21% O_2 where pre-CFC activity was lost in parallel with cell divisions [63].

The better maintenance of HSC was confirmed by Danet et al. in 2003 [65] in 4-day cultures of $Lin^-CD34^+CD38^-$ human bone marrow cells exposed to 1.5% O_2, which resulted in a greater expansion of SRC (Scid Repopulating Cells; a subpopulation of HSC; see Chapter 1), comparing to 20% O_2 cultures. This functional HSC enhancement was paralleled by an enhanced stabilization of HIF-1α, as well as by an up-regulation of expression of angiogenic receptors and secretion of VEGF in culture [65]. The claims that the appropriately low O_2 allows to "push" the balance commitment/self-renewal toward self-renewal [63,64] were confirmed by Eliasson et al. [66] with murine bone marrow LSK cells cultured in 1% O_2. The authors found an increase in 1% O_2 culture of HPP-CFC, CAFC, and Pre-CFC number compared to 20–21% O_2 culture and a decrease of CFC (Committed progenitors) (see Chapter 1, Figure 1.1). Of note, the positive effect on the maintenance of stem cells in very low O_2 concentration could be revealed only in situations with an appropriate stimulation provided in the culture. In that respect, the culture of murine bone marrow lineage-negative cells in 1% O_2 allows a better maintenance of pre-CFC activity than in 20–21% O_2 [67]. This positive effect on pre-CFC is significantly enhanced if IL-6 was added in addition to IL-3 in culture at 1% O_2, while at 20–21%, IL-6 does not exhibit any effect [67].

Combining appropriate stimulation in culture and O_2 concentration, it is possible to improve simultaneously amplification of committed progenitors and the maintenance of HSC. Indeed, such a culture of cord blood $CD34^+$ cells exposed to 3% O_2 allows a full CFC amplification in parallel with enhancing pre-CFC and SRC activity [68]. Similarly, acting on the ex vivo conditions (serum, cytokines), Roy et al. in 2012 [69] obtained in 5% O_2 cultures of cord blood $CD133^+$ cells with better amplification of committed progenitors and maintenance of HSC (SRC). An appropriate O_2 concentration (5%) enhances growth, amplification, and proliferative capacity of committed progenitors (erythroid, granulocyte-monocyte, and megakaryocyte) as well as their response to cytokines relevant for this stage of hematopoiesis [70–72]. It is interesting that Smith and Broxmeyer in 1986 [73] in long-term suspension cultures demonstrated that atmospheric O_2 concentration (20–21%) had a negative impact on the long-term production of committed progenitors in culture. This was much improved if the cultures were exposed to 5% O_2. A posteriori, we can conclude that this approach was, in fact, a first to demonstrate that a physiologically relevant O_2 concentration can maintain ex vivo the pre-CFC, a subpopulation more primitive than committed progenitors (see Chapter 1). Similar results were obtained in stroma-based long-term cultures, providing the information related to long-term culture initiating cells, also a subpopulation more primitive than committed progenitors [74]. A positive effect of low O_2 concentration on pre-CFC and SRC has also been found in cocultures of cord blood $CD34^+$ cells with MStroC (most effective was 1.5% O_2 concentration) [75]. At least in part, this effect can be ascribed to enhanced IL-6 production by stromal cells at 1.5%

O_2 [75]. The same suggestion comes from the results of Zhambalova et al. in 2009 [76], who also found an enhanced secretion of IL-6 in cocultures of hematopoietic cells with human MStroC.

In line with these data, suggesting supporting action of stromal cells exposed to physiologically relevant O_2 concentrations in maintenance of HSC "hypoxic" phenotype, are the results of Kiani et al. in 2014 [77], who evaluated, using a coculture system, the interactive impact of HIF-1α-overexpression in MStroC on HSCs. They found that HIF-1α-overexpression acts via enhanced secretion of SCF (SCF stabilizes HIF1α in HSC, see Chapter 3). Also the addition of exogenous IL-17 in MStroC/CD34$^+$ coculture enhances CD34$^+$ and the total CFC production in the oxygen-dependent manner; the effect is maximal at 3% O_2, and is relatively more pronounced on erythroid progenitors (BFU-E) than on CFU-GM [78]. Thus, the low O_2 concentration indeed appears to approximate better the in vivo microenvironment in cocultures causing the selective modification of hematopoietic cell and MStroC interactions [79].

It is interesting that the BFU-E "hypoxic" feature was evidenced when cord blood CD34$^+$ cells were cultured at 1% O_2 [80]. In liquid cultures exposed to this oxygen concentration, the BFU-E were not only numerically maintained, but also individual progenitor proliferative capacity was enhanced (resulting in very large BFU-E colonies compared to the same type of progenitors in 20% O_2 cultures). It should be stressed, however, that the differentiation of BFU-E through the mature erythroid progenitors (CFU-E) and a "chain" of erythroid precursors is inhibited at 1% O_2 and requires a much higher O_2 concentration [80]. These conclusions were completely confirmed in the course of optimization of ex vivo red blood cell production starting from CD34$^+$ cells [72]. The final result (yield of RBC) was significantly enhanced if the first culture steps, corresponding to the amplification of primitive progenitors, were performed at low O_2 concentrations (1.5% or 5% O_2) and ensuing steps, corresponding to differentiation stages, were performed at 13% O_2 [72]. Indeed, the cell-specific oxygen consumption rate in the similar ex vivo cultures is the highest during the erythroid differentiation onset (Day10-Day11) [81]. These publications [72,80] in fact, point to the low O_2 concentration as a general physiologic regulator of erythropoiesis beyond the EPO-related feedback. Megakaryocytes show a response similar to that of erythroid precursors: while 5% O_2 ensures a higher expression of c-Mpl and IL-3R, and of Mk differentiation-specific surface receptors CD41a, CD42a, and NMDAR1, 20–21% O_2 supported higher expression of the Mk-early and -late-maturation-specific transcription factors GATA-1 (i.e., enhanced maturation [82]). In a similar manner, a higher O_2 concentration enhances the differentiation/maturation of NK cells from their progenitors [83] (the authors claim "differentiation from the HSC," which is not based on the experimental design and the data presented in paper). On the basis of analysis of CD34$^+$CD38$^-$ cells (relatively enriched in HSC but still containing a high proportion of committed progenitors), Fan et al. in 2012 [84] found that the physiologically relevant O_2 concentration (5%) "down regulates the genes driving differentiation." In order to respect the physiological order, we would rather reformulate this statement: hyperoxia (20–21% O_2) up-regulates the genes driving differentiation.

4.3.2 Mesenchymal Stem Cells and Ex Vivo Oxygenation

4.3.2.1 Exposure to Low O_2 Concentrations after Amplification in Atmospheric (20–21%) O_2 Concentrations

Most studies are reporting the results obtained with so-called "Mesenchymal Stem Cells–MSC" previously amplified in cultures of stromal cells in 20% O_2 after one or more passages. This cell population is composed of several functionally distinguishable subsets [85–87] (see Chapter 1). In the further text, we will continue to term this heterogeneous cell population "Mesenchymal Stromal Cells–MStroC" and the term "MSC" will be reserved only for a subpopulation of MStroC exhibiting at least colony-forming capacity, as it has been originally established by Friedenstein in 1970 [88]. Attention should be paid to the interpretation of results issued from these studies since the cultures were exposed to atmospheric O_2 concentrations, which are hyperoxic compared to the physiological environment. Indeed, the relatively long cultivation in these conditions induces a decline in proliferative and differentiation potential of MStroC [89–91], which, in fact, results from their adaptation to high O_2 levels. Thus, the cultured cells become aerobic. However, this adaptation to high O_2 concentrations seems not to be complete since the high O_2 concentration results in some oxidative damage of the genetic material and enhanced genetic alteration-independent senescence [92], which is greatly reduced if they are cultured in physiologically relevant O_2 concentrations as shown for murine [93] and human [94] MStroC. Indeed, the gene expression and bioenergetics data strongly suggest that growth at reduced oxygen tensions (2% and 3% applied in cited studies) favors a natural metabolic state of increased glycolysis and reduced oxidative phosphorylation, which is disturbed at 20% O_2, resulting in abnormally increased levels of oxidative stress [93,94]. These data are in line with the previous study of Ohnishi et al. [95], who found that a short exposure to 1% O_2 of rat MStroC "produced" in 20% O_2 cultures (considered as "MSC" by authors) up-regulates the genes involved in glycolysis and metabolism as well as genes of several molecules involved in cell proliferation and survival [95].

An enhanced expression of energy metabolism-associated genes including GLUT-1, LDH and PDK1 and induced HIF1α expression, associated with enhanced glucose consumption was evidenced in MStroC isolated from human cord blood (passages 3 to 7, in 20% O_2) [96] cultured under low O_2 concentrations (0.5%, 2.5%, 5%). These O_2 concentrations maintained a higher proliferative capacity and significantly reduced damage or cell death. Since these are physiologic properties of MSC in physiologically relevant O_2 concentrations, we do not agree with the authors that these cultures demonstrate "altered metabolism of these human stem cells within the hypoxic environment" [96] (see Chapter 2). These data rather show the negative effect of an inadequately high oxygenation on a cell population that is primarily fitted to low O_2 environment.

In spite of aerobic adaptation of MStroC amplified in 20–21% O_2 cultures, the MSC metabolism seems to be very flexible since several studies demonstrated that the exposure of MStroC to physiologically relevant O_2 concentrations can recover their primitive capacities. The positive effect of low O_2 concentration (5%) was evidenced on the ex vivo amplification of rat bone marrow CFU-F whose number and proliferative capacity was enhanced [97] compared to 20–21%O_2 culture. The adherent stromal cells from 5% O_2 cultures (containing the Mesenchymal Stem Cells), when harvested and loaded into porous ceramic cubes and implanted into syngeneic host animals, produced more bone than cells cultured in 20% oxygen [97]. Of note, the short exposure (24 h) of bone marrow MStroC (passage two; the authors call these cells "MSC") to 1.5% O_2 increases their number and enhances chondrogenic-differentiation potential (while not altering adipogenic and osteogenic ones) [58]. More specifically, a culture in 2% O_2 favors CFU-F expansion, while maintaining MStroC immunophenotype and differentiation potential, which is associated with a rapid proliferation [98]. Similar results were obtained when human bone marrow MStroC cells were maintained in classical cultures under 2% O_2 atmosphere (after being amplified during the "0" passage in 20–21% culture) for up to seven in vitro passages [99]. These cells maintained their multilineage differentiation capabilities over 6 weeks in spite of 30-fold higher expansion fold compared to 20–21% O_2 cultures. In 2% O_2 cultures, human MStroC maintained their growth rates even after reaching confluence, resulting in the formation of multiple cell layers [99].

Another relevant approach [100] showed that the number and proliferative capacity of human bone marrow-derived MSC (CFU-F) was enhanced in three-dimensional (3D) constructs cultured up to 1 month in 2% O_2 (after being amplified during the "0" passage in 20–21% O_2 culture), which is paralleled by a slower proliferation (of note, in these cultures the pericellular O_2 concentrations should be extremely low; see Chapter 2). Upon induction of osteogenic and adipogenic differentiation, the cells previously incubated in 2% cultures showed enhanced expression of differentiation markers. This is consistent with preservation of proliferative and differentiation capacities of MSC in 2% O_2.

In this context a very interesting study of Hung et al. in 2007 [101] should be mentioned, who showed, using the MStroC frozen after the second and third passages in 20% O_2 cultures, that a 20 h exposure to 1% O_2 enhances their engraftment in 2-day-old chicken embryo (as analyzed in 5-day-old chicken embryo). With respect to the nature of this approach we believe that these data, in addition to showing an enhancement of the engraftment capacity concentration (enhanced expression of CX3CR1 and CXCR4) upon exposure to this physiologically relevant O_2, suggests also at least a partial restoring of MSC proliferative capacity. Since the number of clones (composed of a progeny resulting from several divisions of initial MSC (CFU-F)) that responded to adipogenic differentiation is fourfold lower with respect to those that responded to osteogenic differentiation, we can suppose that the different target cells are responders (presumably the clones either growing from committed progenitors or becoming committed progenitors during colony formation). In addition, if analyzed after 14 days it seems that in 1% O_2 a greater proportion of more primitive progenitors (colonies not responding to the induction of differentiation) persist until this time point. Since at day 21 all colonies differentiated in one or the other direction, it is evident that the additional 7 days were necessary to produce the cells (late committed progenitors?) capable of responding to differentiation stimuli in the colonies grown from more primitive progenitors. Upon a proper analysis, even if amplification at 20% O_2 results in a 35% higher number of CFU-F per culture, it turns out that the absolute number of primitive clones is twofold higher in 1% O_2. Similarly, the faster proliferation and better maintenance of undifferentiated characteristics and the differentiation potential of the human Wharton's jelly-derived MStroC was evidenced in 2–3% O_2 compared to 20–21% O_2 cultures from 1 to 9 passages (the passage 0 appears to have been performed in 20–21% O_2) [102], which was associated with a gene expression profile consistent with mesodermal/endothelial fate: several stem cells markers and early mesodermal/endothelial genes such as DESMIN, CD34, and ACTC were up-regulated (the total MStroC population analyzed). Of note, the MStroC cultured in 2–3% O_2 expressed higher mRNA levels of HIFs, Notch receptors, and Notch downstream gene HES1 [102]. Also, Volkmer et al. [103] showed that maintaining constant levels of oxygen (2% O_2) after passage 2 (from passage 0 to 2 the human MStroC were probably amplified in 20% O_2 cultures since they were purchased from a commercial source) improves the osteogenic potential of human MSC and suggest that low oxygen concentrations may preserve the stemness of hMSCs. But the differentiation process itself needs better oxygenation. Using a particular approach—culturing the human MStroC on "nanogratings"—Zhao et al. [104] showed that 21-day culture of the human MStroC (5–15% CFU-F, depending on condition) in 2% O_2 (after 4–5 passages in 20–21 O_2%) maintained a higher viability and differentiation capacity, doubling the CFU-F frequency.

By analyzing population of rapidly dividing MStrC (after passage 6) (mainly Sca-1^+CD44^+ and CD105^+) via loss-of-function and gain-of-function approaches it was found that HIF-1α activity is both necessary and sufficient for induction of VEGFR1 in 1% O_2 culture. Surprisingly, the HIF1α protein levels that the cells issued from 20% O_2 cultures (what authors call "basal conditions"; see Chapter 2) were sufficient to ensure some level of VEGFR1 expression [105]. Of note, the concentration of real MSC in Sca-1^+CD44^+ and CD105^+ cell fraction although certainly inferior to 50%, was obviously sufficient to give a clear signal concerning HIF-1α expression; that is, the results that are in line with those obtained with the hematopoietic cells CD34^+ [106] and Lin$^-$CD$34^+$$38^-$ cells [65] (exposure to 1.5% O_2 also induces VEGF secretion by Lin$^-$CD$34^+$$38^-$ cells [65]). Exposure of MStroC to low O_2 concentration (1%) increases secretion of VEGF, HGF, and basic FGF in an HIF-dependent manner [107]. Since both populations explored in these studies, MStroC and CD34^+CD38^-, represent heterogeneous populations in which the cells exhibiting the "stemness" properties are still a minority (or in any case not more than a half of population), it is not clear if these secretory activities concern the real MSC and HSC, committed progenitors or postprogenitor cells. On the contrary, the "anaerobic/microaerophilic" phenotype is clearly related to the maintenance of "stemness" of MSC. Tamama et al. [107] suggested in 2011 that increase of MSC colony formation related to low O_2 concentration (1%) was not HIF-dependent, which seems not to be in line with the results of Park et al. [108], who showed that constitutive stabilization of HIF-1α exerts a selective influence on CFU-F promoting their self-renewal and proliferative/differentiation capacities without affecting proliferation of the "mass population" (MStroC) [108]. On the contrary, HIF-1α stabilization in MStroC inhibits osteogenic differentiation as well as low O_2 concentration (1%) reversibly decreased osteogenic and adipogenic differentiation [107,108]. In the Tamama et al. study in 2011 [107] the arguments are presented in favor of a scenario in which in low O_2 concentration, the decrease of osteogenic differentiation, is dependent on HIF pathway, whereas decrease of adipogenic differentiation is dependent on the activation of unfolded protein response (UPR), but not HIFs.

The inhibition in low O_2 concentration (2%) of chondrogenesis and osteogenesis in especially designed cultures aimed to stimulate these differentiation processes was evidenced before, in experiments with murine adipose tissue-derived MStroC (total adherent stromal cell population previously produced in 20–21% cultures, cells used after the 2nd passage) [109]. Reported data are in line with the scenario that the higher O_2 concentrations are demanded for differentiation and maturation processes, but are not specific enough to provide the data on the target population that undergoes the differentiations since all results are obtained on "bulk" MStroC. As deduced from Hung et al. in 2007 [101] results on the basis of analysis of the single clones (see above), it is reasonable to suppose that the responders to differentiation are some kind of committed mesenchymal progenitors. All these results are in line with the decreased osteogenic, chondrogenic, and myogenic differentiation of human adipose-tissue MStroC in 2% O_2 (i.e., of their fraction responsive to induction of differentiation) [110]. Of course, the maintenance of the differentiation potential, regularly obtained in physiologically relevant O_2 concentrations, should not be confused with the inhibition of differentiation (upon its induction), also regularly obtained in the same conditions.

In the context of influence of physiologically relevant O_2 concentrations on MStroC differentiation, consider the study of Fink et al. [111], which, in our opinion, represents a "special case." MSC-TERT cells used in this study represent a lineage of human bone marrow stromal cells that was immortalized through the expression of the catalytic subunit of human telomerase and amplified for a large number of passages in 20% O_2 cultures. After previous readaptation of these cells to physiologically relevant O_2 concentration (5%) they were exposed to lower O_2 concentrations (4–1% O_2). It turned out that in the lowest O_2 concentration studied (1%) hMSC-TERT cells (that authors consider as MSC) may acquire adipocyte-mimicking morphology in the absence of true adipogenic conversion. We do not believe that any pertinent conclusion applicable on primary MSC may be derived from these results.

4.3.2.2 Death in Physiologically Relevant O_2 Concentrations and MStroC Population

Zhu et al. [112] claimed in 2006 that "hypoxia" by itself and in association with serum deprivation (3%) induces apoptosis in rat bone marrow MStroC (considered by authors "Mesenchymal Stem Cells"). Note that (1) the authors analyzed the bulk fibroblast-like adherent cells derived from rat bone marrow, a population containing less than 10% of functional MSC (CFU-F), and (2) MStroC are previously amplified in 20–21%O_2. Thus, (1) the variation in apoptosis rate can come from the cells that do not exhibit the functional properties of MSC (see Chapter 1) and (2) it probably impacts the cell fraction that undergoes a hyperaerobic adaptation to nonphysiologically high O_2 concentrations during amplification (see Chapter 2).

Induction of apoptosis in one-half of rat bone marrow MStrC population (previously produced in 20% O_2 cultures) was observed upon exposure to 0.1% O_2; preconditioning in 0.5% O_2 from 24 to 72h decreased the apoptosis rate to ~35% [113]. Similar trends with respect to apoptosis induction as well as a preconditioning effect were obtained by Peterson et al. [114], who used the rat bone marrow MStroC for up to five passages in 20% O_2 cultures before exposing them to 1% O_2. Of course, the apoptosis rate in 1% O_2 was much lower (up to ~5% Annexin$^+$ cells) comparing to one in 0.1% O_2 mentioned previously. If the rat bone marrow MStroC previously amplified in 20% O_2 (residual CD34$^+$ cells removed by immunomagnetic technique) were incubated for 24h in anoxia (0% O_2), 65% cells become apoptotic (Annexin V$^+$) and ~5% dead without apoptotic signs [115]. It is interesting that 48h incubation in 1% O_2 did not induce apoptosis of the murine adipose tissue-derived MStroC, previously amplified in 20% O_2 cultures [101,116] suggesting that the adipose-tissue MStroC might be more flexible from a metabolic viewpoint comparing to the bone marrow or the apoptosis onset that occurred before analysis time point.

To interpret these results properly, it is necessary to consider the functional heterogeneity of MStroC population. They are consistent with the scenario that the cells completely adapted to aerobiosis during the previous culture at 20–21% O_2 (less primitive ones) were impacted by physiologically relevant O_2 concentrations (misleadingly called "hypoxia"), while subpopulations inside MStroC that retained the metabolic flexibility during passages at 20% O_2 (MSC) did not undergo apoptosis. This way, an apparent contradiction related to the short-term exposure to "hypoxia," which induces apoptosis in "MSC," and the long-term exposure, which does not (interpretation of Buravkova et al. [117]), can be cleared out: the less primitive MStroC fraction that undergoes a hyperaerobic adaptation in previous 20–21% cultures does not tolerate lower O_2 concentrations (0–5% O_2) (see top panel of Figure 4.1; adapted from Buravkova et al. [117]) while the more primitive MStroC (actually the fraction corresponding to the MSC and maybe some primitive committed progenitors) retaining their metabolic flexibility (right part of the same figure) do. The first category of cells disappear at the beginning of physiologically relevant O_2 culture (the same effect has a short-term exposure) while the other one continues to proliferate more or less maintaining (depending on the actual O_2 concentration and the other culture conditions) their proliferative and differentiation capacities.

In function of further discussion, it is interesting to evoke the well-established MStroC heterogeneity including more primitive subpopulations [85,87,118–120]. The most potent clones, noticed in early passage cultures, decline in further passages and almost disappear in the late passage cultures [121] (for details see Chapter 1).

4.3.2.3 Culturing MStroC Directly in Physiologically Relevant O_2 Concentrations

When human bone marrow cells without any selection were cultured on fibronectin, directly in a physiologically relevant O_2 concentration (3% O_2), the maintenance of a primitive stem cell population was evidenced, coupled with an extensively high

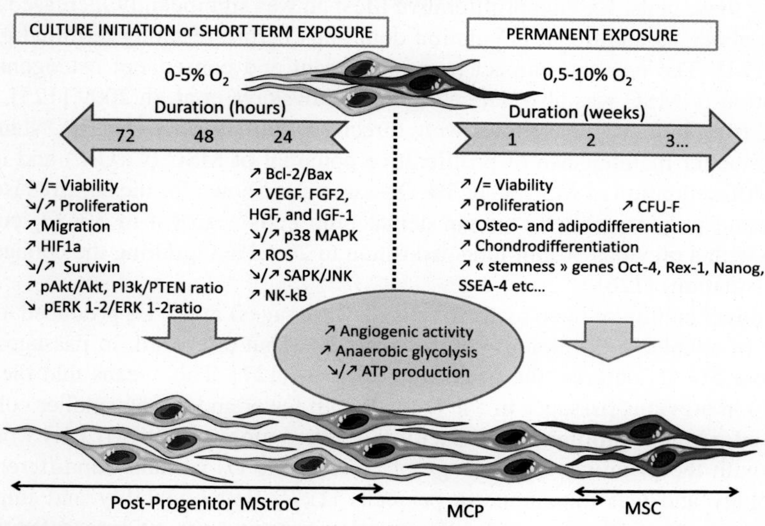

Figure 4.1 The Figure 1 from the review Buravkova et al. [117], modified and adapted as the top part of actual figure is transposed to a simplified heterogeneity model of MStroC (bottom part of actual figure). In our interpretation, the short exposure to a low O_2 concentration or the initial culture period is critical for a fraction of the MStroC, which underwent a full aerobic adaptation during their amplification (passages) in cultures exposed to 20–21% O_2. This fraction is probably composed of postprogenitor cells and relatively more mature committed progenitors (left side of bottom panel), while the primitive progenitors and real MSC are not negatively impacted by low O_2 concentrations. After a short exposure or upon initiation of a culture exposed to a physiologically relevant O_2 concentration, these more primitive cells continue to proliferate in low O_2 concentration without an increase in apoptotic fraction (absence of apoptosis in cultures permanently exposed to low O_2). In other words, at the beginning of a culture exposed to a physiologically relevant O_2 concentration (or a short exposure to the same) of MStroC previously amplified in cultures exposed to high (atmospheric) O_2 concentrations, cells that became strictly aerobic were eliminated, while those still exhibiting anaerobic/microaerophilic properties were selected and amplified in these conditions.

proliferative (>50 doublings) and differentiation potential (bone-forming osteoblasts, cartilage-forming chondrocytes, fat-forming adipocytes, neural cells, pancreatic-islet cells)—these cells were termed "Marrow-Isolated Adult Multilineage Inducible" ("MIAMI") cells [122]. Culture of these cells in 3% O_2 did not increase the rate of cell death and it ensured the maintenance of stemness (high proliferative and differentiation capacity), up-regulated mRNAs for Oct-4, REX-1, telomerase reverse transcriptase, HIF-1-α, as well as the expression of SSEA-4 compared to air O_2 condition [123]. Of note, unlike in 20% O_2, no osteogenic differentiation was detectable in 3% oxygen [123]. Similar results concerning the osteogenic (and adipogenic) differentiation were obtained with bone marrow MStroC by Fehrer et al. [124], who exposed the bone marrow cells directly to 3% O_2 (which was preceded, however, in the course of cell isolation from marrow, with a 2 h incubation with the collagenase in 20–21% O_2). A transfer from 3% O_2 to 20% O_2 cultures enabled the full differentiation of these cells. Ex vivo proliferative lifespan was significantly increased in 3% O_2 with about 10 additional population doublings before reaching terminal growth arrest [124]. The negative impact on commitment and consequent osteogenic differentiation of MSC was also noticed by Grinakovskaya et al. in 2009 [125], when MStroC from human lipoaspirates were directly cultured in 5% O_2. The "standard" feature—better maintenance of proliferative potential of MSC (CFU-F) and inhibition of differentiation of MStroC in 5% O_2—was also shown for the MStroC isolated from dental pulp (so-called "human dental pulp cells") grown in 3% O_2 cultures (though with a previous 30 min-long exposition to 20–21% O_2 during the collagenase-step of isolation) [126].

The direct culture of bone marrow cells (0–2 passages) in 5% O_2 produced a lower number of smaller-size colonies compared to 20%, but afterward, in passage 2, the cells from 5% O_2 outgrew the ones from 20% O_2 [127]. This means that the MSC (CFU-F) in previous passages in 5% O_2, although fewer and giving smaller colonies, conserved a higher proliferative capacity (i.e., were more primitive). This would be in line with the primitive morphology of cells in 5% O_2 cultures (undifferentiated morphology and less mitochondria per cell) [127]. Better viability and improved maintenance of proliferative and differentiation capacity was evidenced for murine and human MStroC derived from fat tissue when directly amplified in 2% O_2 culture with the caveat that the tissue fragments were previously exposed to atmosphere O_2 concentrations during the 2 h enzyme digestion [128,129] as well as 5% O_2 culture for ovine bone marrow MStroC [130]. Data in 2015 [131] obtained with human adipose-tissue MStroC showed that a direct culture in 2% O_2 allows a much better maintenance of their phenotypic and morphologic characteristics as well as of proliferative and differentiation capacities compared to 20–21% O_2 providing also indication that these cells have a low tendency of developing tumors under in situ normoxia (2% O_2). It should be noted that the lower percentages of apoptotic and necrotic cells were evidenced in 2% O_2 cultures [131,132]. It is also worth mentioning that no increase in apoptosis was reported in physiologically relevant O_2 concentrations in these studies.

4.3.2.4 What Can We Learn?

What can be learned about the MSC from ex vivo data obtained with MStroC cultured in physiologically relevant and high (atmospheric) O_2 concentrations? Usual analysis of these articles, which consider as mesenchymal stem cells all the cells adherent to plastic that (1) express specific surface antigens (for human cells CD105, CD73, and CD90) and do not express the other ones (for human cells CD45, CD34, CD14, CD11B, CD79alpha or CD90, HLA-DR) and (2) exhibit capacity (as tested on a bulk population) to undergo osteoblast adipocyte and chondroblast differentiation (as specified by minimal criteria standards of ISCT [133]), points out some contradiction in response of these cells to different oxygen concentrations. We believe that the reason for these "contradictions" is the fact that the functional heterogeneity of so-called "MSC population" (see Chapter 2) is not being taken into account. Indeed, the cells belonging to such a population obtained after one or more passages in cultures exposed to 20% O_2 exhibit a great functional heterogeneity. A majority of them do not have a colony-forming ability while a minority is capable of forming colonies. This minority endowed by colony forming ability was in fact initially defined as "mesenchymal stem cells" and was termed "CFU-F" [88] since it was capable of making colonies composed of fibroblast-like cells. Even this colony-forming cell population is heterogeneous with respect to the individual proliferative and differentiation cell capacities. Roughly, the two types of clonogenic cells can be distinguished: the clones with extremely high proliferative capacities giving very large colonies, and those with much lower proliferative capacities [120]. However, an important point should be stressed: the production and amplification of the so-called "mesenchymal stem cells" that we term "Mesenchymal Stromal Cells" (MSTroC) is usually performed in cultures exposed to atmospheric O_2 concentration (20–21% O_2), which represents a hyperoxic condition (i.e., is far away from physioxia; Chapter 2); thus the MStroC population, obtained this way, underwent an adaptation to hyperoxia. Hence, the functional, metabolic, and other properties of these cells cannot be considered as a physiological standard, neither for a bulk MStroC population, nor for a functionally defined MSC population. This should be taken into account for an appropriate interpretation of the results reviewed above. With that respect, the phenomenon of an onset in MStroC death (mainly by apoptosis) after exposure of cells obtained from passages 1 through 6 can be easily explained: the fraction of cells impacted by the transfer from hyperoxic conditions (wrongly called "normoxia" in most papers) to physiologically relevant O_2 concentrations (1–5%; wrongly called "hypoxia" in most papers) represents the strictly aerobic (i.e., hyperaerobic) cells [43], which are the result of an adaptation to hyperoxic conditions (20–21% O_2) (although it cannot be excluded that some of them are primarily aerobic). This fraction is most probably composed of the cells that do not exhibit stem or progenitor cell properties. Unfortunately, in the papers showing this phenomenon, the authors did not investigate the concentration of Colony Forming Cells in population surviving exposure to a low O_2 concentration. An enhancement of CFU-F frequency in surviving cell fraction would directly confirm this viewpoint.

The other set of data shows that continuous exposure of MStroC to physiologically relevant O_2 concentrations (1–5%) results in enhancement of their proliferative capacities and amplification of MStroC expansion. Considering the previous explanation concerning cell death induction by a short exposure (24–72h) to a low O_2 concentration, it could be assumed that the same phenomenon occurs at the beginning of the cultures where duration is usually 7 or more days. After "elimination" of strictly aerobic cell fraction at the beginning of the culture, the remaining cells (endowed by more or less high proliferative potential, i.e., stem or progenitor cell properties still capable to regain their original microaerophilic or some of them even anaerobic metabolic properties) continued to proliferate, most probably by asymmetric cell divisions. This way, they are producing both primitive mesenchymal stem cells (strictly anaerobic; self-renewal) and committed mesenchymal progenitors (varying from microaerophilic to aerobic intrinsic set-up). The latter could explain a constant apoptotic fraction appearing in some studies during very low O_2 cultures or even in total anoxia.

Without taking into consideration the functional heterogeneity of MStroC (usually considered as MSC), a totally wrong perception of MSC physiology was established. For example, some papers are demonstrating that the differentiation of MStroC was inhibited by physiologically relevant O_2 concentrations, which is presented as another contradiction in response of MSC to hypoxia, contrasting their maintenance of proliferative capacity. But, this phenomenon cannot be understood without taking into consideration the heterogeneity of MStroC population, since only a fraction of MStroC (probably committed progenitors or even more differentiated precursor cells) respond to the differentiation induction in 20% O_2 cultures. Differentiation is an energy-demanding process for which a higher availability of oxygen is required (exactly in the same manner as in the case of hematopoietic progenitors; see the beginning of this chapter). This feature is potentiated by adaptation of progenitors and precursors to high O_2 (20–21%) applied in expansion culture. Thus, the inhibition of differentiation in low O_2 concentrations is an expected and completely logical event, which probably, in most of the cases published and with respect to the stimulation applied, concerns a committed progenitor fraction inside the MStroC population. It could not be excluded, however, that in the case of an appropriate stimulation of both commitment and differentiation, such an inhibitory phenomenon of low O_2 concentration can affect MSC in terms of their commitment [125]. Of course, in the same culture, the proliferative and differentiation capacity of the "real" MSC is maintained, unlike in 20–21% O_2 where they are declining.

There are lots of reasons to not consider MStroC issued from 20–21% O_2 cultures as a physiological standard, which is mainly the case, since these O_2 concentrations are termed "normoxia" (Chapter 2). It was clearly shown that in such a culture, passage by passage, the real MSC population is decreasing with respect to proliferative and differentiation potential while the proportion of nonstem and progenitor cells exhibiting more and more mature morphology increases [89–91]. This is a typical artifact of nonphysiologically elevated O_2 concentration in culture. This is evident since the direct culture of primary bone marrow (or other tissue) cells in culture exposed to physiologically relevant O_2 concentrations reveals the presence of at least two categories of very primitive MSC: one is rapidly proliferating and endowed by extremely elevated proliferative and differentiation potential [122] (referred as Marrow-isolated multilineage inducible cells (MIAMI)) and the second one, composed of the cells proliferating slowly (hence giving smaller colonies at defined time point) but producing the rapidly proliferating high proliferative capacity clones in further passages [127]. It should be stressed that the appearance of these latter colony-forming cells was usually interpreted as "inhibition" of MSC proliferation. As it could be noticed the primitive MSC (we are not talking about bulk MStroC population) exhibit a functional heterogeneity that seems to be analogous to the one described for HSCs. Following this reasoning, it is even possible that a very primitive category of MSC exhibiting weak or null colony forming capacity exists. These hypothetic cells should have a high in vivo capacity in terms of regeneration of mesenchymal tissues. Even if in numerous experimental set-ups it was demonstrated that MStroCs enhance the in vivo regeneration of various tissues, the existence of such a category of MSC has not been established until now.

Setting apart the absence of a straightforward and simple in vivo model, the major problem of this approach is, again, the heterogeneity of MStroC, since the majority of these cells is represented by postprogenitor cell populations endowed by a high secretory capacity. Indeed, the secretion of cytokines and growth factors after transplantation of MStroC is recognized as a major mechanism enhancing the tissue regeneration upon their in vivo injection. These conceptual problems cannot be solved without selection of functionally homogeneous primitive MSC (elimination of poststem and progenitor cell populations) followed by an injection in vivo (various models of tissue regeneration can be exploited).

Taken together, it is evident that the low O_2 cultures (i.e., cultures in physiologically relevant concentrations) revealed the real functional hierarchy in the mesenchymal stem cell compartment, which seems to be analogous to the one already established for HSCs. The habit of naming as "mesenchymal stem cells" a population of stromal cells that is, from a functional point of view, very heterogeneous, impairs an appropriate interpretation of a large quantity of experimental data, adding to the confusion in this field that began with ascribing to MSC the properties of poststem and committed progenitor cell populations that make up a majority of the MStroC population. The necessary minimum for "adjusting" the interpretation of data is to at least respect Friedenstein's original definition of MSC. The other significant problem is the fact that nonphysiologically elevated O_2 cultures are considered as a physiological standard ("normoxia"). It is perfectly clear that in these conditions, the MStroC metabolism is altered [43]. In spite of this, the terms "alteration" and "altered" are used for the cell phenomena in physiologically relevant O_2 cultures. A reversed perception is established for the meaning of the term "adaptation." Instead of using this term for the situation when primarily anaerobic and microaerophilic MSC were cultured in nonphysiologically elevated O_2 concentrations ("hyperoxic adaptation"), it is usually used to describe the primary properties of these cells as glycolitic metabolism, low number and activity of mitochondria, and so on ("adaptation to hypoxia").

Also, in some papers, the concentration of 5% O_2 was considered as physiologically relevant, and 1% O_2 as "hypoxia." On the basis of all the data published so far it can be considered that both of these concentrations, as well as the other O_2 concentrations in the range between 1% and 5% O_2, are physiological in different microenvironment niches. Furthermore, in this range O_2 concentrations regulate balance of self-renewal/commitment of the real MSC and, at the same time, modulate the differentiation of committed mesenchymal progenitors. These progenitors demand, in turn, for full differentiation, a better oxygenation—an appropriate concentration between 5% and 13% associated with an appropriate stimulation.

4.3.3 Neuronal Stem Cells

As already mentioned at the beginning of this chapter for different categories of embryonic cells, the proliferation of rat "neural mesencephalic precursors" was promoted and apoptosis was reduced when cells were grown in 3% O_2, resulting in a greater number of "precursors" compared to "standard" 20–21% O_2 [134]. In addition, in 3% O_2 the percentage of neurons of dopaminergic phenotype increased to 56% compared with 18% in 20–21% O_2, resulting, altogether, in a ninefold net increase in dopamine neuron yield. The similar conclusion was derived from the culture in 5% O_2 of cells isolated from rat embryonic day 14.5 sciatic nerve behaving as stem cells [135] and in 3% O_2 (NCS isolated from rat embryonic mesencephalon), with an involvement of HIF-1α in elaboration of this effect [136]. Also the generation of dopamine neurons from long-term cultures of human fetal mesencephalic precursor cells exposed to 3% O_2 resulted in higher cellular yields than from cultures exposed to 20–21% O_2 [137]. It was also shown that the 20–21% O_2, unlike 5% O_2, induces apoptosis in a greater extent in multipotent cortical NCS and oligodendrocyte progenitors than in committed neuronal progenitors [138]. The physiologically relevant O_2 concentration (5%) also maintains the undifferentiated state an immortalized line with the NSC properties derived from hNSC cultures (parental cells) isolated and propagated from the diencephalic and telencephalic brain regions of a human fetus at 10.5 weeks [139]. Using the same cell line it has been demonstrated that the actual O_2 concentration (1%, 2.5%, 5%, and 20%) influences the balance between quiescence and proliferation and self-renewal and differentiation [140].

Adult neural stem cells (NSC) represent a heterogeneous population of proliferative, multipotent cells that can be isolated from specific regions of the central nervous system (CNS). Cultured NSC stem cells are endowed with the capacity to self-renew and differentiate into neurons, astrocytes, and oligodendrocytes in predictable proportions [141]. Stem cells have been isolated from many regions of the embryonic nervous system [142]. Initially, they have been found in the rostrolateral region of the lateral ventricle of the adult mouse (actively proliferating cells) [143] and in adult striatum. Adult neural stem cells have now been found in the two principal adult neurogenic regions, the hippocampus and the subventricular zone (SVZ), and in some nonneurogenic regions, including the spinal cord [144] (for review see [145]).

The publications on embryonic neuronal cells cultured in O_2 concentrations lower than atmospheric ones were followed soon after by the first report concerning the behavior of adult neuronal stem cells in culture exposed to physiologically relevant O_2 concentrations [146]. This approach is directly inspired by the data concerning the physiological oxygenation of brain tissue: the O_2 concentration varies from 0.5% in the midbrain to about 8% at pia mater [147,148] (reviewed in [149]). The "neurospheres" generated from adult forebrain were grown in a low O_2 concentration (30 mmHg = 3.9%) dissolved O_2 obtained after short flushing of a hermetic chamber with 5% CO_2 in nitrogen (0% O_2). This physiologically relevant O_2 concentration greatly enhanced the frequency of neurospheres as well as the number of neurons per sphere in parallel with the cell divisions [146] clearly showing the negative impact of nonphysiologically high atmosphere (20–21% O_2 concentration) on neuronal stem cells.

Note that the similar phenomenon was obtained on murine bone marrow hematopoietic pre-CFU-C cells in culture exposed to 0.9% O_2 [64]. This effect of low O_2 concentration was mimicked by Erythropoietin (EPO), which promoted the production of neuronal progenitors at the expense of multipotent ones [146]. Such promotion was blocked by coadministration of an EPO-neutralizing antibody. It became obvious that even in the adult organism EPO is an important factor for neuronal stem and progenitor cells (i.e., not as erythropoiesis-specific as previously considered) [137]. However, EPO is not the only factor, since Notch signaling as well as the other stemness-related pathways are implicated in maintenance of the undifferentiated (primitive) state of NCS as well as of other stem cell types (reviewed in [150]). Pistollato et al. [151] showed in 2007 that ex vivo expansion of human postnatal brain $CD133^+$ $Nestin^+$ cells was enhanced at 5% oxygen, while raising oxygen tension to 20% depleted undifferentiated cells and promoted astrocyte differentiation even in the presence of mitogens. When undifferentiated cells were expanded at 5% oxygen and then differentiated at 20% oxygen, oligodendrocyte maturation was further enhanced 2.5-fold. These results indicate that physiologically relevant O_2 concentrations act in favor of maintenance of an undifferentiated state while the higher O_2 concentrations induce exhaustion of these cells by enhanced differentiation (i.e., the same phenomenon evidenced in other stem cell types).

Furthermore, if the human embryonic cell (hEC; this term is used for the ex vivo propagated blastocyst internal mass cells) -derived neurospheres were cultured in O_2 concentration as low as 0.1% for 12 h, a significantly up-regulated HIF-1α and HIF-2α were evidenced, while producing a biphasic response within HIF targets, including erythropoietin, vascular endothelial growth factor, and Bcl-2 family members, during hypoxia and subsequent reoxygenation [152]. This cytoprotective phenotype resulted in a 50% increase in both total and neural precursor cell survival after either hydrogen peroxide insult or oxygen-glucose deprivation.

Altogether, a growing amount of evidence shows that culturing embryonic and adult NSCs in physiologically relevant O_2 concentrations not only increases cell proliferation and survival but also stimulates the dopaminergic differentiation capacity of the cells [153–159]. Furthermore the key to achieve "passagable" NSC without loss of identity; that is, maintaining their primitive properties is the combination of both absence of EGF and propagation in physiological levels (3%) of O_2 (termed appropriately "physiological normoxia" by authors) [160].

4.3.4 Muscle Satellite Cells

Satellite cells (SC) represent a distinct lineage of myogenic high proliferative capacity cells with a "satellite" position in relation to skeletal muscular fibers [161]. "Satellite cells" were considered unipotent stem cells, nonproliferating in the steady state and exhibiting the ability of the postnatal maintenance and repair of muscle [162]. However, some data suggest that they can activate the proangiogenic [163] and adipogenic [164,165] differentiation program. In vivo SC would be exposed to oxygen concentrations of ~2–5% [164]. In respect to the proliferative activity, differentiation and proliferative capacities as well as metabolic properties SC exhibit an intrinsic heterogeneity [166,167].

Soon after the first papers suggesting that physiologically relevant O_2 concentrations ex vivo facilitate a self-renewal effect on HSCs [61,63], publications appeared demonstrating similar effects with lower O_2 concentrations on SC: (1) both SC proliferation and survival of mature fibers increased in 6% O_2 versus nonphysiologic 20–21% O_2 cultures (of note, the adipogenic differentiation program is activated only in 20–21% O_2 cultures) [165] and (2) 3% O_2 cultures enhanced the individual SC clones proliferative activity (a remarkable increase in the percentage of large-sized colonies) in parallel with allowing continued cell cycle progression, resulting, ultimately, in the enhanced in vitro replicative life span [168]. In line with these data, Liu et al. [169] showed in 2012 that 1% O_2 culture of murine muscle SC promotes self-renewal divisions without affecting the overall proliferation of primary myoblasts but also diminishes the differentiated state of satellite cell-derived primary myoblasts. The experiments with the bovine muscle SC showed that a low O_2 concentration (1%, termed "hypoxia" by the authors) stimulates the proliferation of satellite cells but also promotes their myogenic differentiation [170]. Similarly, a low O_2 concentration (2%) promotes a promyogenic shift in murine muscle SC, but, one part of these cells maintained the primitiveness in this condition [171]. The expression HIF1α in SC precedes their activation and seems to be related to the proangiogenic differentiation program switch on [163]. In the context of the SC heterogeneity, and in view of the specific differentiation ways related to the anaerobic/microaerophilic metabolism, these apparently opposite simultaneous responses (maintaining of primitiveness and differentiation) may not be contradictory. Namely, SCs are comprised of at least two distinct, although not irreversibly committed, populations of cells distinguishable for prominent differences in basal biological features such as proliferation, metabolism, and differentiation [166]. Indeed, both the high proliferative (HP) and low proliferative (LP) (highly committed) SC clones were greatly enhanced in 2% compared to 20–21% O_2 cultures [167]. In addition, the stimulation in ex vivo culture may influence the predominant exhibition of one or the other effect on these two SC target populations.

It should be noted that in tissue hypoxia (see Chapter 2), which occurs after long-term exposure of organism to lowered O_2 concentrations in atmosphere (high altitudes), has a negative impact on SC: they exhibit a significantly lower ability to regenerate skeletal muscle, compared to before this high-altitude expedition [172]. This is another reason to not use the term "hypoxia" for physiologically relevant O_2 concentrations applied to the ex vivo cultures.

4.4 Conclusions

From the studies on the stem/progenitor cell systems reviewed in this chapter, after taking into consideration (1) the heterogeneity of cell populations studied, (2) type of stimulation of culture, and (3) applying a correction of the viewpoint (i.e., considering the atmospheric O_2 concentrations as not physiologic but the increased ones), it can be concluded that the O_2 concentration to which the culture is exposed has a great impact on the self-renewal, commitment, and differentiation processes. Although the real dissolved O_2 concentrations in cultures exposed to atmospheric O_2 concentrations are lower than 20–21% O_2, they still remain in most of the cultures, higher comparing to the physiologically relevant values (see Chapter 2) provoking metabolic alterations [43], increasing the ROS concentration with consequent lipid, protein, and nucleic acid damages [25,173] and activating senescence and death

Oxygen level-related response of MStroC Subpopulations

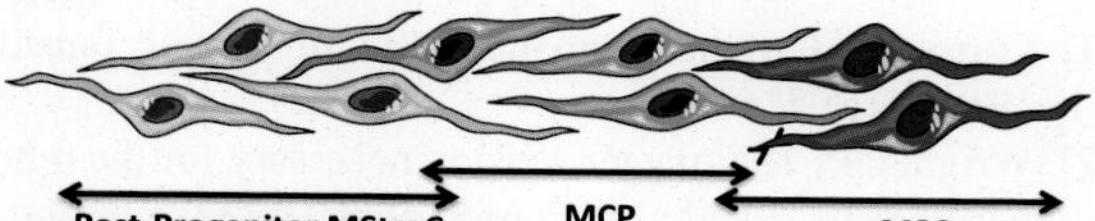

		Post-Progenitor MStroC	MCP	MSC
0-0,9 %	apoptose	+++	-	-
	Self-renewal	NA	+	++
	commitment	NA	+	+
	differentiation	-	+/-	NA
1-5 %	apoptose	++	-	-
	Self-renewal	NA	+	+++
	commitment	NA	NA	++
	differentiation	-	+/-	NA
5-13 %	apoptose	-	-	-
	Self-renewal	NA	-	++
	commitment	NA	NA	++
	differentiation	++	++	NA
20-21 %	apoptose	-	-	-
	Self-renewal	NA	-	+
	commitment	NA	NA	+++
	differentiation	+++	+++	NA

Figure 4.2 The cell stage-dependant response to the culture O_2 concentration with respect to the heterogeneity of MStroC.

[174]. On the other hand, the repeatedly obtained results showing that the stemness is conserved if the cultures are exposed to the lower O_2 concentrations (0–5%), where the dissolved O_2 level is much lower than in the gas phase, reaching even anoxia in some cases, can only stress the anaerobic/facultative anaerobic nature of stem cell populations.

It is evident that, if properly stimulated, the primitive stem cell subsets respond to the physiologically relevant O_2 concentrations by maintaining their proliferative capacities, either setting the self-renewal/commitment balance toward self-renewal or by maintaining the quiescent state (Figure 4.2). Increase in O_2 concentration from the lowest values (stem cells niche ones) toward the intermediary physiologically relevant values is premising for the commitment of stem cells (if appropriate stimulation exists), which is paralleled by their maintenance, decline, or total exhaustion in function of actual oxygenation level. The response of committed progenitors is rather different: the very low O_2 concentrations induce the quiescence of these cells, the intermediate O_2 concentrations maintain the proliferative progenitors with a decreased differentiation rate, and the higher ones enhance the differentiation process. Some committed progenitors whose differentiation specifically demands reduced oxygenation (i.e. endothelial) are stimulated to differentiate in low O_2 concentrations. Altogether the ex vivo data suggest that O_2 requirement and sensitivity to low oxygen condition depend on the developmental stage of stem cells and the way and level of commitment of progenitor cells.

References

[1] Carrel A. On the permanent life of tissues outside of the organism. J Exp Med 1912;15(5):516–28.

[2] Wright GP. The oxygen tension necessary for the mitosis of certain embryonic and neoplastic cells. J Pathol Bacteriol 1928;31:735–52.

[3] Morriss GM, New DA. Effect of oxygen concentration on morphogenesis of cranial neural folds and neural crest in cultured rat embryos. J Embryol Exp Morphol 1979;54:17–35.

[4] Hashimoto S, Minami N, Takakura R, Yamada M, Imai H, Kashima N. Low oxygen tension during in vitro maturation is beneficial for supporting the subsequent development of bovine cumulus-oocyte complexes. Mol Reprod Dev 2000;57(4):353–60.

[5] Karagenc L, Sertkaya Z, Ciray N, Ulug U, Bahceci M. Impact of oxygen concentration on embryonic development of mouse zygotes. Reprod Biomed Online 2004;9(4):409–17.

[6] Thompson JG, Simpson AC, Pugh PA, Donnelly PE, Tervit HR. Effect of oxygen concentration on in-vitro development of preimplantation sheep and cattle embryos. J Reprod Fertil 1990;89(2):573–8.

[7] Batt PA, Gardner DK, Cameron AW. Oxygen concentration and protein source affect the development of preimplantation goat embryos in vitro. Reprod Fertil Dev 1991;3(5):601–7.

[8] Rho GJ, S B, Kim DS, Son WJ, Cho SR, Kim JG, et al. Influence of in vitro oxygen concentrations on preimplantation embryo development, gene expression and production of Hanwoo calves following embryo transfer. Mol Reprod Dev 2007;74(4):486–96.

[9] Dumoulin JC, Meijers CJ, Bras M, Coonen E, Geraedts JP, Evers JL. Effect of oxygen concentration on human in-vitro fertilization and embryo culture. Hum Reprod 1999;14(2):465–9.

[10] Catt JW, Henman M. Toxic effects of oxygen on human embryo development. Hum Reprod 2000;15(Suppl. 2):199–206.

[11] García J, Sepúlveda S, Noriega-Hoces L. Beneficial effect of reduced oxygen concentration with transfer of blastocysts in IVF patients older than 40 years old. Health 2010;2:1010–7.

[12] Thomson JA, Itskovitz-Eldor J, Shapiro SS, Waknitz MA, Swiergiel JJ, Marshall VS, et al. Embryonic stem cell lines derived from human blastocysts. Science 1998;282(5391):1145–7.

[13] Adelman DM, Maltepe E, Simon MC. Multilineage embryonic hematopoiesis requires hypoxic ARNT activity. Genes Dev 1999;13(19):2478–83.

[14] Powers DE, Millman JR, Huang RB, Colton CK. Effects of oxygen on mouse embryonic stem cell growth, phenotype retention, and cellular energetics. Biotechnol Bioeng 2008;101(2):241–54.

[15] Mathieu J, Zhang Z, Nelson A, Lamba DA, Reh TA, Ware C, et al. Hypoxia induces re-entry of committed cells into pluripotency. Stem Cells 2013;31(9):1737–48.

[16] Ezashi T, Das P, Roberts RM. Low O_2 tensions and the prevention of differentiation of hES cells. Proc Natl Acad Sci USA 2005;102(13):4783–8.

[17] Forsyth NR, Musio A, Vezzoni P, Simpson AH, Noble BS, McWhir J. Physiologic oxygen enhances human embryonic stem cell clonal recovery and reduces chromosomal abnormalities. Cloning Stem Cells 2006;8(1):16–23.

[18] Forsyth NR, Kay A, Hampson K, Downing A, Talbot R, McWhir J. Transcriptome alterations due to physiological normoxic (2% O_2) culture of human embryonic stem cells. Regen Med 2008;3(6):817–33.

[19] Forristal CE, Wright KL, Hanley NA, Oreffo RO, Houghton FD. Hypoxia inducible factors regulate pluripotency and proliferation in human embryonic stem cells cultured at reduced oxygen tensions. Reproduction 2010;139(1):85–97.
[20] Lengner CJ, Gimelbrant AA, Erwin JA, Cheng AW, Guenther MG, Welstead GG, et al. Derivation of pre-X inactivation human embryonic stem cells under physiological oxygen concentrations. Cell 2010;141(5):872–83.
[21] Narva E, Pursiheimo JP, Laiho A, Rahkonen N, Emani MR, Viitala M, et al. Continuous hypoxic culturing of human embryonic stem cells enhances SSEA-3 and MYC levels. PLoS One 2013;8(11):e78847.
[22] Petruzzelli R, Christensen DR, Parry KL, Sanchez-Elsner T, Houghton FD. HIF-2alpha regulates NANOG expression in human embryonic stem cells following hypoxia and reoxygenation through the interaction with an Oct-Sox cis regulatory element. PLoS One 2014;9(10):e108309.
[23] Prado-Lopez S, Conesa A, Arminan A, Martinez-Losa M, Escobedo-Lucea C, Gandia C, et al. Hypoxia promotes efficient differentiation of human embryonic stem cells to functional endothelium. Stem Cells 2010;28(3):407–18.
[24] Koay EJ, Athanasiou KA. Hypoxic chondrogenic differentiation of human embryonic stem cells enhances cartilage protein synthesis and biomechanical functionality. Osteoarthritis Cartilage 2008;16(12):1450–6.
[25] Saretzki G, Armstrong L, Leake A, Lako M, von Zglinicki T. Stress defense in murine embryonic stem cells is superior to that of various differentiated murine cells. Stem Cells 2004;22(6):962–71.
[26] Takahashi K, Yamanaka S. Induction of pluripotent stem cells from mouse embryonic and adult fibroblast cultures by defined factors. Cell 2006;126(4):663–76.
[27] Yoshida Y, Takahashi K, Okita K, Ichisaka T, Yamanaka S. Hypoxia enhances the generation of induced pluripotent stem cells. Cell Stem Cell 2009;5(3):237–41.
[28] Shimada H, Hashimoto Y, Nakada A, Shigeno K, Nakamura T. Accelerated generation of human induced pluripotent stem cells with retroviral transduction and chemical inhibitors under physiological hypoxia. Biochem Biophys Res Commun 2012;417(2):659–64.
[29] Muchkaeva IA, Dashinimaev EB, Artyuhov AS, Myagkova EP, Vorotelyak EA, Yegorov YY, et al. Generation of iPS cells from human hair follicle dermal papilla cells. Acta Naturae 2014;6(1):45–53.
[30] Guo CW, Kawakatsu M, Idemitsu M, Urata Y, Goto S, Ono Y, et al. Culture under low physiological oxygen conditions improves the stemness and quality of induced pluripotent stem cells. J Cell Physiol 2013;228(11):2159–66.
[31] Page RL, Ambady S, Holmes WF, Vilner L, Kole D, Kashpur O, et al. Induction of stem cell gene expression in adult human fibroblasts without transgenes. Cloning Stem Cells 2009;11(3):417–26.
[32] Liu SP, Fu RH, Wu DC, Hsu CY, Chang CH, Lee W, et al. Mouse-induced pluripotent stem cells generated under hypoxic conditions in the absence of viral infection and oncogenic factors and used for ischemic stroke therapy. Stem Cells Dev 2014;23(4):421–33.
[33] Sugimoto K, Yoshizawa Y, Yamada S, Igawa K, Hayashi Y, Ishizaki H. Effects of hypoxia on pluripotency in murine iPS cells. Microsc Res Tech 2013;76(10):1084–92.
[34] Cooper PD, Burt AM, Wilson JN. Critical effect of oxygen tension on rate of growth of animal cells in continuous suspended culture. Nature 1958;182(4648):1508–9.
[35] Balin AK, Goodman BP, Rasmussen H, Cristofalo VJ. The effect of oxygen tension on the growth and metabolism of WI-38 cells. J Cell Physiol 1976;89(2):235–49.
[36] Balin AK, Fisher AJ, Carter DM. Oxygen modulates growth of human cells at physiologic partial pressures. J Exp Med 1984;160(1):152–66.

[37] Zwartouw HT, Westwood JC. Factors affecting growth and glycolysis in tissue culture. Br J Exp Pathol 1958;39(5):529–39.
[38] Clark ME. Growth and morphology of adult mouse fibroblasts under anaerobic conditions and at limiting oxygen tensions. Exp Cell Res 1964;36:548–60.
[39] Kilburn DG, Lilly MD, Self DA, Webb FC. The effect of dissolved oxygen partial pressure on the growth and carbohydrate metabolism of mouse LS cells. J Cell Sci 1969;4(1):25–37.
[40] Richter A, Sanford KK, Evans VJ. Influence of oxygen and culture media on plating efficiency of some mammalian tissue cells. J Natl Cancer Inst 1972;49(6):1705–12.
[41] Packer L, Fuehr K. Low oxygen concentration extends the lifespan of cultured human diploid cells. Nature 1977;267(5610):423–5.
[42] Roy S, Khanna S, Bickerstaff AA, Subramanian SV, Atalay M, Bierl M, et al. Oxygen sensing by primary cardiac fibroblasts: a key role of p21(Waf1/Cip1/Sdi1). Circ Res 2003;92(3):264–71.
[43] Gnaiger E. Oxygen conformance of cellular respiration. A perspective of mitochondrial physiology. Adv Exp Med Biol 2003;543:39–55.
[44] Till JE, Mc CE. A direct measurement of the radiation sensitivity of normal mouse bone marrow cells. Radiat Res 1961;14:213–22.
[45] Rosendaal M, Hodgson GS, Bradley TR. Organization of haemopoietic stem cells: the generation-age hypothesis. Cell Tissue Kinet 1979;12(1):17–29.
[46] Taylor WG, Camalier RF. Modulation of epithelial cell proliferation in culture by dissolved oxygen. J Cell Physiol 1982;111(1):21–7.
[47] Bath C. Human corneal epithelial subpopulations: oxygen dependent ex vivo expansion and transcriptional profiling. Acta Ophthalmol 2013;91(Thesis 4):1–34.
[48] Magnussen JD. The influence of oxygen tension on the production of erythrocytes in vitro. Acta Pharmacol Toxicol (Copenh) 1949;5(2):153–63.
[49] Shibata T. Studies on erythropoiesis. II. In vitro studies on red cell proliferation under varied oxygen tension. Acta Med Okayama 1964;18:179–88.
[50] Pulvertaft RJ. The effect of reduced oxygen tension on platelet formation in vitro. J Clin Pathol 1958;11(6):535–42.
[51] Bradley TR, Hodgson GS, Rosendaal M. The effect of oxygen tension on haemopoietic and fibroblast cell proliferation in vitro. J Cell Physiol 1978;97(3 Pt 2 Suppl. 1):517–22.
[52] Bradley TR, Hodgson GS. Detection of primitive macrophage progenitor cells in mouse bone marrow. Blood 1979;54(6):1446–50.
[53] Hodgson GS, Bradley TR, Radley JM. In vitro production of CFU-S and cells with erythropoiesis repopulating ability by 5-fluorouracil treated mouse bone marrow. Int J Cell Cloning 1983;1(1):49–56.
[54] Rich IN, Kubanek B. The effect of reduced oxygen tension on colony formation of erythropoietic cells in vitro. Br J Haematol 1982;52(4):579–88.
[55] Maeda H, Hotta T, Yamada H. Enhanced colony formation of human hemopoietic stem cells in reduced oxygen tension. Exp Hematol 1986;14(10):930–4.
[56] Katahira J, Mizoguchi H. Improvement of culture conditions for human megakaryocytic and pluripotent progenitor cells by low oxygen tension. Int J Cell Cloning 1987;5(5):412–20.
[57] Ishikawa Y, Ito T. Kinetics of hemopoietic stem cells in a hypoxic culture. Eur J Haematol 1988;40(2):126–9.
[58] Martin-Rendon E, Hale SJ, Ryan D, Baban D, Forde SP, Roubelakis M, et al. Transcriptional profiling of human cord blood CD133^{+} and cultured bone marrow mesenchymal stem cells in response to hypoxia. Stem Cells 2007;25(4):1003–12.

[59] Moussavi-Harami F, Duwayri Y, Martin JA, Moussavi-Harami F, Buckwalter JA. Oxygen effects on senescence in chondrocytes and mesenchymal stem cells: consequences for tissue engineering. Iowa Orthop J 2004;24:15–20.
[60] Cipolleschi MG, Dello Sbarba P, Olivotto M. The role of hypoxia in the maintenance of hematopoietic stem cells. Blood 1993;82(7):2031–7.
[61] Ivanovic Z, Bartolozzi B, Bernabei PA, Cipolleschi MG, Rovida E, Milenkovic P, et al. Incubation of murine bone marrow cells in hypoxia ensures the maintenance of marrow-repopulating ability together with the expansion of committed progenitors. Br J Haematol 2000;108(2):424–9.
[62] Cipolleschi MG, Rovida E, Ivanovic Z, Praloran V, Olivotto M, Dello Sbarba P. The expansion of murine bone marrow cells preincubated in hypoxia as an in vitro indicator of their marrow-repopulating ability. Leukemia 2000;14(4):735–9.
[63] Ivanovic Z, Dello Sbarba P, Trimoreau F, Faucher JL, Praloran V. Primitive human HPCs are better maintained and expanded in vitro at 1 percent oxygen than at 20 percent. Transfusion 2000;40(12):1482–8.
[64] Ivanovic Z, Belloc F, Faucher JL, Cipolleschi MG, Praloran V, Dello Sbarba P. Hypoxia maintains and interleukin-3 reduces the pre-colony-forming cell potential of dividing CD34(+) murine bone marrow cells. Exp Hematol 2002;30(1):67–73.
[65] Danet GH, Pan Y, Luongo JL, Bonnet DA, Simon MC. Expansion of human SCID-repopulating cells under hypoxic conditions. J Clin Invest 2003;112(1):126–35.
[66] Eliasson P, Karlsson R, Jönsson JI. Hypoxia expands primitive progenitors cells from mouse bone marrow during in vitro culture and preserves the colony-forming ability. In: Koka PS, editor. Stem cell research compendium, vol. 1. Nova Science Publishers; 2008. p. 329–42.
[67] Kovacevic-Filipovic M, Petakov M, Hermitte F, Debeissat C, Krstic A, Jovcic G, et al. Interleukin-6 (IL-6) and low O(2) concentration (1%) synergize to improve the maintenance of hematopoietic stem cells (pre-CFC). J Cell Physiol 2007;212(1):68–75.
[68] Ivanovic Z, Hermitte F, Brunet de la Grange P, Dazey B, Belloc F, Lacombe F, et al. Simultaneous maintenance of human cord blood SCID-repopulating cells and expansion of committed progenitors at low O_2 concentration (3%). Stem Cells 2004;22(5):716–24.
[69] Roy S, Tripathy M, Mathur N, Jain A, Mukhopadhyay A. Hypoxia improves expansion potential of human cord blood-derived hematopoietic stem cells and marrow repopulation efficiency. Eur J Haematol 2012;88(5):396–405.
[70] LaIuppa JA, Papoutsakis ET, Miller WM. Oxygen tension alters the effects of cytokines on the megakaryocyte, erythrocyte, and granulocyte lineages. Exp Hematol 1998;26(9):835–43.
[71] Mostafa SS, Miller WM, Papoutsakis ET. Oxygen tension influences the differentiation, maturation and apoptosis of human megakaryocytes. Br J Haematol 2000;111(3):879–89.
[72] Vlaski M, Lafarge X, Chevaleyre J, Duchez P, Boiron JM, Ivanovic Z. Low oxygen concentration as a general physiologic regulator of erythropoiesis beyond the EPO-related downstream tuning and a tool for the optimization of red blood cell production ex vivo. Exp Hematol 2009;37(5):573–84.
[73] Smith S, Broxmeyer HE. The influence of oxygen tension on the long-term growth in vitro of haematopoietic progenitor cells from human cord blood. Br J Haematol 1986;63(1):29–34.
[74] Koller MR, Bender JG, Miller WM, Papoutsakis ET. Reduced oxygen tension increases hematopoiesis in long-term culture of human stem and progenitor cells from cord blood and bone marrow. Exp Hematol 1992;20(2):264–70.

[75] Hammoud M, Vlaski M, Duchez P, Chevaleyre J, Lafarge X, Boiron JM, et al. Combination of low O(2) concentration and mesenchymal stromal cells during culture of cord blood CD34(+) cells improves the maintenance and proliferative capacity of hematopoietic stem cells. J Cell Physiol 2012;227(6):2750–8.
[76] Zhambalova AP, Darevskaya AN, Kabaeva NV, Romanov YA, Buravkova LB. Specific interaction of cultured human mesenchymal and hemopoietic stem cells under conditions of reduced oxygen content. Bull Exp Biol Med 2009;147(4):525–30.
[77] Kiani AA, Abdi J, Halabian R, Roudkenar MH, Amirizadeh N, Soleiman Soltanpour M, et al. Over expression of HIF-1alpha in human mesenchymal stem cells increases their supportive functions for hematopoietic stem cells in an experimental co-culture model. Hematology 2014;19(2):85–98.
[78] Krstic A, Vlaski M, Hammoud M, Chevaleyre J, Duchez P, Jovcic G, et al. Low O_2 concentrations enhance the positive effect of IL-17 on the maintenance of erythroid progenitors during co-culture of $CD34^+$ and mesenchymal stem cells. Eur Cytokine Netw 2009;20(1):10–6.
[79] Jing D, Wobus M, Poitz DM, Bornhauser M, Ehninger G, Ordemann R. Oxygen tension plays a critical role in the hematopoietic microenvironment in vitro. Haematologica 2012;97(3):331–9.
[80] Cipolleschi MG, D'Ippolito G, Bernabei PA, Caporale R, Nannini R, Mariani M, et al. Severe hypoxia enhances the formation of erythroid bursts from human cord blood cells and the maintenance of BFU-E in vitro. Exp Hematol 1997;25(11):1187–94.
[81] Browne SM, Daud H, Murphy WG, Al-Rubeai M. Measuring dissolved oxygen to track erythroid differentiation of hematopoietic progenitor cells in culture. J Biotechnol 2014;187:135–8.
[82] Mostafa SS, Papoutsakis ET, Miller WM. Oxygen tension modulates the expression of cytokine receptors, transcription factors, and lineage-specific markers in cultured human megakaryocytes. Exp Hematol 2001;29(7):873–83.
[83] Yun S, Lee SH, Yoon SR, Myung PK, Choi I. Oxygen tension regulates NK cells differentiation from hematopoietic stem cells in vitro. Immunol Lett 2011;137(1–2):70–7.
[84] Fan J, Cai H, Li Q, Du Z, Tan W. The effects of ROS-mediating oxygen tension on human CD34(+)CD38(−) cells induced into mature dendritic cells. J Biotechnol 2012;158(3):104–11.
[85] Colter DC, Sekiya I, Prockop DJ. Identification of a subpopulation of rapidly self-renewing and multipotential adult stem cells in colonies of human marrow stromal cells. Proc Natl Acad Sci USA 2001;98(14):7841–5.
[86] Prockop DJ, Sekiya I, Colter DC. Isolation and characterization of rapidly self-renewing stem cells from cultures of human marrow stromal cells. Cytotherapy 2001;3(5):393–6.
[87] Smith JR, Pochampally R, Perry A, Hsu SC, Prockop DJ. Isolation of a highly clonogenic and multipotential subfraction of adult stem cells from bone marrow stroma. Stem Cells 2004;22(5):823–31.
[88] Friedenstein AJ, Chailakhjan RK, Lalykina KS. The development of fibroblast colonies in monolayer cultures of guinea-pig bone marrow and spleen cells. Cell Tissue Kinet 1970;3(4):393–403.
[89] Vacanti V, Kong E, Suzuki G, Sato K, Canty JM, Lee T. Phenotypic changes of adult porcine mesenchymal stem cells induced by prolonged passaging in culture. J Cell Physiol 2005;205(2):194–201.
[90] Larson BL, Ylostalo J, Lee RH, Gregory C, Prockop DJ. Sox11 is expressed in early progenitor human multipotent stromal cells and decreases with extensive expansion of the cells. Tissue Eng Part A 2010;16(11):3385–94.

[91] Larson BL, Ylostalo J, Prockop DJ. Human multipotent stromal cells undergo sharp transition from division to development in culture. Stem Cells 2008;26(1):193–201.
[92] Tarte K, Gaillard J, Lataillade JJ, Fouillard L, Becker M, Mossafa H, et al. Clinical-grade production of human mesenchymal stromal cells: occurrence of aneuploidy without transformation. Blood 2010;115(8):1549–53.
[93] Fan G, Wen L, Li M, Li C, Luo B, Wang F, et al. Isolation of mouse mesenchymal stem cells with normal ploidy from bone marrows by reducing oxidative stress in combination with extracellular matrix. BMC Cell Biol 2011;12:30.
[94] Estrada JC, Albo C, Benguria A, Dopazo A, Lopez-Romero P, Carrera-Quintanar L, et al. Culture of human mesenchymal stem cells at low oxygen tension improves growth and genetic stability by activating glycolysis. Cell Death Differ 2012;19(5):743–55.
[95] Ohnishi S, Yasuda T, Kitamura S, Nagaya N. Effect of hypoxia on gene expression of bone marrow-derived mesenchymal stem cells and mononuclear cells. Stem Cells 2007;25(5):1166–77.
[96] Lavrentieva A, Majore I, Kasper C, Hass R. Effects of hypoxic culture conditions on umbilical cord-derived human mesenchymal stem cells. Cell Commun Signal 2010;8:18.
[97] Lennon DP, Edmison JM, Caplan AI. Cultivation of rat marrow-derived mesenchymal stem cells in reduced oxygen tension: effects on in vitro and in vivo osteochondrogenesis. J Cell Physiol 2001;187(3):345–55.
[98] Dos Santos F, Andrade PZ, Boura JS, Abecasis MM, da Silva CL, Cabral JM. Ex vivo expansion of human mesenchymal stem cells: a more effective cell proliferation kinetics and metabolism under hypoxia. J Cell Physiol 2010;223(1):27–35.
[99] Grayson WL, Zhao F, Bunnell B, Ma T. Hypoxia enhances proliferation and tissue formation of human mesenchymal stem cells. Biochem Biophys Res Commun 2007;358(3):948–53.
[100] Grayson WL, Zhao F, Izadpanah R, Bunnell B, Ma T. Effects of hypoxia on human mesenchymal stem cell expansion and plasticity in 3D constructs. J Cell Physiol 2006;207(2):331–9.
[101] Hung SC, Pochampally RR, Hsu SC, Sanchez C, Chen SC, Spees J, et al. Short-term exposure of multipotent stromal cells to low oxygen increases their expression of CX3CR1 and CXCR4 and their engraftment in vivo. PLoS One 2007;2(5):e416.
[102] Nekanti U, Dastidar S, Venugopal P, Totey S, Ta M. Increased proliferation and analysis of differential gene expression in human Wharton's jelly-derived mesenchymal stromal cells under hypoxia. Int J Biol Sci 2010;6(5):499–512.
[103] Volkmer E, Kallukalam BC, Maertz J, Otto S, Drosse I, Polzer H, et al. Hypoxic preconditioning of human mesenchymal stem cells overcomes hypoxia-induced inhibition of osteogenic differentiation. Tissue Eng Part A 2010;16(1):153–64.
[104] Zhao F, Veldhuis JJ, Duan Y, Yang Y, Christoforou N, Ma T, et al. Low oxygen tension and synthetic nanogratings improve the uniformity and stemness of human mesenchymal stem cell layer. Mol Ther 2010;18(5):1010–8.
[105] Okuyama H, Krishnamachary B, Zhou YF, Nagasawa H, Bosch-Marce M, Semenza GL. Expression of vascular endothelial growth factor receptor 1 in bone marrow-derived mesenchymal cells is dependent on hypoxia-inducible factor 1. J Biol Chem 2006;281(22):15554–63.
[106] Piccoli C, D'Aprile A, Ripoli M, Scrima R, Boffoli D, Tabilio A, et al. The hypoxia-inducible factor is stabilized in circulating hematopoietic stem cells under normoxic conditions. FEBS Lett 2007;581(16):3111–9.
[107] Tamama K, Kawasaki H, Kerpedjieva SS, Guan J, Ganju RK, Sen CK. Differential roles of hypoxia inducible factor subunits in multipotential stromal cells under hypoxic condition. J Cell Biochem 2011;112(3):804–17.

[108] Park IH, Kim KH, Choi HK, Shim JS, Whang SY, Hahn SJ, et al. Constitutive stabilization of hypoxia-inducible factor alpha selectively promotes the self-renewal of mesenchymal progenitors and maintains mesenchymal stromal cells in an undifferentiated state. Exp Mol Med 2013;45:e44.
[109] Malladi P, Xu Y, Chiou M, Giaccia AJ, Longaker MT. Effect of reduced oxygen tension on chondrogenesis and osteogenesis in adipose-derived mesenchymal cells. Am J Physiol Cell Physiol 2006;290(4):C1139–46.
[110] Lee JH, Kemp DM. Human adipose-derived stem cells display myogenic potential and perturbed function in hypoxic conditions. Biochem Biophys Res Commun 2006;341(3):882–8.
[111] Fink T, Abildtrup L, Fogd K, Abdallah BM, Kassem M, Ebbesen P, et al. Induction of adipocyte-like phenotype in human mesenchymal stem cells by hypoxia. Stem Cells 2004;22(7):1346–55.
[112] Zhu W, Chen J, Cong X, Hu S, Chen X. Hypoxia and serum deprivation-induced apoptosis in mesenchymal stem cells. Stem Cells 2006;24(2):416–25.
[113] Chacko SM, Ahmed S, Selvendiran K, Kuppusamy ML, Khan M, Kuppusamy P. Hypoxic preconditioning induces the expression of prosurvival and proangiogenic markers in mesenchymal stem cells. Am J Physiol Cell Physiol 2010;299(6):C1562–70.
[114] Peterson KM, Aly A, Lerman A, Lerman LO, Rodriguez-Porcel M. Improved survival of mesenchymal stromal cell after hypoxia preconditioning: role of oxidative stress. Life Sci 2011;88(1–2):65–73.
[115] Chang W, Song BW, Lim S, Song H, Shim CY, Cha MJ, et al. Mesenchymal stem cells pretreated with delivered Hph-1-Hsp70 protein are protected from hypoxia-mediated cell death and rescue heart functions from myocardial injury. Stem Cells 2009;27(9):2283–92.
[116] Efimenko A, Starostina E, Kalinina N, Stolzing A. Angiogenic properties of aged adipose derived mesenchymal stem cells after hypoxic conditioning. J Transl Med 2011;9:10.
[117] Buravkova LB, Andreeva ER, Gogvadze V, Zhivotovsky B. Mesenchymal stem cells and hypoxia: where are we? Mitochondrion 2014;19(Pt A):105–12.
[118] Colter DC, Class R, DiGirolamo CM, Prockop DJ. Rapid expansion of recycling stem cells in cultures of plastic-adherent cells from human bone marrow. Proc Natl Acad Sci USA 2000;97(7):3213–8.
[119] Conget PA, Allers C, Minguell JJ. Identification of a discrete population of human bone marrow-derived mesenchymal cells exhibiting properties of uncommitted progenitors. J Hematother Stem Cell Res 2001;10(6):749–58.
[120] Mabuchi Y, Morikawa S, Harada S, Niibe K, Suzuki S, Renault-Mihara F, et al. LNGFR(+)THY-1(+)VCAM-1(hi+) cells reveal functionally distinct subpopulations in mesenchymal stem cells. Stem Cell Rep 2013;1(2):152–65.
[121] Mets T, Verdonk G. In vitro aging of human bone marrow derived stromal cells. Mech Ageing Dev 1981;16(1):81–9.
[122] D'Ippolito G, Diabira S, Howard GA, Menei P, Roos BA, Schiller PC. Marrow-isolated adult multilineage inducible (MIAMI) cells, a unique population of postnatal young and old human cells with extensive expansion and differentiation potential. J Cell Sci 2004;117(Pt 14):2971–81.
[123] D'Ippolito G, Diabira S, Howard GA, Roos BA, Schiller PC. Low oxygen tension inhibits osteogenic differentiation and enhances stemness of human MIAMI cells. Bone 2006;39(3):513–22.
[124] Fehrer C, Brunauer R, Laschober G, Unterluggauer H, Reitinger S, Kloss F, et al. Reduced oxygen tension attenuates differentiation capacity of human mesenchymal stem cells and prolongs their lifespan. Aging Cell 2007;6(6):745–57.

[125] Grinakovskaya OS, Andreeva ER, Buravkova LB, Rylova YV, Kosovsky GY. Low level of O_2 inhibits commitment of cultured mesenchymal stromal precursor cells from the adipose tissue in response to osteogenic stimuli. Bull Exp Biol Med 2009;147(6):760–3.

[126] Iida K, Takeda-Kawaguchi T, Tezuka Y, Kunisada T, Shibata T, Tezuka K. Hypoxia enhances colony formation and proliferation but inhibits differentiation of human dental pulp cells. Arch Oral Biol 2010;55(9):648–54.

[127] Basciano L, Nemos C, Foliguet B, de Isla N, de Carvalho M, Tran N, et al. Long term culture of mesenchymal stem cells in hypoxia promotes a genetic program maintaining their undifferentiated and multipotent status. BMC Cell Biol 2011;12:12.

[128] Valorani MG, Germani A, Otto WR, Harper L, Biddle A, Khoo CP, et al. Hypoxia increases Sca-1/CD44 co-expression in murine mesenchymal stem cells and enhances their adipogenic differentiation potential. Cell Tissue Res 2010;341(1):111–20.

[129] Valorani MG, Montelatici E, Germani A, Biddle A, D'Alessandro D, Strollo R, et al. Pre-culturing human adipose tissue mesenchymal stem cells under hypoxia increases their adipogenic and osteogenic differentiation potentials. Cell Prolif 2012;45(3):225–38.

[130] Zscharnack M, Poesel C, Galle J, Bader A. Low oxygen expansion improves subsequent chondrogenesis of ovine bone-marrow-derived mesenchymal stem cells in collagen type I hydrogel. Cells Tissues Organs 2009;190(2):81–93.

[131] Choi JR, Pingguan-Murphy B, Wan Abas WA, Yong KW, Poon CT, Noor Azmi MA, et al. In situ normoxia enhances survival and proliferation rate of human adipose tissue-derived stromal cells without increasing the risk of tumourigenesis. PLoS One 2015;10(1):e0115034.

[132] Choi JR, Pingguan-Murphy B, Wan Abas WA, Noor Azmi MA, Omar SZ, Chua KH, et al. Impact of low oxygen tension on stemness, proliferation and differentiation potential of human adipose-derived stem cells. Biochem Biophys Res Commun 2014;448(2):218–24.

[133] Dominici M, Le Blanc K, Mueller I, Slaper-Cortenbach I, Marini F, Krause D, et al. Minimal criteria for defining multipotent mesenchymal stromal cells. The International Society for Cellular Therapy position statement. Cytotherapy 2006;8(4):315–7.

[134] Studer L, Csete M, Lee SH, Kabbani N, Walikonis J, Wold B, et al. Enhanced proliferation, survival, and dopaminergic differentiation of CNS precursors in lowered oxygen. J Neurosci 2000;20(19):7377–83.

[135] Morrison SJ, Csete M, Groves AK, Melega W, Wold B, Anderson DJ. Culture in reduced levels of oxygen promotes clonogenic sympathoadrenal differentiation by isolated neural crest stem cells. J Neurosci 2000;20(19):7370–6.

[136] Zhang CP, Zhu LL, Zhao T, Zhao H, Huang X, Ma X, et al. Characteristics of neural stem cells expanded in lowered oxygen and the potential role of hypoxia-inducible factor-1Alpha. Neurosignals 2006;15(5):259–65.

[137] Storch A, Paul G, Csete M, Boehm BO, Carvey PM, Kupsch A, et al. Long-term proliferation and dopaminergic differentiation of human mesencephalic neural precursor cells. Exp Neurol 2001;170(2):317–25.

[138] Chen HL, Pistollato F, Hoeppner DJ, Ni HT, McKay RD, Panchision DM. Oxygen tension regulates survival and fate of mouse central nervous system precursors at multiple levels. Stem Cells 2007;25(9):2291–301.

[139] De Filippis L, Lamorte G, Snyder EY, Malgaroli A, Vescovi AL. A novel, immortal, and multipotent human neural stem cell line generating functional neurons and oligodendrocytes. Stem Cells 2007;25(9):2312–21.

[140] Santilli G, Lamorte G, Carlessi L, Ferrari D, Rota Nodari L, Binda E, et al. Mild hypoxia enhances proliferation and multipotency of human neural stem cells. PLoS One 2010;5(1):e8575.

[141] Vescovi AL, Parati EA, Gritti A, Poulin P, Ferrario M, Wanke E, et al. Isolation and cloning of multipotential stem cells from the embryonic human CNS and establishment of transplantable human neural stem cell lines by epigenetic stimulation. Exp Neurol 1999;156(1):71–83.
[142] Temple S. The development of neural stem cells. Nature 2001;414(6859):112–7.
[143] Morshead CM, van der Kooy D. Postmitotic death is the fate of constitutively proliferating cells in the subependymal layer of the adult mouse brain. J Neurosci 1992;12(1):249–56.
[144] Reynolds BA, Weiss S. Generation of neurons and astrocytes from isolated cells of the adult mammalian central nervous system. Science 1992;255(5052):1707–10.
[145] Lim DA, Alvarez-Buylla A. Adult neural stem cells stake their ground. Trends Neurosci 2014;37(10):563–71.
[146] Shingo T, Sorokan ST, Shimazaki T, Weiss S. Erythropoietin regulates the in vitro and in vivo production of neuronal progenitors by mammalian forebrain neural stem cells. J Neurosci 2001;21(24):9733–43.
[147] Dings J, Meixensberger J, Jager A, Roosen K. Clinical experience with 118 brain tissue oxygen partial pressure catheter probes. Neurosurgery 1998;43(5):1082–95.
[148] Jorgensen JR, Juliusson B, Henriksen KF, Hansen C, Knudsen S, Petersen TN, et al. Identification of novel genes regulated in the developing human ventral mesencephalon. Exp Neurol 2006;198(2):427–37.
[149] Csete M. Oxygen in the cultivation of stem cells. Ann N Y Acad Sci 2005;1049:1–8.
[150] Gustafsson MV, Zheng X, Pereira T, Gradin K, Jin S, Lundkvist J, et al. Hypoxia requires notch signaling to maintain the undifferentiated cell state. Dev cell 2005;9(5):617–28.
[151] Pistollato F, Chen HL, Schwartz PH, Basso G, Panchision DM. Oxygen tension controls the expansion of human CNS precursors and the generation of astrocytes and oligodendrocytes. Mol Cell Neurosci 2007;35(3):424–35.
[152] Francis KR, Wei L. Human embryonic stem cell neural differentiation and enhanced cell survival promoted by hypoxic preconditioning. Cell Death Dis 2010;1:e22.
[153] Jensen P, Gramsbergen JB, Zimmer J, Widmer HR, Meyer M. Enhanced proliferation and dopaminergic differentiation of ventral mesencephalic precursor cells by synergistic effect of FGF2 and reduced oxygen tension. Exp Cell Res 2011;317(12):1649–62.
[154] Krabbe C, Bak ST, Jensen P, von Linstow C, Martinez Serrano A, Hansen C, et al. Influence of oxygen tension on dopaminergic differentiation of human fetal stem cells of midbrain and forebrain origin. PLoS One 2014;9(5):e96465.
[155] Krabbe C, Courtois E, Jensen P, Jorgensen JR, Zimmer J, Martinez-Serrano A, et al. Enhanced dopaminergic differentiation of human neural stem cells by synergistic effect of Bcl-xL and reduced oxygen tension. J Neurochem 2009;110(6):1908–20.
[156] Liu S, Tian Z, Yin F, Zhao Q, Fan M. Generation of dopaminergic neurons from human fetal mesencephalic progenitors after co-culture with striatal-conditioned media and exposure to lowered oxygen. Brain Res Bull 2009;80(1–2):62–8.
[157] Stacpoole SR, Bilican B, Webber DJ, Luzhynskaya A, He XL, Compston A, et al. Efficient derivation of NPCs, spinal motor neurons and midbrain dopaminergic neurons from hESCs at 3% oxygen. Nat Protoc 2011;6(8):1229–40.
[158] Stacpoole SR, Webber DJ, Bilican B, Compston A, Chandran S, Franklin RJ. Neural precursor cells cultured at physiologically relevant oxygen tensions have a survival advantage following transplantation. Stem Cells Transl Med 2013;2(6):464–72.
[159] Milosevic J, Schwarz SC, Krohn K, Poppe M, Storch A, Schwarz J. Low atmospheric oxygen avoids maturation, senescence and cell death of murine mesencephalic neural precursors. J Neurochem 2005;92(4):718–29.

[160] Bilican B, Livesey MR, Haghi G, Qiu J, Burr K, Siller R, et al. Physiological normoxia and absence of EGF is required for the long-term propagation of anterior neural precursors from human pluripotent cells. PLoS One 2014;9(1):e85932.
[161] Mauro A. Satellite cell of skeletal muscle fibers. J Biophys Biochem Cytol 1961;9:493–5.
[162] Schultz E, Gibson MC, Champion T. Satellite cells are mitotically quiescent in mature mouse muscle: an EM and radioautographic study. J Exp Zool 1978;206(3):451–6.
[163] Rhoads RP, Johnson RM, Rathbone CR, Liu X, Temm-Grove C, Sheehan SM, et al. Satellite cell-mediated angiogenesis in vitro coincides with a functional hypoxia-inducible factor pathway. Am J Physiol Cell Physiol 2009;296(6):C1321–8.
[164] Redshaw Z, Loughna PT. Oxygen concentration modulates the differentiation of muscle stem cells toward myogenic and adipogenic fates. Differentiation 2012;84(2): 193–202.
[165] Csete M, Walikonis J, Slawny N, Wei Y, Korsnes S, Doyle JC, et al. Oxygen-mediated regulation of skeletal muscle satellite cell proliferation and adipogenesis in culture. J Cell Physiol 2001;189(2):189–96.
[166] Rossi CA, Pozzobon M, Ditadi A, Archacka K, Gastaldello A, Sanna M, et al. Clonal characterization of rat muscle satellite cells: proliferation, metabolism and differentiation define an intrinsic heterogeneity. PLoS One 2010;5(1):e8523.
[167] Urbani L, Piccoli M, Franzin C, Pozzobon M, De Coppi P. Hypoxia increases mouse satellite cell clone proliferation maintaining both in vitro and in vivo heterogeneity and myogenic potential. PLoS One 2012;7(11):e49860.
[168] Chakravarthy MV, Spangenburg EE, Booth FW. Culture in low levels of oxygen enhances in vitro proliferation potential of satellite cells from old skeletal muscles. Cell Mol Life Sci 2001;58(8):1150–8.
[169] Liu W, Wen Y, Bi P, Lai X, Liu XS, Liu X, et al. Hypoxia promotes satellite cell self-renewal and enhances the efficiency of myoblast transplantation. Development 2012;139(16):2857–65.
[170] Kook SH, Son YO, Lee KY, Lee HJ, Chung WT, Choi KC, et al. Hypoxia affects positively the proliferation of bovine satellite cells and their myogenic differentiation through up-regulation of MyoD. Cell Biol Int 2008;32(8):871–8.
[171] Koning M, Werker PM, van Luyn MJ, Harmsen MC. Hypoxia promotes proliferation of human myogenic satellite cells: a potential benefactor in tissue engineering of skeletal muscle. Tissue Eng Part A 2011;17(13–14):1747–58.
[172] Mancinelli R, Pietrangelo T, La Rovere R, Toniolo L, Fano G, Reggiani C, et al. Cellular and molecular responses of human skeletal muscle exposed to hypoxic environment. J Biol Regul Homeost Agents 2011;25(4):635–45.
[173] Lee HC, Wei YH. Mitochondrial biogenesis and mitochondrial DNA maintenance of mammalian cells under oxidative stress. Int J Biochem Cell Biol 2005;37(4):822–34.
[174] Parrinello S, Samper E, Krtolica A, Goldstein J, Melov S, Campisi J. Oxygen sensitivity severely limits the replicative lifespan of murine fibroblasts. Nat Cell Biol 2003;5(8):741–7.

Quiescence/Proliferation Issue and Stem Cell Niche

Chapter Outline

5.1 Embryo and Embryonic Cells

As already mentioned, neurogenesis through the whole embryonic development of mice takes place in very poorly oxygenated areas in which the neural stem and progenitor cells actively proliferate [1]. It is interesting that the development of sheep and cattle embryos can occur in a simple defined medium without somatic cell support even in anoxia (0% O_2) at the extent comparable to the one at 21% O_2 while under lowered O_2 concentrations (2–12%) it is even more efficient [2]. These data, suggesting that during embryonic development system-specific stem cells can proliferate either at very low O_2 concentrations or even in anoxia, seems to be in line with the experimental findings showing that the human embryonic stem cells (ESCs) continue to proliferate when the respiratory chain (complex III) is blocked. Furthermore, this respiratory blockade results in enhancement of their pluripotency [3]. Perturbation of mitochondrial function by PTPMT1 deficiency (knockout mice), which alters mitochondrial metabolism, increases LSK $CD150^+CD48^-Flk2^-$ pool (the cell population considered by the authors as "HSC") by ~40-fold [4].

5.2 Stem Cells in "Hypoxic Niche"—Quiescence and/or Slow Proliferation; Case of Hematopoietic Stem Cells

5.2.1 Arguments in Favor of HSC Proliferation

The first study that related "hypoxia" to the stem cell niche [5] showed that the preclonogenic elements (cells more primitive than committed progenitors) continued to

Anaerobiosis and Stemness. http://dx.doi.org/10.1016/B978-0-12-800540-8.00005-3

proliferate and to generate committed progenitors in situations when the respiratory chain was blocked by excess of pyruvate, while the committed progenitors showed a proliferative arrest. These data suggested some anaerobic/microaerophilic aspects of primitive hematopoietic progenitor/stem cells. When the same group exposed the mouse bone marrow cells to 1% O_2 atmosphere in culture, they found that more primitive the hematopoietic cell subset is more "resistant" to low O_2 concentration [6]. Since it is considered that the primitive stem cells (MRA) are in the G_0 phase of the cycle, the low-energy demands in quiescence have been proposed as an explanation for this "resistance." However, when the cells were stimulated by a better "suited" cytokine cocktail, it was obvious that the ex vivo amplification of committed progenitors (CFC) under 1% O_2 occurs in parallel with full maintenance of MRA cells (HSC), which is possible only if the MRA self-renew at some extent (asymmetric model is probable) [7]. This is the first publication that related HSC self-renewal to low O_2 concentrations. Applying an ex vivo essay ("Pre-CFC" or culture repopulating ability (CRA)) analog to in vivo MRA one, it was shown that a substantial proportion of, if not all, cells endowed with Pre-CFC activity (which is maintained at 1% or 0.9% O_2) is in the active cell cycle since inactivated by 5-Fluorouracil (5FU) [8,9]. Furthermore, after sorting of the cells on the basis of their proliferative history, it was shown that in 0.9% O_2 cultures, the pre-CFC activity was conserved in all cell fractions (0, 1, and >1 cell divisions) but it was tenfold more pronounced after one or more divisions compared to nondivided cells. If compared to air (21% O_2) cultures, the mean pre-CFC activity after 0.9% O_2 culture per one $CD34^+$ cell was 10-, 100-, and 500-fold higher for fractions divided 0, 1, and >1 times, respectively. This is the formal proof that the maintenance of high proliferative capacity of a subpopulation of primitive progenitors/HSC occurs in parallel with the cell divisions; that is, that a fraction of these cells self-renew [9].

A laconic statement that "Hypoxia maintains HSC ex vivo by suppressing cell cycle and inducing quiescence" was frequently made, referring to the experimental data that do not present enough specific results to deduce it. That was the case [10,11] with the study that demonstrated that cord blood $CD34^+$ cells can reenter in G_0 after being divided ex vivo in a culture under 0.1% O_2 and stimulated by IL-3 [12], which shows in fact, that, after the first division, one-half of $CD34^+$ cells reenter in G_0 but another half continues to actively proliferate (G1, S2G2M). Since committed progenitors (Colony Forming Cells–SCF) represent one-half of cord blood $CD34^+$ cells and the "HSC" (exhibiting in vivo repopulating ability) less than 2% [13] (let's suppose, for the sake of argument, that this fraction may be at most 10%), we actually cannot know to which fraction of cells these HSC belong: to the one that reentered in G_0 or the one continuing to proliferate. It is also possible that this HSC minority is dispersed in two fractions. On the other hand, having in mind the low percentage of real HSC in CB $CD34^+$ populations, it is evident that out of ~50% of cells reentering in G_0, a substantial proportion belongs to the committed progenitors. This point should be kept in mind as well as the fact that about one-half of CB $CD34^+$ cells can proliferate in nearly anoxic conditions (0.1% O_2) and that at least some HSC can be in this fraction. An additional argument for the claim that the HSC may proliferate in nearly anoxic conditions are the data obtained with FDCP-Mix cell line [14] showing that under 0.1% O_2

two subpopulations of cells can be distinguished inside this cell line: the first (majority of cells) composed of more mature cells that enter in G_0 and the second (minority), composed of immature cells with high proliferative potential that continue to proliferate slowly. This subpopulation slowly proliferating at 0.1% O_2 is endowed by "serial culture-repopulating capacity"; that is, it exhibits some stem cell properties [14].

Another argument that HSC proliferate in a "hypoxic" niche can be found in the Foxo3/Cited2 studies. Foxo3a is essential for the maintenance of the hematopoietic stem cells (HSCs) since its loss leads to a defect in HSC self-renewal capacity that impairs the maintenance of their in vivo pool during aging [15]. Foxo3, a direct target gene of HIF-1α, is related to the metabolic regulation in a low O_2 environment [16]. It is activated in response to hypoxic stress and inhibits HIF-1-induced apoptosis via regulation of Cited2 [17]. Taking into account that the most recent data show that Cited2 is required for the maintenance of glycolytic metabolism in adult HSCs [18], the described features may be operating in the context of physioxia of HSC niche (low O_2 environment). This scenario implies that HSC proliferation is compatible with the "hypoxic" niche and glycolytic metabolic profile since the self-renewal cannot operate without the proliferation of cells.

Nevertheless, it is generally accepted today that in adult tissues, a reservoir of primitive stem cells is being maintained in the state called "quiescence"—the G_0 phase of cycle. The dominant opinion today is that the quiescence of stem cells is related to their position in hypoxic niches and their "hypoxic" phenotype [19].

5.2.2 Arguments in Favor of Quiescence of HSC

Indeed, there are solid data demonstrating that the "HSC hypoxic niche" cellular and biochemical factors increase the proportion of phenotypically defined cell populations enriched in HSC.

In xenotransplantation experiments it was shown that a significant proportion of human cord blood $Lin^-CD34^+CD38^-$ cells enter in G_0 phase and acquire a "hypoxic" phenotype in the recipient mouse bone marrow [20]. The similar effect was found in a culture exposed to 1% O_2 on SLK $FLT3^-CD34^-$, which authors consider "LT-HSC" [21].

The data demonstrates that Tpo/MPL signaling plays a role in maintaining the $LSKCD34^-MPL^+$ cells in quiescence in the osteoblastic bone marrow niche [22]. In fact, total LSK $CD34^-$ cell population, mainly quiescent in adult BM (considered by authors as long-term (LT-HSC)) as well as half of the LSK $CD34^+$ population expresses the Tpo receptor Mpl; administration of an anti-Mpl neutralizing antibody, AMM2, "suppressed" the quiescence of LSK $CD34^-$ cells (decreased the proportion of cells in G_0 from 90.4% to 69.9%) and enabled HSC engraftment without irradiation [23]. As mentioned above, Tpo, which was shown to stabilize HIF-1α transcripts [24] and cell cycle quiescence (including LSK $CD34^-$ population that authors consider as "HSC"), seems to be maintained by appropriate HIF-1α levels [14]. In HIF-1α-deficient mice, the G_0 LSK $CD34^-$ cell proportion decreases from 83.6% to 65% (mean values) [11]. Du et al. [25] demonstrated that Cited2, a negative regulator of HIF-1, regulates LSK $CD34^-$ cells (considered as "HSC" by authors) quiescence through both HIF-1–dependent and HIF-1–independent pathways. The deletion of Cited2 resulted in a decrease of G_0 cells proportion (as detected by Pyronin/Hoechst 33342 technique) from 28.5% to 11.5% and from 73.8% to 49.5%, respectively, for LSK and LSKCD34$^-$ cells [25], which is partially rescued by the additional deletion of HIF-1α in *Cited2*-knockout BM.

Based on the above, the TPO/MPL mechanism appears to be a plausible explanation for the hypoxia-mimicking effect of the stem cell niche leading to the quiescence of phenotypically defined cell populations considered "HSC." The similar conclusion is provided in experiments with the pharmacological stabilization of HIF-1α [26] increasing proportion in G_0 phase from 62.6% to 89.9% of LSK48$^-$ cells considered "HSC" (but also of LKS48$^+$ cells (from 47.7% to 65.4%) considered as "progenitors"). Conversely, if HIF-1α and HIF-2α are down-regulated by the deletion of Meis1 [27] the proportion of G_0 cells in LSK Flk2$^-$CD34$^-$ population (considered LT-HSC by authors) decreases significantly although not in a large proportion (from 83% to 72%).

As mentioned earlier, TGF under a low O_2 concentration (1%) (or with HIF overexpression) enhances cycle arrest of CD34$^+$/CD38$^-$ cells as estimated on the basis of Pyronin/Hoechst 33258 cell cycle analysis (quiescent cell proportion from 7.3% to ~13%). Furthermore, CD34$^+$ cells cultured in 1% O_2 secreted high levels of latent TGFβ, suggesting an auto- or paracrine role of TGFβ in the regulation of quiescence, which would, however, not be direct [28]. Again, CD34$^+$/CD38$^-$ population, although more enriched in HSC comparing to CD34$^+$ population, is still composed mainly of the committed progenitors (see Chapter 1).

Most of these studies show an increase in G_0-phase cell proportion (various phenotypically defined cell populations more or less enriched in HSC) in parallel with the enhancement of functional HSC activity evidenced by the in vivo repopulation capacity and vice versa, their decrease in parallel with the diminution of functional HSC activity. However, it should be stressed that G_0-phase fraction never reached 100% in observed populations and that these cell populations, even enriched in HSC, do not represent the pure HSC populations (real HSC are still relative minority) and a nonnegligible proportion is composed of multipotent, bipotent, and unipotent progenitors. Thus, it is misleading to identify these cell populations by the term "HSC," which should be restricted to the cells shown to exhibit stem cell properties with respect to the "referent" definition of HSC, which is the functional one (see Chapter 1). *Therefore, on the basis of these studies, the quiescence can be "attributed" neither exclusively to HSC nor to all HSC.* In addition, the range of changes in "induction of quiescence" (e.g., 62.6–89.9%) or "suppression of quiescence" (83.6–65%; 83–72%; or 90.4–69.9%) are far from an exclusive "all or nothing" effect. So, even if we admit that subpopulations of LT-HSC were predominantly in the quiescent state and had robust reconstitution activity of hematopoiesis, it cannot be excluded that a proportion of them proliferate within the stem cell niche.

5.2.3 At Least a Fraction of HSC Proliferates in Steady State

Taking into account these remarks, in our opinion the data concerning the quiescence of phenotypically defined populations enriched in HSC cited above are not incompatible with the concept of HSC slow cycling low-energy self-renewal divisions in a "hypoxic" niche [29–31]. In that respect it is plausible to cite the study of Cheshier et al. [32], who found that at any time point, 25% of LSK Thy1.1$^+$ cells (considered by authors as "LT-HSC") were in G_1SG_2M phases of the cell cycle. They specified (literal citation): "About 50% of LT-HSC incorporated BrdU by 6 days and >90% incorporated BrdUrd by 30 days. *By 6 months, 99% of LT-HSC had incorporated BrdU.* We calculated that approximately 8% of LT-HSC asynchronously entered the cell cycle per day. Nested reverse transcription–PCR analysis revealed cyclin D2 expression in a high proportion of LT-HSC. Although "75% of LT-HSC are quiescent in G_0 at any time-point, all HSC are being recruited into the cycle regularly such that 99% of LT-HSC divide on average every 57 days" [32]. These data do not seem to be in substantial contradiction with the recent modeling studies showing that Bromodeoxyuridine Label-Retaining Cells (BrdU LRC). BrdU LRC can reflect slow activation of quiescent cells and their possible return into quiescence after division [33] and that the BrdU LRC phenomenon concerns mainly the slowly proliferating cells that cycle about once every 149–193 days [34]. It seems that HSCs are not stochastically entering the cell cycle but reversibly switch from dormancy to self-renewal upon stimulation [35].

5.3 Mesenchymal Stem Cell Case

Some interesting data were obtained ex vivo with mesenchymal stem cells from rat bone marrow (the authors consider adherent cells after 1st, 2nd, 3rd, and 4th passage defined by the phenotype: CD90$^+$CD54$^+$CD44$^+$CD29$^+$CD45$^-$CD11b$^-$ as "mesenchymal stem cells") [36]. It was shown that these cells proliferate and amplify in anoxia (0% O_2) although with a higher proportion of apoptotic cells compared to 5% O_2 and air (20–21% O_2). These results are consistent with the scenario that in functionally heterogeneous mesenchymal cell population some cells exhibit an anaerobic character but some are O_2-dependent (probably a fraction lacking stem cell properties).

5.4 Conclusions

On the basis of all presented arguments, the usual simplified notion of HSC that are quiescent in the steady state should not be taken rigidly as a dogma, since at least some HSC are undergoing slow proliferation in the "hypoxic" niche. These cells are activated and recruited to replace the descendant cell populations either in the steady-state conditions (to compensate for the physiological loss of mature cells with limited life span) or to regenerate cell compartments lost during an insult. Response of HSC and HPC on low O_2 concentrations is presented in Figure 5.1 as well as the steady-state situation (in frame).

Furthermore, the ability to proliferate is crucial for existence of the self-renewal phenomenon of HSC. Our proposal that low-energy (anaerobic/microaerophilic) proliferation is the key event of the self-renewal as low availability of energy may

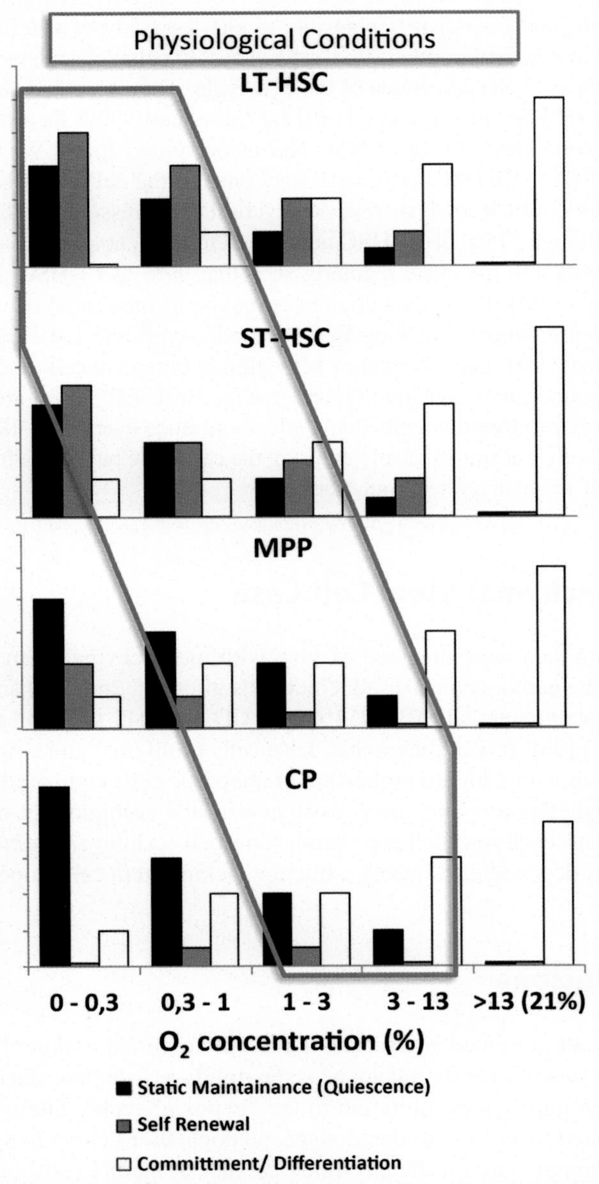

Figure 5.1 Response of different stem and progenitor cell populations to the O_2 levels. Only functionally defined populations are considered (see as presented in Chapter 1, Figure 1.1): LTC-HSC, long-term repopulating hematopoietic stem cells; ST-HSC, short-term repopulating hematopoietic stem cells; MPP, multipotent hematopoietic progenitors; CP, committed hematopoietic progenitors. The physiological state is given in the frame.

prevent cells from commitment and subsequent differentiation, which require much more energy than simple cell division uncoupled with differentiation [14] will be discussed in further chapters. The ex vivo experimental findings of a simultaneous production of committed progenitors [5] or amplification of nonfunctional cells bearing phenotype associated to the HSC [4] in anaerobiosis could be explained by an imperfect blockade of the cell respiration, extremely low O_2 quantities available and/or a suboptimal and a partial stimulation in culture. These factors may result in commitment and differentiation after HSC division and consequent exhaustion of a functional HSC pool in parallel with retaining a phenotype associated with HSC ("dissociation phenotype/function").

References

[1] Lee YM, Jeong CH, Koo SY, Son MJ, Song HS, Bae SK, et al. Determination of hypoxic region by hypoxia marker in developing mouse embryos in vivo: a possible signal for vessel development. Dev Dyn 2001;220:175–86.

[2] Thompson JG, Simpson AC, Pugh PA, Donnelly PE, Tervit HR. Effect of oxygen concentration on in-vitro development of preimplantation sheep and cattle embryos. J Reprod Fertil 1990;89:573–8.

[3] Varum S, Momcilović O, Castro C, Ben-Yehudah A, Ramalho-Santos J, Navara CS. Enhancement of human embryonic stem cell pluripotency through inhibition of the mitochondrial respiratory chain. Stem Cell Res 2009;3:142–56.

[4] Yu WM, Liu X, Shen J, Jovanovic O, Pohl EE, Gerson SL, et al. Metabolic regulation by the mitochondrial phosphatase PTPMT1 is required for hematopoietic stem cell differentiation. Cell Stem Cell 2013;12:62–74.

[5] Dello Sbarba P, Cipolleschi MG, Olivotto M. Hemopoietic progenitor cells are sensitive to the cytostatic effect of pyruvate. Exp Hematol 1987;15:137–42.

[6] Cipolleschi MG, Dello Sbarba P, Olivotto M. The role of hypoxia in the maintenance of hematopoietic stem cells. Blood 1993;82:2031–7.

[7] Ivanović Z, Bartolozzi B, Bernabei PA, Cipolleschi MG, Rovida E, Milenković P, et al. Incubation of murine bone marrow cells in hypoxia ensures the maintenance of marrow-repopulating ability together with the expansion of committed progenitors. Br J Haematol 2000;108:424–9.

[8] Cipolleschi MG, Rovida E, Ivanovic Z, Praloran V, Olivotto M, Dello Sbarba P. The expansion of murine bone marrow cells preincubated in hypoxia as an in vitro indicator of their marrow-repopulating ability. Leukemia 2000;14:735–9.

[9] Ivanovic Z, Belloc F, Faucher JL, Cipolleschi MG, Praloran V, Dello Sbarba P. Hypoxia maintains and interleukin-3 reduces the pre-colony-forming cell potential of dividing CD34(+) murine bone marrow cells. Exp Hematol 2002;30:67–73.

[10] Iriuchishima H, Takubo K, Matsuoka S, Onoyama I, Nakayama KI, Nojima Y, et al. Ex vivo maintenance of hematopoietic stem cells by quiescence induction through Fbxw7α overexpression. Blood 2011;117:2373–7.

[11] Takubo K, Goda N, Yamada W, Iriuchishima H, Ikeda E, Kubota Y, et al. Regulation of the HIF-1alpha level is essential for hematopoietic stem cells. Cell Stem Cell 2010;7:391–402.

[12] Hermitte F, Brunet de la Grange P, Belloc F, Praloran V, Ivanovic Z. Very low O_2 concentration (0.1%) favors G_0 return of dividing $CD34^+$ cells. Stem Cells 2006;24:65–73.

[13] Miller PH, Cheung AM, Beer PA, Knapp DJ, Dhillon K, Rabu G, et al. Enhanced normal short-term human myelopoiesis in mice engineered to express human-specific myeloid growth factors. Blood 2013;121:e1–4.
[14] Guitart AV, Debeissat C, Hermitte F, Villacreces A, Ivanovic Z, Boeuf H, et al. Very low oxygen concentration (0.1%) reveals two FDCP-Mix cell subpopulations that differ by their cell cycling, differentiation and p27KIP1 expression. Cell Death Differ 2011;18:174–82.
[15] Miyamoto K, Araki KY, Naka K, Arai F, Takubo K, Yamazaki S, et al. Foxo3a is essential for maintenance of the hematopoietic stem cell pool. Cell Stem Cell 2007;1:101–12.
[16] Jensen KS, Binderup T, Jensen KT, Therkelsen I, Borup R, Nilsson E, et al. FoxO3A promotes metabolic adaptation to hypoxia by antagonizing Myc function. EMBO J 2011;30:4554–70.
[17] Bakker WJ, Harris IS, Mak TW. FOXO3a is activated in response to hypoxic stress and inhibits HIF1-induced apoptosis via regulation of CITED2. Mol Cell 2007;28:941–53.
[18] Du J, Li Q, Tang F, Puchowitz MA, Fujioka H, Dunwoodie SL, et al. Cited2 is required for the maintenance of glycolytic metabolism in adult hematopoietic stem cells. Stem Cells Dev 2014;23:83–94.
[19] Arai F, Suda T. Quiescent stem cells in the niche. StemBook [Internet] 2008. Cambridge (MA): Harvard Stem Cell Institute.
[20] Shima H, Takubo K, Tago N, Iwasaki H, Arai F, Takahashi T, et al. Acquisition of G_0 state by CD34-positive cord blood cells after bone marrow transplantation. Exp Hematol 2010;38:1231–40.
[21] Eliasson P, Rehn M, Hammar P, Larsson P, Sirenko O, Flippin LA, et al. Hypoxia mediates low cell-cycle activity and increases the proportion of long-term-reconstituting hematopoietic stem cells during in vitro culture. Exp Hematol 2010;38:301–10. e2.
[22] Yoshihara H, Arai F, Hosokawa K, Hagiwara T, Takubo K, Nakamura Y, et al. Thrombopoietin/MPL signaling regulates hematopoietic stem cell quiescence and interaction with the osteoblastic niche. Cell Stem Cell 2007;1:685–97.
[23] Arai F, Yoshihara H, Hosokawa K, Nakamura Y, Gomei Y, Iwasaki H, et al. Niche regulation of hematopoietic stem cells in the endosteum. Ann N Y Acad Sci 2009;1176:36–46.
[24] Yoshida K, Kirito K, Yongzhen H, Ozawa K, Kaushansky K, Komatsu N. Thrombopoietin (TPO) regulates HIF-1alpha levels through generation of mitochondrial reactive oxygen species. Int J Hematol 2008;88(1):43–51.
[25] Du J, Chen Y, Li Q, Han X, Cheng C, Wang Z, et al. HIF-1α deletion partially rescues defects of hematopoietic stem cell quiescence caused by Cited2 deficiency. Blood 2012;119:2789–98.
[26] Forristal CE, Winkler IG, Nowlan B, Barbier V, Walkinshaw G, Levesque JP. Pharmacologic stabilization of HIF-1α increases hematopoietic stem cell quiescence in vivo and accelerates blood recovery after severe irradiation. Blood 2013;121:759–69.
[27] Kocabas F, Zheng J, Thet S, Copeland NG, Jenkins NA, DeBerardinis RJ, et al. Meis1 regulates the metabolic phenotype and oxidant defense of hematopoietic stem cells. Blood 2012;120:4963–72.
[28] Wierenga AT, Vellenga E, Schuringa JJ. Convergence of hypoxia and TGFβ pathways on cell cycle regulation in human hematopoietic stem/progenitor cells. PLoS One 2014;9:e93494.
[29] Ivanovic Z. Hypoxia or in situ normoxia: the stem cell paradigm. J Cell Physiol 2009;219:271–5.
[30] Ivanovic Z. Respect the anaerobic nature of stem cells to exploit their potential in regenerative medicine. Regen Med 2013;8:677–80.

[31] Guitart AV, Hammoud M, Dello Sbarba P, Ivanovic Z, Praloran V. Slow-cycling/quiescence balance of hematopoietic stem cells is related to physiological gradient of oxygen. Exp Hematol 2010;38:847–51.
[32] Cheshier SH, Morrison SJ, Liao X, Weissman IL. In vivo proliferation and cell cycle kinetics of long-term self-renewing hematopoietic stem cells. Proc Natl Acad Sci USA 1999;96:3120–5.
[33] Glauche I, Moore K, Thielecke L, Horn K, Loeffler M, Roeder I. Stem cell proliferation and quiescence–two sides of the same coin. PLoS Comput Biol 2009;5:e1000447.
[34] van der Wath RC, Wilson A, Laurenti E, Trumpp A, Liò P. Estimating dormant and active hematopoietic stem cell kinetics through extensive modeling of bromodeoxyuridine label-retaining cell dynamics. PLoS One 2009;4:e6972.
[35] Wilson A, Laurenti E, Oser G, van der Wath RC, Blanco-Bose W, Jaworski M, et al. Hematopoietic stem cells reversibly switch from dormancy to self-renewal during homeostasis and repair. Cell 2008;135:1118–29.
[36] Anokhina EB, Buravkova LB, Galchuk SV. Resistance of rat bone marrow mesenchymal stromal precursor cells to anoxia in vitro. Bull Exp Biol Med 2009;148:148–51.

Metabolic Peculiarities of the Stem Cell Entity: Energetic Metabolism and Oxidative Status

Chapter Outline

6.1 Embryonic Stem Cells

6.1.1 Energetic Profile during Embryonic Development

Energetic metabolism during mammalian embryonic development undergoes dramatic changes.

The earliest stage of the development refers to the preimplantation embryo, in which each individual blastomere is totipotent. During this period, oxidative metabolism is the preferential energetic production pathway based on the uptake of pyruvate, lactate, amino acids, and triglyceride-derived fatty acids [2–5]. Interestingly, true primordial germ cell oocytes and spermatozoid also use pyruvate secreted by

Anaerobiosis and Stemness. http://dx.doi.org/10.1016/B978-0-12-800540-8.00006-5

the ovarian follicle or lactate secreted by the Sertoli cells [6]. Actually, early embryo development is inhibited by the high concentration of glucose [2]. Energy production of early embryo relies on abundant maternal mitochondria inherited from the oocyte. Despite their functional capacity to produce ATP, mitochondria of oocytes and newly fertilized eggs are structurally undeveloped, consisting of spheric structures with truncated cristae that predominantly reside near the nucleus [7]. Since preimplantation embryo is developed in a low O_2 environment, it would be tempting to test if these mitochondria work in anaerobic mode as it was evidenced for some somatic cells [8,9].

This stage is characterized by high ATP total and ATP/ADP ratio that decreases with the development. This ATP serves as a potentially allosteric inhibitor of glycolysis (by inhibition of the rate-limiting glycolytic enzyme phosphofructokinase-1, PFK1) and participates in the ATP-dependent ion transport into the embryo [10]. Also, metabolism of the early embryo is marked by the high uptake of the bicarbonate that could serve as a carbon source for the anaplerosis, nucleotide synthesis, and gluconeogenesis [11,12].

In the next stage, the morula, blastomeres are segregated into the pluripotent inner cell mass (ICM) and trophectoderm, which accelerates metabolic activity and significantly increases growth. Glucose uptake, glucose transporter type 1 (GLUT1), and GLUT3 expression, activities of glycolytic enzymes (hexokinase (HK) and PFK1 enzymes), gradually increase. As a result, flux into the pentose phosphate pathway (PPP) for nucleotide synthesis also increases. These changes led to a bivalent energetic pattern, a mixture of glycolysis and oxidative phosphorylation (OXPHOS). Increase of the mitochondrial activity observed at that stage seems to be largely due to the mitochondria of trophectoderm while mitochondria in the ICM have a lower membrane potential [7].

Anaerobic energetic profile is accelerated further in the blastocyst stage where glucose uptake exceeds that of pyruvate or lactate and is predominantly metabolized through glycolysis [10]. Importantly, implanting blastocyst is developed in the low O_2 microenvironment [13–16] that favors glycolytic shift. Precisely, in the rodent uterus O_2 concentrations is 3.5–5% during the period of the late blastocyst and the implantation [17] while in the human uterus O_2 concentration is about 2.5% during early pregnancy.

Later embryonic development is marked by mitochondrial replication, maturation into tubular cristae-dense structures, and cytosolic deployment that enables reinitiation of oxidative metabolism and progressive decline in glycolysis [7,10]. This coincides with increase in the molecular oxygen 8.6% (60 mm Hg) [18] that is sensed by the embryo, after placentation.

Described environmental and developmental changes have determined the functional features of the different types of the embryonic stem cells (ESCs) currently used in the experimental studies. Mouse and human ESCs are isolated from ICM of preimplantation embryos [19–21] while epiblast stem cells (EpiSCs) represent cells from the postimplantation epiblast, a later stage in development [22]. After isolation, ESCs can be cultured in vitro indefinitely using either a feeder layer of fibroblast cells or an artificial substrate such as matrigel with proper supplementation of necessary growth factors [23,24].

ESCs and EpiSCs are pluripotent, yet display distinct features in terms of gene expression, epigenetic modifications, and developmental capacity following blastocyst injection. In contrast to mESCs, though isolated from the same tissue, hESCs are similar to EpiSCs based on transcriptional and protein expression profiles and their epigenetic state. Actually, it was suggested that compared to mESCs, hESCs represent a more developed stage that is close to the EpiSCs stage [25]. Thus, pluripotency should not be considered as a single defined state but as a collection of several intermediate stages, with similarities and differences in measurable characteristics relating to gene expression and cellular phenotype, metabolic and energetic pattern [16].

6.1.2 Energy Production in ESCs

ESCs maintain rapid proliferation due to a shortened cell cycle (with a shortened G_1 and an extended S phase), as compared with their somatic counterparts and adult stem cells [26]. Therefore, we might expect that ESCs fuel their energy demand mostly by the glycolysis and exhibit anabolic metabolism [27] as is the case with other highly proliferating biological systems (i.e., cancer cells that preferentially utilize glycolysis even when O_2 is available, or many unicellular organisms that utilize fermentation).

Indeed, in vitro measuring showed that ESCs produce more lactate, consume more glucose, exhibit a lower oxygen consumption rate, have a lower mitochondrial membrane potential and increased levels of glycolytic and PPP enzymes than differentiated cells [28–32]. In addition, Zhang et al. demonstrated that uncoupling protein 2 (UCP2) a mitochondrial inner membrane protein is more highly expressed in the pluripotent stem cells than in fibroblasts [25]. Using $^{13}C_6$-labeled glucose, the authors confirmed that ESCs use UCP2 to increase the glycolytic flux and shunt pyruvate away from oxidation in the mitochondria for use in the PPP [25].

This activation of glycolysis accompanied by high PPP activity is thus the preferred metabolic state of rapidly proliferating cells due to several reasons. First, even if glycolysis is a less efficient way for energy production (2 ATP per molecule of glucose versus 36 ATP generated in the OXPHOS upon complete oxidation of the one molecule of glucose), it still ensures generation of biomass (macromolecules) as well as ATP needed for cell divisions and production of the viable daughter cells [27]. Some glucose is converted into acetyl-CoA for fatty acid synthesis, into glycolytic intermediaries for the nonessential amino acid synthesis, or PPP. The major functions of the PPP are to protect cells against oxidative stress by producing NADPH, which is necessary for the regeneration of glutathione [33] and to act as an alternative pathway to produce ribose-5-phosphate and NADPH for the synthesis of DNA, RNA, and fatty acids (Figure 6.1). Therefore, converting all the glucose to CO_2 via OXPHOS in mitochondria for maximization of ATP production is not sustainable for the proliferating cells. Second, mitochondrial OXPHOS is the major cellular source of the reactive oxygen species (ROS) [34]. Thus, glycolytic energetic orientation preserves the proliferating cells of the ROS excess that causes apoptosis/senescence, DNA damage, and cell cycle arrest [34].

Although not efficient in the energy production as OXPHOS, glycolysis can produce ATP at a rapid rate; in the presence of excess glucose, glycolysis can produce a

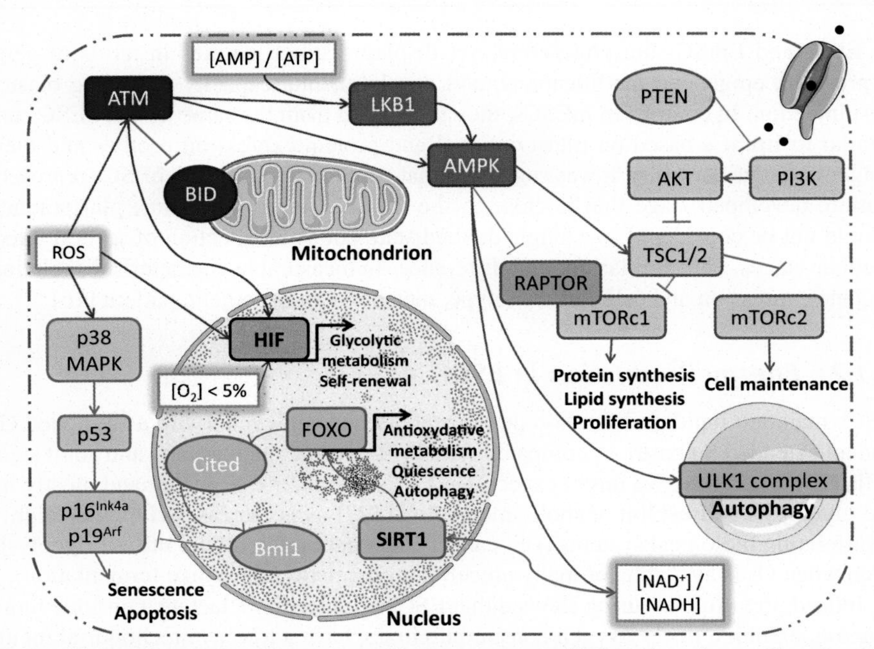

Figure 6.1 Biochemical pathways contributing to stem cells function. Stem cells primarily generate cellular energy via glycolysis. Importantly, metabolism of glucose through glycolysis also generate intermediaries that serve as substrates in different biosynthetic pathways: PPP, phospholipid, and nonessential amino acid synthesis (e.g., serine). PPP provide the substrate for the nucleotide synthesis and NADPH implicated in the oxidative stress protection. PDK is highly activated in the stem cell and inhibits PDH. This reaction promotes pyruvate conversation to lactate and inhibits pyruvate transformation to Ac-CoA that enters in TCA. Production of lactate from pyruvate enables NAD^+ regeneration that ensures glycolytic flux. Even that mitochondrial function is diminished stem cell exhibit TCA cycle activity that serves to generate metabolite for fatty acid and amino acid synthesis. Stem cells are express enzyme isoforms (e.g., HK1, PKM2) and metabolites (e.g., F1,6 BP) that favor utilization of glycolysis as an energy source. In addition, stem cells critically depend on PML activity that controls PGC-1α/PPAR-δ transcription complex, which targets genes promoting FAO program. Glutamine metabolism is important energetic pathway for the proliferating cells. Deamidation of glutamine generates glutamate via GLS. Glutamate is converted via transamination to TCA intermediaries such as alpha-ketoglutarate and oxidative stress protector NADPH. Also, pluripotent stem cells are critically dependant on threonine metabolism. Threonine contributes to the synthesis of glycine and epigenetic modulator SAM. Amino acids serine, glycine, and threonine are critical for the sphingolipid and folate metabolism. G6P, glucose 6-phosphate; F6P, fructose 6-phosphate; F16BP, fructose 1,6-bisphosphate; GA3P, glyceraldehyde-3 phosphate; 1,3, BPG, 1,3, bisphosphoglycerate; PEP, phosphoenolpyruvate; PKM2, pyruvate kinase isoform M2; PPP, pentose phosphate pathway; RIBO-5-P, ribose-5-phosphate; PDK, pyruvate dehydrogenase kinase; PDH, pyruvate dehydrogenase; NADPH, nicotinamide adenine dinucleotide phosphate reduced; NAD^+, oxidized nicotinamide adenine dinucleotide phosphate; Ac-CoA, acetyl-CoA; HK1, hexokinase-1; pyruvate kinase isozymes M2; PML, promyelocytic leukemia protein; PPPR, peroxisome proliferator-activated receptor; PGC-1α, peroxisome proliferator-activated receptor gamma coactivator 1-alpha; FAO, fatty acid oxidation; GLS, glutaminase; OA, oxaloacetate; SAM, S-adenosylmethionine.

larger percentage of ATP than OXPHOS up to 100-fold compared with mitochondrial ATP generation [35,36].

So, glycolysis-based anabolic requirements seem to be critical for the energetic profile of the pluripotent cells. Interestingly, hESCs are markedly glycolytic even in the ambient air [25,37], at concentrations of O_2 much higher than those existing during embryonic development [17] exhibiting the Warburg effect as a cancer cell [38]. This indicates that, in this particular case (cells propagated for a long time ex vivo in atmospheric O_2 concentration), a "pluripotent stage-specific metabolic program" is relatively less sensitive to the O_2 in the culture condition [39], and anabolic abovementioned requirements seem to be crucial in determining the energetic profile of ESCs.

Recent studies using ESCs in standard laboratory conditions (20% O_2/5–7% CO_2) demonstrated that while mESCs, which are considered the most primitive ESCs [40], are metabolically bivalent (they can switch between OXPHOS or glycolysis on demand), EpiSC and hESC are almost exclusively glycolytic. Recently, Takashima et al. showed that standard laboratory hESCs are primed or more mature than naïve or ground-state hESCs [41]. Namely, they demonstrated that transient ectopic exposure of NANOG and KLF4 reprogram primed hESC into naïve hESCs. These induced hESCs have metabolic activity similar to mESCs or preimplantation mouse embryo [41].

Altogether, we could consider that a significant relationship exists between metabolic phenotype and pluripotent developmental stage correlating with developmental metabolic energetic changes demonstrated in vivo.

Even if they are mostly glycolytic, ESCs possesses functional mitochondria, which are immature compared to those in more differentiated cells, with lower transcription of electron transport chain (ETC) components, respiration, and ETC activity [25].

EpiSCs contain more morphologically mature mitochondria than mESCs but are obligatory glycolytic due to low cytochrome C expression [16]. Moreover, hESC have a similar mitochondrial mass and oxygen consumption rate as fibroblasts when normalized to protein content, but their mitochondria consume rather than produce ATP [25,28].

It has been suggested that mitochondrial oxidation operates both to recycle NAD^+ to keep glycolytic flux and to generate intermediaries for the fatty acids and amino acids synthesis [42] (Figure 6.1). Thus, mitochondrial function does contribute to the maintenance of pluripotent cells in nascent, yet in responsive state to meet demands of the differentiation [30,43,44]. Hence, ESCs with high mitochondrial potential (indicator of the mitochondrial activity) form teratomas more efficiently than ESCs with lower mitochondrial potential [45]. Accordingly, blastocyst with mitochondrial enzyme deficiency will show defects in development [46].

Energetic metabolism modulation could divert the pluripotent stage of the ESC. Stimulation of the glycolysis in ESCs by low O_2 concentration [47,48], inhibition of the mitochondrial respiration [49], and supplementation of the insulin [50] sustain the stemness, while inhibition of the glycolysis blocks proliferation and promotes cell death [32]. Also, in vitro supplementation with the cell permeable α-ketoglutarate, an intermediate in TCA, directly supports ESC self-renewal while cell permeable succinate promotes differentiation [51] by the transcriptional and epigenetic regulation of the pluripotent gene expression.

Further, functional mitochondrial respiration is critical for differentiation of the mESC or hESC to somatic cells [30,52–54]. Differentiation of the ESC is accompanied by a decline in glycolysis and an increase in expression of the mtDNA transcription factors, mitochondrial DNA (mtDNA) replication factors, components of the fatty acid oxidation, enzymes of the tricarboxylic acid (TCA) cycle, and subunits of the ETC along with mitochondrial biogenesis [30,55–57]. However, another study showed that differentiation of ESCs to neural stem cells is accompanied by increased glycolysis and decreased oxidative phosphorylation, although glycolysis declines upon terminal differentiation to neurons [58].

Astonishingly, these energetic changes could be reversed if somatic cells were reprogrammed to divert in pluripotent cells (induced pluripotent cells, iPSC) [59]. Reprogramming is associated with a major bioenergetic restructuring to facilitate a conversion from somatic mitochondrial oxidation to a glycolysis-dependent pluripotency state. This comprises mitochondrial reconstruction from a mature cristae-rich morphology in somatic cells to more immature spherical and cristae-poor structures, an increase in glucose utilization and lactate production, and the expression of glycolytic enzymes [60]. Importantly, all these changes precede expression of the pluripotent markers, further confirming metabolic determination of the pluripotency [31].

Also, stimulating glycolysis by elevating medium glucose levels increases reprogramming efficiency, whereas inhibition of glycolysis reduces reprogramming, in agreement with a recent report [31,61].

6.1.3 Other Energetic Pathways Important for the Energetic Homeostasis in ESC

Fatty acids oxidation (FAO) seems to be critical in the generation of energy in mESCs [62] as inhibition of the rate-limiting enzyme of the FAO carnitine-O palmitoyl transferase leads to ATP depletion and decreased resistance to nutrient deprivation. The importance of the FAO for the pluripotence stage metabolic homeostasis seems not to be due to OXPHOS ATP production, as ESCs have low oxidative metabolism and mitochondrial activity. However, FAO is the important source of acetyl-CoA and NADPH. Acetyl-CoA reacts with oxaloacetate to form citrate, which can enter two metabolic chain reactions that produce cytosolic NADPH. Also, acetyl-CoA could be used as a substrate for the histone acetylation pattern that enables epigenetic regulation of the stem cell functions (see paragraph“Metabolic-based Epigenetic Control of the Transcription Program in Stem Cells”).

Yanes et al. [63] found that ESCs have a lipid profile that differs from the differentiated cells and that ESCs are characterized by an accumulation of unsaturated metabolites. This unsaturated metabolome renders ESCs reactive and sensitive to oxidation reactions associated with differentiation [63]. mESCs are characterized by a unique amino acid catabolism. Using systematic depletion of all 20 amino acids in tissue culture media, Wang et al. found that mouse ESCs are critically dependent on threonine [64]. Catabolism of the threonine is critical for the nucleotide synthesis and rapid proliferation of mESC. Threonine-depleted media cause a reduction in self-renewal gene expression in mESCs, and cell cycle arrest [65]. mESCs have extremely high levels in threonine catabolizing enzyme L-threonine dehydratase (TDH) relative to differentiated cells. TDH increases the synthesis of S-adenosyl methionine (SAM), which is critical for the epigenetic control of pluripotency [66].

Glutamine (Gln) metabolism is an important energetic pathway for the proliferating cells [67]. This mitochondrial pathway provides TCA intermediaries such as alpha-ketoglutarate and oxidative stress protector NADPH (Figure 6.1). In contrast to the clear contributions of Gln metabolism to cancer development and progression, its role in stem cell homeostasis is still elusive [67].

6.1.4 Nutrient-Sensing Pathways That Coordinate Energy Metabolism with Stem Cell Function

Interplay between energetic pathways has an essential role in the determination of the stem cell fate [68]. Using nutrient-sensing pathways, including the insulin-forkhead box O factors (FOXO) pathway, mammalian target of rapamycin (mTOR), AMP-activated protein kinase (AMPK) stem cells coordinate the metabolic process in order to maintain cellular homeostasis [69].

6.1.4.1 mTOR

In response to several factors such as O_2, nutrients (amino acids, glucose) and growth factors, mammalian target of rapamycin (mTOR) control cellular energetics by inducing protein and lipid synthesis [70] glycolysis, mitochondrial activity and biogenesis, and also inhibition of autophagy [71,72] (Figure 6.2). It exists in two distinct complexes, mTOR complex 1 (mTORSC1) and mTOR complex 2 (mTORC2) with overlapping functions [68]. Different growth factors stimulate phosphatidyl inositol 3kinase (PI3K)-AKT kinase pathway that activates mTORC1 through inhibition of the mTOR inhibitory proteins, the tuberous sclerosis complex 1/2 (TSC1/2) [73–75].

When the nutrients are abundant, activated PI3/AKT-mTORC1 pathway represses FOXO that is responsible for pro-autophagic gene response. Thus in the starvation, anabolic processes are inhibited and autophagic response is activated in order to ensure cellular energetic balance.

Disruption of the complex following treatment with the mTOR pharmacologic inhibitor rapamycin lowered mitochondrial membrane potential, oxygen consumption, and ATP synthetic capacity [71,76]. Inhibition of mTOR by rapamycin or by knockdown in mouse and human ESCs reduces the expression of pluripotency markers and impairs the self-renewal of ESCs in vitro [45,77]. Furthermore, deletion of mTOR inhibits proliferation in the early mouse embryo, causes defects in cell size and lethality between E6.5 and 7.5 [78]. Somewhat surprisingly, mTOR activity appears to promote ESC differentiation. For example, the transition from ESCs to more differentiated cell types is associated with increased activation of mTORC1 signaling [79,80]. Furthermore, overexpression of a constitutively active form of S6 kinase 1, a substrate of mTORC1, promotes differentiation of ESC [79]. Interestingly, inhibition of mTOR by rapamycin enhances reprogramming efficiency of mouse fibroblast and generation of iPSC [71,81].

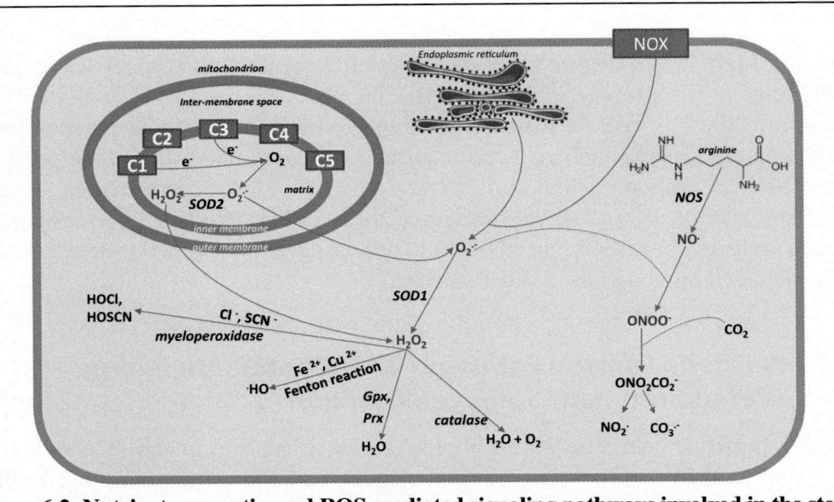

Figure 6.2 Nutrient, energetic, and ROS-mediated signaling pathways involved in the stem cell homeostasis. LKB/AMPK pathway senses cellular energetic status. When cellular energetic status is low (AMP/ATP ratio is high), LKB activates AMPK by phosphorylation, which in turn activates glucose uptake and inhibits mTOR in order to activate catabolic and inhibits anabolic energy consuming pathways. In addition, by direct activation of the serine/threonine-protein kinase ULK1, AMPK controls autophagic response. mTOR exists in two distinct complexes, mTORC1 and mTORC2, with overlapping functions. mTOR is associated with its binding partner, Raptor. PI3K/AKT pathway is activated by insulin or other growth factors. Different growth factors stimulate PI3K-AKT kinase pathway that activates mTORC1 through inhibition of the mTOR inhibitory proteins, TSC1/2. Also, glucose and amino acids can stimulate mTOR. PTEN is negative regulator of PI3K/AKT and consecutively of mTOR activity. In response to elevated ROS, ATM protein kinase, a critical genomic stability regulator, is activated. ATM-mediated phosphorylation of BID suppresses mitochondrial ROS production that preserve redox stem cell homeostasis. Also, ATM blocks mTOR activation through activation of the LKB/AMPK pathway, which further decreases mitochondrial ROS production. HIF-1α is stabilized in response to low O_2 concentration, increased level of ROS, or by activated ATM. In this situation, it forms with its transcription partner HIF-1β, functional HIF-1 transcription factor that induces expression of numerous genes involved in the glycolytic metabolism. But when the cell could not cope with oxidative stress, elevated ROS leads to p38 MAPK and tumor suppressor p53 activation, which in turn triggers up-regulation of the CDK inhibitors $p16^{Ink4a}$ and $p19^{Arf}$, block of cell cycle, senescence, and premature exhaustion of stem cells. Further, SIRT regulates energetic metabolism in order to adapt it to maintain cellular homeostasis according to oxidative cell status. SIRT1 senses when the NAD^+ levels become high (oxidative environment or low cellular energetic status) and activates expression of FOXO transcription factors. These transcription factors are the critical mediator against oxidative stress by activating the transcription of antioxidative defense genes and induction of autophagy. Also, FOXOs activate transcription of Cited2 and polycomb protein Bmi1, which inhibit expression of cyclin-dependent kinase inhibitors $p16^{Ink4a}$ and $p19^{Arf}$. In the low energetic state, activated AMPK stimulate STR1 by activating expression of NAD^+. Also, FOXO activity is positively regulated by AMPK. AMP, ATP, adenosine monophosphate, or triphosphate; LKB, liver kinase B1 AMPK, AMP-activated protein kinase; PI3K, phosphatidylinositol 3-kinase; mTOR mammalian target of rapamycin; TSC1/2, tuberous sclerosis complex 1/2; PTEN, phosphatase, and tensin homolog; ATM, ataxia telangiectasia mutated kinase; BID, BH3-interacting domain death agonist; HIF, hypoxia-inducible factor; MAPK, mitogen-activated protein kinase; SIRT1, Sirtuin 1; FOXO, forkhead homeobox type O; Cited2, Cbp/p300-interacting transactivator 2; Bmi1, B cell-specific Moloney murine leukemia virus integration site 1.

Thus, mTOR activity does not exhibit exclusively self-renewing or differentiation-inducing outcomes. Regarding its physiological effects, it seems more adequate to consider mTOR signaling as critical to adapt cellular energetic to particular environmental and functional circumstances whether it is promotion of the self-renewal or of the differentiation process.

6.1.4.2 AMPK and LKB1

AMPK activated protein kinase (AMPK) is a cellular energetic sensor that is activated when the ratio AMP/ATP is high (low energetic state). In this situation, AMP is fixed to the subunit of the enzyme and provokes its conformation change that allows its activation by the LKB1 (Liver kinase B1). Activated AMPK inhibits anabolic pathways (pathways that consume ATP), such as lipogenesis and mTOR-regulated protein synthesis, while simultaneously activating catabolic pathways such as FAO and glucose uptake [82], autophagy, and mitochondrial biogenesis for efficient ATP generation [83]. AMPK mediates control of autophagy indirectly via inhibition of the mTOR or directly by the activation of the autophagy-related proteins [84]. Also, AMPK is implicated in gene expression and cell cycle regulation [85,86]. An activation of AMPK in vitro maintains mESCs self-renewal and pluripotency [87]. Interestingly, chemical AMPK activators impede fibroblast reprogramming and decrease of generation of iPS in mouse and human, via inhibition of Oct4 expression [88]. Accordingly, RNA interference knockdown of LKB1 in hESCs and inhibition of AMPK result in up-regulation of pluripotency genes of Oct4 and Nanog, but in down regulation of differentiation markers [89].

6.1.4.3 FOXOs

Exhibiting multiple roles in the stem cell's metabolic homeostasis, the forkhead box protein O (FOXO) family of transcription factors has been shown to be essential for both adult and ESCs [90]. FOXO proteins activate the expression of genes that encode enzymes required for the detoxification of ROS and promote quiescence [91–93]. FOXO activity is positively regulated by AMPK and negatively regulated by PI3K pathway signaling, as AKT (activated by PI3K) phosphorylates FOXO proteins to cause its translocation out of the nucleus. Also, FOXO proteins are implicated in the regulation of the cell cycle, DNA repair and apoptosis [94], and LKB1 transcription [95]. FOXO family members are important for the maintenance of the pluripotency of ESCs [96], which is partly explained by the regulation of OCT4 and SOX2 expression [97].

6.2 Adult Stem Cells

In contrast to ESCs, somatic stem cells divide rapidly during fetal development to support organogenesis but are often quiescent or slowly cycling in adult tissues, entering the cell cycle periodically to maintain tissue homeostasis [66].

Various stem cells, including hematopoietic stem cells (HSCs), mesenchymal stem cells (MSCs), and neural stem cells (NSCs), reside in low oxygenated niche [67] (see Chapters 3–5). Although experimental evidence indicates that low oxygen tensions contribute to the maintenance of an undifferentiated state, influence proliferation, and cell fate commitment and energetic profile [18,48,98], the glycolytic profile in stem cells is a result of the intrinsic metabolic set-up.

It is suggested that adult stem cells exhibit self-renewal and multipotency mainly employing asymmetric cell division. This implies that adult stem cells should possess an energetic pattern that ensures quantitatively and qualitatively two opposite demands: quiescence (slow cycling) and self-renewal in low oxygenated niche or lineage specification toward more energy requiring progenitors. For this reason, adult stem cells should exhibit huge metabolic plasticity that enables energetic reprogramming on the basis of the functional demands.

6.2.1 Energetic Profile of HSCs

Recent direct measurements revealed that O_2 saturation in bone marrow is low (1.3–4%). Sophisticated microscopic system evidenced bone marrow O_2 decreasing gradient, from endosteal to deep perisinusoidal regions [99] where the most primitive HSCs were localized [100]. Along this gradient reside the different functional categories of the HSCs. Generally, slow cycling HSCs possess long-term reconstitution ability (LT-HSCs) upon transplantation, whereas actively cycling HSCs exhibit only short–term reconstitution (ST-HSCs). LT-HSCs give rise to ST-HSCs, which in turn produce lineage-specific progenitors [98] (see Chapter 1). These functional distinctions in the primitive hematopoietic cells correlate with metabolic/energetic particularities.

It should be noted that in further text "HSCs" stand for "population enriched true HSCs," containing multipotent and committed progenitors cells as well (see Chapter 1).

We could logically suppose that HSCs residing in low O_2 niche rely heavily on the anaerobic glycolysis to maintain their quiescent state [67]. As expected, in vitro studies showed that mouse LT-HSCs are more glycolytic because they consume less oxygen, have lower mitochondrial potential [101], and exhibit increased levels of glycolytic metabolites and enzymes in comparison with ST-HSCs, restricted progenitors, or differentiated cells [36,102]. Glycolysis is regulated by the activity of pyruvate dehydrogenase kinase (PDK) enzymes, which inactivate pyruvate dehydrogenase, therefore promoting glycolytic production of lactate from pyruvate rather than allowing pyruvate to enter the TCA cycle [103]. Combined deletion of PDK2 and PDK4 results in mild defects in HSC reconstituting ability and an increase in proliferation, suggesting that persistent pyruvate dehydrogenase activation impairs HSC function [36]. Using metabolome analysis, the same authors found that fructose 1,6-bisphosphate (F1,6BP) is enriched in HSCs compared to committed progenitors. F1,6BP allosterically activates pyruvate kinase and the ensuing pyruvate levels were also drastically elevated exclusively in primitive hematopoietic cells (Figure 6.1). This adaptation occurs to maximize glycolytic flux for this ATP-producing step of glycolysis. Also, TCA cycle-related metabolites such as 2-ketoglutarate, acetyl-CoA, and succinyl-CoA were not detected, supporting the role of a diminished TCA cycle in HSC-enriched population [36].

A proteomic study of the transition from HSC-enriched population Lin$^-$Sca-1$^+$c-kit$^+$ (LSK) to myeloid progenitors showed that hexokinase isoform HK1 expression is higher in LSK populations whereas HK2 and HK3 are more expressed in the myeloid progenitors [104]. HK1 is associated with catabolism and utilization of the glycolysis as the source of the energy in slowly cycling cells, whereas HK2 and HK3 are associated with the anabolic roles [105] in fast proliferating committed progenitors [6].

However, the HSC-enriched population also has functional mitochondria that enable a rapid switch to mitochondrial OXPHOS to meet the robust energy demands associated with differentiation [106]. Moreover, mitochondria in these primitive hematopoietic cells are relatively inactive [101]. In contrast, when HSCs move toward lineage-committed progenitors, mitochondrial content and associated ROS production increase [67].

Mitochondrial metabolism is also important for the differentiation of HSCs. Conditional deletion of PTPMT1, a mitochondrial phosphatase that dephosphorylates phosphatidylinositol phosphates in the mitochondrial membrane, results in a striking accumulation of LSK cells and severe anemia caused by a block in differentiation [106]. This is accompanied by decreased oxygen consumption by hematopoietic progenitors, suggesting impaired mitochondrial respiration. Chemical inhibitor of the PTPMT1 induces reprogramming of cellular metabolism from OXPHOS to glycolysis leading to the preservation of the LT-HSC activity in culture (ex vivo) in part through hyperactivation of the AMPK [107].

Furthermore, a recent study provided direct evidence that fatty acid metabolism and promyelocytic leukemia protein (PML)/peroxisome proliferator-activated receptor (PPAR)-δ signaling have a crucial role for functioning of HSCs. Ito et al. identified that PML controls activation of the peroxisome proliferator-activated receptor-γ coactivator (PGC)-1α/PPAR-δ transcription complex, which stimulates expression of the genes promoting FAO [108]. Inhibition of mitochondrial FAO reduces both numbers and reconstitution capacity of HSCs, whereas agonists of the FAO pathway improved HSC maintenance. Also, FAO is operational in LSK stem cells but undetected in differentiated cells [109]. As it was suggested for ESCs, importance of FAO for the HSCs lies in generation of the intermediaries enabling epigenetic control of the stem cell fate and oxidative stress protection [67].

As in pluripotent stem cells, different signaling pathways are involved in the regulation of the self-renewal/differentiation and quiescence (slow cycling)/proliferation balance of somatic stem cells.

6.2.1.1 mTOR

It plays dual roles in HSC maintenance. TSC1 deletion (a model of constitutive mTORC1 activation) in HSCs leads to the loss of quiescence through increased mitochondrial biogenesis and production of ROS [110]. Furthermore, mTOR activation in HSCs by inflammatory cytokines leads to defective hematopoiesis [111]. Also, mTOR stimulation through phosphatase and tensin homolog deleted on chromosome 10 (PTEN) deletion depletes HSCs [112–114]. These effects were attenuated with rapamycin treatment. mTOR hyperactivation is associated with increased numbers of mitochondria and higher ROS levels in adult HSCs [110]. Constitutively active AKT signaling that activates mTOR causes accelerated proliferation accompanied with differentiation and consequently exhaustion of HSCs [115].

But inhibition of mTORC1 complex [116], through ablation of its binding partner Raptor, leads to defective hematopoiesis and reduced ability of HSCs to reconstitute irradiated mice.

Ultimately, mTOR signaling is directly responsible for HSC aging; rapamycin treatment in aged mice increases lifespan. The summation of these studies indicates that mTOR is essential in HSCs maintenance. But although maintaining lower TOR activity is beneficial for HSCs, a complete loss of mTOR activity is detrimental.

Further, importance of mTORC1-regulated autophagy for HSC maintenance was evidenced. LSK deficient in the proteins implicated in the autophagic response, autophagy related 7 (Atg7), lost the hematopoietic reconstitution ability in immunodeficient mice [117].

6.2.1.2 AMPK and LKB1

It was suggested that AMPK functions in HSCs influence mitochondrial homeostasis. HSCs with inactive AMPK have decreased levels of ATP and mitochondrial DNA (mtDNA) [118]. Numbers of HSCs in the bone marrow are substantially reduced several months after AMPK activity is abrogated [118] suggesting that the regulation of mitochondrial homeostasis by AMPK contributes to the control of HSC proliferation. However, AMPK activity is not necessary for HSC reconstitutive capacity in bone marrow transplantation assays [118].

In contrast, it seems that LKB1 has crucial roles in regulating HSC function [119,120]. Conditional deletion of LKB1 in the mouse hematopoietic system leads to a loss of quiescence in HSCs and to a transient increase in the number of hematopoietic progenitors, which is followed by a decline in the number of hematopoietic cells in the bone marrow and progressive pancytopenia. The effects of LKB1 on HSCs are likely mediated via regulation of mitochondrial biogenesis, as LKB1 deletion leads to down regulation of PGC–PPAR pathway coactivators and their target genes, which have critical role in the mitochondrial biogenesis and function [119].

6.2.1.3 FOXOs

In addition to the role in the regulation of redox homeostasis, FOXOs protect HSCs from metabolic stress during both starvation and cytokine withdrawal via induction of autophagy in mouse HSCs, which allows avoidance of an energy crisis [121]. Deletion of FOXO family members leads to the premature depletion of these adult stem cells, over proliferation of committed progenitors, increase in the differentiated effector cells, and decrease in the hematopoietic reconstitutive capacity (i.e., the number of real HSC) [91,92,122]. The physiological role of FOXOs in the hematopoietic cells is partly explained by FOXO-mediated upregulation of expression of genes that are involved in cell cycle arrest, including p27KIP1, p57KIP2, and cyclin G2 (encoded by Cdkn1b, Cdkn1c, and Ccng2, respectively) [91].

6.2.2 Other Adult Stem Cells

6.2.2.1 Neural Stem Cells

Neural stem cells (NSCs) in the dentate gyrus of the hippocampus are thought to reside in a low oxygen environment, given their staining with the hypoxia marker pimonidazole and poor vascularization of this anatomic structure [123]. NSCs generate new neurons that are necessary for the cognitive function. Proliferative NSCs and progenitor cells (NPCs) have high glycolytic activity [124] but require a high level of lipid synthesis for proper proliferation [125]. Also, neurogenesis from ESCs is stimulated by the eicosanoid pathway and fatty acid metabolism [63].

Considering the regulatory metabolic pathways, mTOR-regulated activity is implicated in NSCs/NPCs proliferation, differentiation, and migration [126].

AMPK activity controls NSCs/NPCs proliferation. For example, in the developing mouse brain, abrogating normal AMPK activity leads to defective mitosis of NPCs and abnormal brain development [127]. This possible effect is consistent with the identification of a role for AMPK substrate phosphorylation in mitosis [85].

FOXO proteins also have a crucial role in NSC fate decisions [128,129]. Adult *Foxo3*$^{-/-}$ mice have fewer NSCs than their wild-type counterparts, and cultured *Foxo3*$^{-/-}$ NSCs exhibit decreased self-renewal and an impaired ability to generate different neural lineages. Gene expression profiling suggests that FOXO3 regulates the NSC pool by inducing a gene expression program that preserves quiescence, prevents premature differentiation, and controls oxygen metabolism.

6.2.2.2 Muscle Stem Cells

Regeneration of the skeletal muscle is enabled by the muscle stem cells (MuSCs) and satellite cells (SCs) that are localized, as HSCs, in an anatomically specialized location SC niche [130,131]. The majority of SCs exist in a quiescent/slowly cycling state. In response to injury or trauma, the SCs become actively cycling cells that simultaneously proliferate and differentiate toward myogenic lineage. One proportion of SCs returns to G_0, quiescent state, ensuring the stem cell pool. After several rounds of proliferation, SCs exit the cell cycle and undergo differentiation and fusion to form immature myotubes, which mature to myofibers [132].

Energy demands and metabolic status of SCs are an area that is still elusive. SCs contain small and scarce mitochondria, which appear to be tightly packed around the nucleus. Differentiation and fusion into myotubes/myofibers is associated with a dramatic increase of aerobic metabolism. Surprisingly, interventions that promote a shift in skeletal muscle metabolism from glycolysis to OXPHOS, such as chronic low-frequency stimulation of the peroneal nerve (a model of endurance exercise training), have been observed to lead to an increase in SC number [133]. Further, caloric restriction induces an increase in total SC number, increased SC mitochondrial abundance and OXPHOS activity, and an increased proliferative capacity of SCs associated with a significant increase in the transplant efficiency of these cells [134]. The elevated SC mitochondrial activity has been proposed as the likely mechanism for the observed increase in SC transplant efficiency

[135]. In contrast, Rocheteau et al. identified SCs showing the lowest level of mitochondrial activity as the population most capable of replenishing the SC pool [136,137].

Altogether, these results suggest that SCs refer to heterogeneous population exhibiting different metabolic properties. The heterogeneity and incapacity to isolate pure SC population could be an explanation for the metabolic difference observed in the different studies.

It is interesting that the number of SCs can be influenced by their microenvironment, fibers to which they are attached, with more SCs being associated with fibers that are predominantly oxidative (slow, type I fibers) than with fibers that rely primarily on glycolysis (fast, type II fibers) [132,138,139].

Gene expression analysis indicated the importance of the lipid metabolism for the SC function, as it is the case for the HSCs/HPCs since the genes for lipid transporter activity were also enriched in quiescent SCs [140].

In contrast to the SCs, the energetic metabolism of the muscle progenitors and precursors are well characterized. Utilizing the C2C12 myogenic cell line, it was demonstrated that proliferating myoblasts rely on the glycolysis, with an increase in both mitochondrial density and OXPHOS activity following differentiation [141,142]. Upon differentiation into myotubes, glycolysis increases as well as OXPHOS. The latter is mediated partly due to expression of the highly active M1 isoform of glycolytic enzyme pyruvate kinase (PKM1) that directs the pyruvate into TCA [6]. Also, differentiation is associated with Sirtuin 1, AMPK, and FoxO3A down regulation [135,143]. Increased availability of amino acids strongly stimulates muscle protein synthesis and reparation is critically regulated by PI3K-AKT-mTOR pathway [144–146]. Furthermore, AMPK-FOXO axis functions as the energetic homeostasis salvation mechanism in fibroblasts and skeletal myotubes. Under glucose deprivation AMPK phosphorylates FOXO3, leading to its mitochondrial accumulation and the expression of OXPHOS machinery [147] as a recovery mechanism to sustain energy metabolism during starvation.

6.2.2.3 Mesenchymal Stem Cells

Routinely in laboratory practice heterogeneous population of the mesenchymal stromal cells (MStroC) containing stem cells (MSCs) is expanded at 20% of O_2, which is much higher than in the physiological niche (see Chapters 1 and 4). Even this expansion in ambient air induces higher expression of the glycolytic enzymes and lower levels of the OXPHOS proteins in MStroC than in more differentiated osteoblasts [148]. This indicates that glycolytic-based anabolic shift is an intrinsic property of primitive cells that could be potentiated in vitro when MStroC culture is maintained in low O_2 concentration. This treatment ameliorates MSCs functional properties: colony forming capacity, proliferative capacity, long-term self-renewal, and ability for the osteogenic, chondrogenic, or adipogenic differentiation [149,150] as well as a decrease in senescence and ROS production. Interestingly, it was demonstrated that osteogenesis and adipogenesis of MSCs are associated with mitochondrial biogenesis and an increase of OXPHOS, while chondrogenesis required reduced O_2 consumption and mitochondrial activity [151]. Also, MSCs show extreme metabolic flexibility, surviving in the ischemic condition for several days (work in progress).

6.3 Oxidative Status of the Stem Cells

Stem cells' fate is tightly regulated by the level of O_2 present in their niche [152] (see Chapters 3–5). This is driven by numerous oxidation/reduction reactions that govern stem cells' energetic metabolism and gene expression. But utilization of O_2 as an integral part of biochemical pathways gives rise to continuous generation of metabolic by-products, ROS and reactive nitrogen species (RNS). In this manner, the aerobes develop physiological mechanisms that enable ROS/RNS to be either detoxified or utilized as intracellular or intercellular signaling messengers. Thus, balance between ROS and RNS production and neutralization by the cell antioxidative defense molecule determines cell oxidative status.

6.3.1 ROS and RNS: Origins and Features

ROS and RNS are not single entities but represent a broad range of chemically distinct reactive species with diverse biological reactivity [153]. There are several intracellular sources of ROS: mitochondria ETC, the membrane-bound NADPH oxidase (NOX) complex, endoplasmic reticulum, oxidoreductase enzymes, and metal catalyzed oxidation [34,153] (Figure 6.3). It was estimated that on average 0.1% to 0.2% of the O_2 consumed by the mitochondria is transformed in ROS [154] but it depends greatly on the cell metabolic pattern.

Leaked electrons from ETC can be captured by O_2, forming anion superoxide (O_2^-) which, along with nitric oxide (NO•), initiates production of ROS and RNS. O_2^- can be rapidly converted into hydrogen peroxide (H_2O_2) by superoxide dismutase (SOD) or it can react with NO• and form peroxynitrite ($ONOO^-$).

H_2O_2 can be converted into hydroxyl radical (•HO) in the presence of metal (Fe^{2+} or Cu^{2+}) ions (Fenton reaction). In the presence of CO_2, which can be formed in the multiple intracellular enzymatic reactions, $ONOO^-$ forms RNS called nitrosoperoxocarboxylate, ($ONO_2CO_2^-$), which decomposes rapidly into nitrogen dioxide radical (NO•) and carbonate radical (CO_3^-). The vast majority of the biological reactions of ($ONO_2CO_2^-$) including those with oxidant sensitive probes are mediated by these radicals [155]. In the presence of chloride and thiocyanate, myeloperoxidase catalyzes H_2O_2 conversion into oxidants hypochlorous (HOCl) or hypothiocyanous acid (HOSCN), respectively [156,157]. The cellular content of the abovementioned oxidants is a result of their generation and capture by a battery of anti-oxidant enzymes (catalase, glutathione peroxidase, glutathione reductases, peroxiredoxins, thioredoxins, thioredoxin reductases, methionine sulphoxide reductases, etc.) [153].

Different ROS and RNS have distinct chemical properties, which result from different reactivity, half-life, and lipid solubility [158]. As signaling molecules they

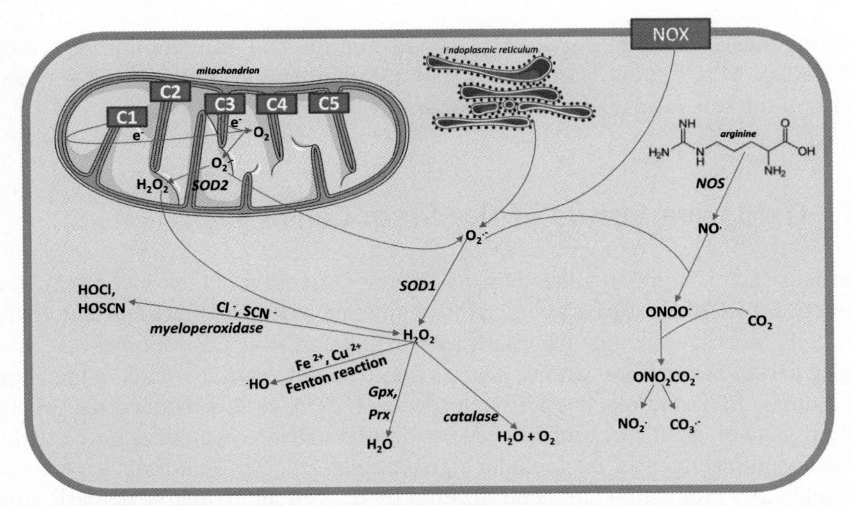

Figure 6.3 Schematic illustration of the cellular ROS and RNS network. Leaked electrons from mitochondria electron transport chain (ETC) can be captured by O_2, forming anion superoxide (O_2^-), which, together with nitric oxide (NO•) initiates production of ROS and RNS. O_2^- can be generated by the membrane-bound NADPH oxidase (NOX) complex, endoplasmic reticulum, or mitochondrial ETC. Once formed, O_2^- is rapidly converted into hydrogen peroxide (H_2O_2) by superoxide dismutase (SOD) or reacts with NO• to form peroxynitrite ($ONOO^-$). H_2O_2 can be converted into hydroxyl radical (•HO) in the presence of metal (Fe^{2+}or Cu^{2+}) ions in the Fenton reaction. In the presence of CO_2, $ONOO^-$ forms the RNS nitrosoperoxocarboxylate, ($ONO_2CO_2^-$), which decompensates rapidly into nitrogen dioxide radical (NO_2) and carbonate radical (CO_3^-). In the presence of chloride and thiocyanate, myeloperoxidase catalyzes the conversion of H_2O_2 to the oxidants hypochlorous (HOCl) or hypothiocyanous acid (HOSCN), respectively. Catalase, glutathione peroxidase (Gpx) or peroxiredoxins (Prx), decompose H_2O_2 to H_2O. C1–5; respiratory chain (C) complexes 1 to 5. ROS and RNS suggested to be involved in HSC fate are highlighted in red.

induce oxidation, nitrosylation, glutathionylation of the cysteine-depending proteins (mediated preferentially by H_2O_2 or NO•, respectively), oxidation of the metalloproteins containing iron-sulfur cluster (mediated by O_2^-), or protein hydroxylation (mediated by the •HO) [159]. These chemical modifications cause allosteric or conformational change of enzymes, which triggers distinct signaling events [158,160]. Also, the role of the RNS in the stem cell biology has been established [161–163]. However, compared with the work on ROS, there are considerably fewer studies that address RNS-mediated effects on stem cells and the signaling networks in which they participate [164] are less abundant.

6.3.2 ROS and RNS Rheostat

Metabolism of the stem cell is adapted to minimize ROS generation. As described in this chapter, stem cell survival is highly dependent on glycolysis, anabolic pathways, and scarce mitochondrial activity. Furthermore, HSC-enriched population or ESCs showed importantly higher expression of antioxidative defense genes compared to progenitors cells [102,165]. At low oxygen concentration, like the one presented in stem cell niche, mitochondria can still work albeit with less efficiency [166]. This minimal mitochondrial activity is required for the anaplerosis and for the production of basal quantity ROS required for the stem cell maintenance [167,168]. The appropriate/physiological ROS production is required to assure functions of HSCs/HPCs, mobilization, survival, differentiation, proliferation, and response to growth factors [34,167,169–173]. Furthermore, ROS modify gene expression and key metabolic regulators activity determining specific metabolic pattern and cell fate decision [159]. The physiological level of ROS stimulates MStroC proliferation [174] or ESC self-renewal. It was demonstrated that ROS directly modulate expression of the Oct4, required for the maintenance of the ESC pluripotency. But, compared to the adult stem cells, ESCs are less sensitive to ROS-mediated DNA damage and senescence due to the efficient capability to repair DNA double-strand breaks [175] that could represent an adaptation to aerobic conditions during numerous passages of ESC lines ex vivo.

Mitochondrial abundance and respiration are increased in the committed progenitors, which is associated with increased ROS production [101,167,176]. This augmentation is required for the efficient stem cell differentiation. For example, differentiation of ESCs toward the cardiac lineage or MStroCs toward adipocytes, chondrocytes, or osteoblasts depends on the NOX or mitochondria produced-ROS [177–181].

Interestingly, reprogramming the parental somatic cells that are associated with mitochondrial regression, ETC down regulation, and glycolysis enzyme up-regulation interconnects increased ROS production and DNA damage [182] that could be diminished in the low O_2 environment or in the presence of the anti-oxidant [47,182].

In contrast, oxidative stress appears when ROS production overcomes antioxidative defense activity. The excess of ROS is associated with DNA damage, lipid peroxidation, senescence, and apoptosis [183–185]. Reduction in the long-term bone marrow reconstitution capacity (thus, reduction in real HSC number) evidenced by the serial transplantations is caused by the elevation of the ROS levels and accumulation of the DNA damage in HSCs/HPCs, which leads to cell aging and HSC pool exhaustion [108,172,186]. In ESCs increased ROS production induces transient cell cycle arrest and apoptosis [187]. Also, a high level of ROS attenuates MSC function (adhesion, length of the telomeres) [174].

Altogether these data give evidence that physiological ROS and RNS production are needed for stem cell maintenance, and that increased ROS and RNS production associated with oxidative stress provoke stem cell functional exhaustion.

6.3.3 ROS-Modified Nutrient-Sensing Pathways and Redox Sensors Implicated in Stem Cell Functions

ROS as signaling molecules modulate nutrient-sensing pathways or activity of redox sensors to direct metabolic flux [188–190] in order to maintain stem cell redox homeostasis that predispose stem cell survival and function.

Recently several redox sensors have been shown to control stem cell functions.

Ataxia telangiectasia mutated (ATM) is a key regulator of the cell cycle checkpoint in the response of the DNA damage. It is directly activated by the ROS-induced cysteine residues oxidation [191,192], which in turn stimulates antioxidative cell response. Among several mechanisms, this response implicates ATM-mediated suppression of pro-apoptotic protein BID that is recognized as a mitochondrial ROS inducer [193] (Figure 6.2). Ito and coworkers were the first to demonstrate that in $Atm^{-/-}$ mice, an increase in the levels of ROS coincides with the loss of NSCs or HSCs from the phenotypically defined LSK population [108,194]. The underlying mechanisms were shown to involve ROS-mediated activation of p38 mitogen-activated protein kinase (MAPK), which up-regulates the cyclin-dependent kinase inhibitors, $p16^{Ink4a}$ and $p19^{Arf}$ [186]. Also, oxidation of p38 MAPK is implicated in the regulation of the self-renewal of the mESC and NSCs [194,195] as well as in the balance between MSCs proliferation and differentiation [196].

In response to oxidative stress, activated ATM-dependent inhibition of mTOR maintains the quiescence and function of HSCs by repressing ROS production and mitogenesis [110,191]. It was suggested that mTOR can be activated directly in response to oxidative stress [197]. Overactivity of mTOR in $Tsc1^{-/-}$ mice increases mitochondrial biogenesis and ROS production; this provokes attenuated HSCs function [110]. But in response to the oxidative stress, ROS abrogates AKT kinase function, which is manifested in the inhibition of the mTOR and consecutive decrease in the ROS cellular content. It correlates with HSCs arrest in quiescent state and with the block in differentiation [198]. In contrast, ROS-mediated activation of the mTOR seems to have a positive effect on the self-renewal and proliferation on the NSCs/NPCs [199].

Further, ROS-induced activation of tumor suppressor p53 triggers apoptosis or expression of the antioxidative defense enzymes in the primitive hematopoietic cells and ESC [191,200].

Key metabolic mediators of HSCs, hypoxia inducible factors (HIF-1 and HIF-2) (see Chapter 7), have also been implicated in regulation of the stem cell redox status [36,101,201]. ROS-stabilized HIF-1 [202] induces adaptive metabolic responses that ensure redox homeostasis [203] such as inhibition of the mTOR signaling and mitochondrial activity [203] and activation of glycolytic metabolism [103]. Also, HIF-2 stimulates the expression of genes encoding antioxidant enzymes such as SOD in mice [204]. Depletion of the α subunit of HIF-2 (HIF-2α) leads to reduced expression of genes with antioxidant functions, such as heme oxygenase 1 (HMOX1) [205]. In correlation, HIF-2α loss induces increased cellular ROS and activation of p53 that in turn promote antioxidative response [195,206]. Also, genetic ablation HIF-2α causes activation of the OXPHOS and an increase in ROS associated with ensuing of the quiescence and self-renewal of HSCs [206]. Transcription factor MEIS1 controls HIFs. Loss of MEIS1 results in down regulation of both HIF-1α and HIF-2α, leading to the stimulation of the mitochondrial metabolism, ROS production, and apoptosis of HSCs that is reversible in the presence of antioxidants [207].

FOXO transcription factors activate the expression of genes that encode enzymes required for the detoxification of ROS [208]. Knockout mice for FOXO demonstrate increased cycling and impaired long-term repopulating potential of primitive hematopoietic cells [91,92]. FOXO-mediated ROS production regulation is critical for NSCs maintenance. Inhibition of FOXO leads to an increase in the ROS content that primes NSC differentiation [129]. FOXO-mediated regulation of ROS in HSCs is mediated partly by the regulation of ATM expression [122]. Treatment of $Atm^{-/-}$ and *Foxo*-deficient mice with of *N*-acetyl cysteine (NAC), scavenger of HOCl and ·HO [209], reversed ROS-induced defects. Several factors are known to coordinate the balance between stem cell self-renewal and commitment with mitochondrial function. These include members of the polycomb family of proteins, such as B lymphoma Mo-MLV insertion region 1 homolog (Bmi1) and PR-domain-containing 16 (Prdm-16), and regulators of the DNA damage response, for example apurinic/apyrimidinic (AP) endonuclease1/redox factor-1 and redox regulator nuclear factor erythroid-2-related factor 2 (Nrf2) [34,210–212].

NAD^+/NADH ratio is a measure of cellular redox status and is important for the maintenance of the glycolytic flux. High NAD^+/NADH ratio, a state associated with low-energy as well as oxidative stress [213], activates the histone deacetylase Sirtuin1 (SIRT1) (one of seven mammalian Class III deacetylases called Sirtuins, SIRT). In turn, activated SIRT control expression of the OXPHOS enzymes and PGC1, key regulator of the mitochondrial biogenesis [214,215]. Thus, SIRT regulates energetic metabolism in order to adapt it to maintain cellular homeostasis according to oxidative cell status. This also implicates SIRT1 crosstalk with other metabolic regulators (e.g., FOXO3, p53), repair of DNA damage [216,217], and control of apoptosis [218,219]. In the low energetic state, activated AMPK stimulates STR1 by activating expression of NAD^+ [220,221]. Loss of the SIRT1 results in the aging, defective lineage specification, accumulation of ROS and DNA damage in HSCs and HPCs [222]. Also, activated SIRT1 promotes proliferation of adult rat MuSCs [143] and represses their differentiation into myocytes [135]. Indeed, a reduction in NAD^+/NADH, a state inhibitory for SIRT1 activity, is associated with mouse muscle cell differentiation. In the ESCs, SIRT1 contributed to pluripotent stem cell function since it is highly expressed in human and mouse ESCs, and its expression declines during differentiation [223]. Yet, under basal conditions, mouse ESCs that lack SIRT1 show no obvious defects [224].

In opposition, as widely present cofactor of the many metabolism-regulated enzymes, decrease in the NAD^+/NADH ratio below the physiological level could cause an important decrease in enzymatic activity (reductive stress), resulting in an altered metabolic homeostasis [225].

It is interesting to note that while adult stem cells are highly sensitive to oxidative stress, progenitors seem to be more resistant even if the content of the ROS is higher [108,167]. Atm deletion is deleterious for the HSC activity, but proliferation and differentiation of the progenitors is not impaired. Also, in the study of Yahata et al., actively cycling progenitors (Lin-$CD34^+$ $CD38^+$) accumulate much less ROS or DNA damage than a relatively quiescent, more primitive population with the highest proportion of HSCs [176].

What the underlying mechanism is of progenitors ROS resistance is still an open question. More efficient DNA repair mechanism, purge of damaged cells, antiapoptotic activity of mitochondria, and so on? These hypotheses seem to be worth testing.

However, even though progenitors can proliferate and differentiate in an ROS relatively rich environment, their lifespan is limited accordingly to the ROS-based theory of aging [226]. Conversely, stem cells cannot cope with oxidative stress, but they are, at least partially, "immortal." So, could we think that living with O_2 enables us to evolve, to differentiate, to expand, but deprives us from the *jeunesse éternelle*?

6.4 Technical Limitations in the Stem Cell Metabolic Studies

Exploring energetic metabolism and oxidative status of the stem cells is limited to a certain extent to technical impediments that should be considered in order to achieve equivocal and adequate interpretation of data. First, data obtained from the experiment in vitro should be interpreted cautiously because somatic stem cell properties can be profoundly changed in culture [159]. Second, due to the inability to obtain pure population of stem cells, metabolic studies are performed in the heterogeneous population consisting in a large proportion of the committed progenitors cells in which the stem cells remain a minority but the results were usually ascribed to them. For example, commonly used population for examining murine HSC activity comprised of cells that do not express surface markers presented on the mature blood cells (Lineage neg Lin^-, "L"), expressing markers stem cell antigen (Sca-1, "S"), c-kit (CD117, "K") (LSK). Moreover, only less than 10% of LSK have HSC potential based on the abovementioned assay [227] and the majority of cells are multipotent progenitors [228] (see Chapter 1). Third, considering redox biology of the stem cells, the characterization and identification of oxidant species responsible for redox-mediated stem cell fate determination and underlying signaling pathways remains ambiguous. This is partly due to the lack of suitable methodologies and chemical probes that allow unequivocal distinction between different ROS/RNS and their localization, kinetics, and quantity.

Neglecting these points might be both at the origin of the misinterpretations of the results and the source of inaccurate conclusions in stem cell physiology studies.

References

[1] McKnight SL. On getting there from here. Science 2010;330(6009):1338–9.
[2] Brinster RL, Troike DE. Requirements for blastocyst development in vitro. J Anim Sci 1979;49(Suppl. 2):26–34.
[3] Martin KL, Leese HJ. Role of glucose in mouse preimplantation embryo development. Mol Reprod Dev 1995;40(4):436–43.
[4] Jansen S, Pantaleon M, Kaye PL. Characterization and regulation of monocarboxylate cotransporters Slc16a7 and Slc16a3 in preimplantation mouse embryos. Biol Reprod 2008;79(1):84–92.
[5] Leese HJ. Metabolism of the preimplantation embryo: 40 years on. Reproduction 2012;143(4):417–27.

[6] Shyh-Chang N, Daley GQ, Cantley LC. Stem cell metabolism in tissue development and aging. Development 2013;140(12):2535–47.
[7] Van Blerkom J. Mitochondria in early mammalian development. Seminars Cell Dev Biol 2009;20(3):354–64.
[8] Tomitsuka E, Kita K, Esumi H. An anticancer agent, pyrvinium pamoate inhibits the NADH-fumarate reductase system–a unique mitochondrial energy metabolism in tumour microenvironments. J Biochem 2012;152(2):171–83.
[9] Weinberg JM, Venkatachalam MA, Roeser NF, Nissim I. Mitochondrial dysfunction during hypoxia/reoxygenation and its correction by anaerobic metabolism of citric acid cycle intermediates. Proc Natl Acad Sci USA 2000;97(6):2826–31.
[10] Johnson MT, Mahmood S, Patel MS. Intermediary metabolism and energetics during murine early embryogenesis. J Biol Chem 2003;278(34):31457–60.
[11] Graves CN, Biggers JD. Carbon dioxide fixation by mouse embryos prior to implantation. Science 1970;167(3924):1506–8.
[12] Brinster RL. Parental glucose phosphate isomerase activity in three-day mouse embryos. Biochem Genet 1973;9(2):187–91.
[13] Leese HJ, Barton AM. Pyruvate and glucose uptake by mouse ova and preimplantation embryos. J Reprod Fertil 1984;72(1):9–13.
[14] Houghton FD, Thompson JG, Kennedy CJ, Leese HJ. Oxygen consumption and energy metabolism of the early mouse embryo. Mol Reprod Dev 1996;44(4):476–85.
[15] Facucho-Oliveira JM, St John JC. The relationship between pluripotency and mitochondrial DNA proliferation during early embryo development and embryonic stem cell differentiation. Stem Cell Rev 2009;5(2):140–58.
[16] Zhou W, Choi M, Margineantu D, Margaretha L, Hesson J, Cavanaugh C, et al. HIF-1alpha induced switch from bivalent to exclusively glycolytic metabolism during ESC-to-EpiSC/hESC transition. EMBO J 2012;31(9):2103–16.
[17] Fischer B, Bavister BD. Oxygen tension in the oviduct and uterus of rhesus monkeys, hamsters and rabbits. J Reprod Fertil 1993;99(2):673–9.
[18] Simon MC, Keith B. The role of oxygen availability in embryonic development and stem cell function. Nat Rev Mol Cell Biol 2008;9(4):285–96.
[19] Evans MJ, Kaufman MH. Establishment in culture of pluripotential cells from mouse embryos. Nature 1981;292(5819):154–6.
[20] Brook FA, Gardner RL. The origin and efficient derivation of embryonic stem cells in the mouse. Proc Natl Acad Sci USA 1997;94(11):5709–12.
[21] Thomson JA, Itskovitz-Eldor J, Shapiro SS, Waknitz MA, Swiergiel JJ, Marshall VS, et al. Embryonic stem cell lines derived from human blastocysts. Science 1998;282(5391): 1145–7.
[22] Tesar PJ, Chenoweth JG, Brook FA, Davies TJ, Evans EP, Mack DL, et al. New cell lines from mouse epiblast share defining features with human embryonic stem cells. Nature 2007;448(7150):196–9.
[23] Stojkovic P, Lako M, Stewart R, Przyborski S, Armstrong L, Evans J, et al. An autogeneic feeder cell system that efficiently supports growth of undifferentiated human embryonic stem cells. Stem Cells 2005;23(3):306–14.
[24] Wang G, Zhang H, Zhao Y, Li J, Cai J, Wang P, et al. Noggin and bFGF cooperate to maintain the pluripotency of human embryonic stem cells in the absence of feeder layers. Biochem Biophys Res Commun 2005;330(3):934–42.
[25] Zhang J, Khvorostov I, Hong JS, Oktay Y, Vergnes L, Nuebel E, et al. UCP2 regulates energy metabolism and differentiation potential of human pluripotent stem cells. EMBO J 2011;30(24):4860–73.

[26] Becker KA, Ghule PN, Therrien JA, Lian JB, Stein JL, van Wijnen AJ, et al. Self-renewal of human embryonic stem cells is supported by a shortened G1 cell cycle phase. J Cell Physiol 2006;209(3):883–93.
[27] Vander Heiden MG, Cantley LC, Thompson CB. Understanding the Warburg effect: the metabolic requirements of cell proliferation. Science 2009;324(5930):1029–33.
[28] Varum S, Rodrigues AS, Moura MB, Momcilovic O, Easley CA, Ramalho-Santos J, et al. Energy metabolism in human pluripotent stem cells and their differentiated counterparts. PloS One 2011;6(6):e20914.
[29] Cho YM, Kwon S, Pak YK, Seol HW, Choi YM, Park do J, et al. Dynamic changes in mitochondrial biogenesis and antioxidant enzymes during the spontaneous differentiation of human embryonic stem cells. Biochem Biophys Res Commun 2006;348(4):1472–8.
[30] Chung S, Dzeja PP, Faustino RS, Perez-Terzic C, Behfar A, Terzic A. Mitochondrial oxidative metabolism is required for the cardiac differentiation of stem cells. Nat Clin Pract Cardiovasc Med 2007;4(Suppl. 1):S60–7.
[31] Folmes CD, Nelson TJ, Martinez-Fernandez A, Arrell DK, Lindor JZ, Dzeja PP, et al. Somatic oxidative bioenergetics transitions into pluripotency-dependent glycolysis to facilitate nuclear reprogramming. Cell Metab 2011;14(2):264–71.
[32] Kondoh H, Lleonart ME, Nakashima Y, Yokode M, Tanaka M, Bernard D, et al. A high glycolytic flux supports the proliferative potential of murine embryonic stem cells. Antioxidants Redox Signal 2007;9(3):293–9.
[33] Carracedo A, Cantley LC, Pandolfi PP. Cancer metabolism: fatty acid oxidation in the limelight. Nat Rev Cancer 2013;13(4):227–32.
[34] Wang K, Zhang T, Dong Q, Nice EC, Huang C, Wei Y. Redox homeostasis: the linchpin in stem cell self-renewal and differentiation. Cell Death Dis 2013;4:e537.
[35] Guppy M, Greiner E, Brand K. The role of the Crabtree effect and an endogenous fuel in the energy metabolism of resting and proliferating thymocytes. Eur J Biochem/FEBS 1993;212(1):95–9.
[36] Aad G, Abajyan T, Abbott B, Abdallah J, Abdel Khalek S, Abdelalim AA, et al. Search for dark matter candidates and large extra dimensions in events with a photon and missing transverse momentum in pp collision data at sqrt[s]=7 TeV with the ATLAS detector. Phys Rev Lett 2013;110(1):011802.
[37] Zhang WC, Shyh-Chang N, Yang H, Rai A, Umashankar S, Ma S, et al. Glycine decarboxylase activity drives non-small cell lung cancer tumor-initiating cells and tumorigenesis. Cell 2012;148(1–2):259–72.
[38] Warburg O. On the origin of cancer cells. Science 1956;123(3191):309–14.
[39] Teslaa T, Teitell MA. Pluripotent stem cell energy metabolism: an update. EMBO J 2015;34(2):138–53.
[40] Nichols J, Smith A. Naive and primed pluripotent states. Cell Stem Cell 2009;4(6):487–92.
[41] Takashima Y, Guo G, Loos R, Nichols J, Ficz G, Krueger F, et al. Resetting transcription factor control circuitry toward ground-state pluripotency in human. Cell 2014;158(6):1254–69.
[42] Shyh-Chang N, Zheng Y, Locasale JW, Cantley LC. Human pluripotent stem cells decouple respiration from energy production. EMBO J 2011;30(24):4851–2.
[43] Prigione A, Adjaye J. Modulation of mitochondrial biogenesis and bioenergetic metabolism upon in vitro and in vivo differentiation of human ES and iPS cells. Int J Dev Biol 2010;54(11–12):1729–41.
[44] Armstrong L, Tilgner K, Saretzki G, Atkinson SP, Stojkovic M, Moreno R, et al. Human induced pluripotent stem cell lines show stress defense mechanisms and mitochondrial regulation similar to those of human embryonic stem cells. Stem Cells 2010;28(4):661–73.

[45] Schieke SM, Ma M, Cao L, McCoy Jr JP, Liu C, Hensel NF, et al. Mitochondrial metabolism modulates differentiation and teratoma formation capacity in mouse embryonic stem cells. J Biol Chem 2008;283(42):28506–12.

[46] Johnson MT, Yang HS, Magnuson T, Patel MS. Targeted disruption of the murine dihydrolipoamide dehydrogenase gene (Dld) results in perigastrulation lethality. Proc Natl Acad Sci USA 1997;94(26):14512–7.

[47] Ezashi T, Das P, Roberts RM. Low O_2 tensions and the prevention of differentiation of hES cells. Proc Natl Acad Sci USA 2005;102(13):4783–8.

[48] Mohyeldin A, Garzon-Muvdi T, Quinones-Hinojosa A. Oxygen in stem cell biology: a critical component of the stem cell niche. Cell Stem Cell 2010;7(2):150–61.

[49] Varum S, Momcilovic O, Castro C, Ben-Yehudah A, Ramalho-Santos J, Navara CS. Enhancement of human embryonic stem cell pluripotency through inhibition of the mitochondrial respiratory chain. Stem Cell Res 2009;3(2–3):142–56.

[50] Chen G, Gulbranson DR, Hou Z, Bolin JM, Ruotti V, Probasco MD, et al. Chemically defined conditions for human iPSC derivation and culture. Nat Methods 2011;8(5): 424–9.

[51] Carey BW, Finley LW, Cross JR, Allis CD, Thompson CB. Intracellular alpha-ketoglutarate maintains the pluripotency of embryonic stem cells. Nature 2015;518(7539):413–6.

[52] Song JY, Park R, Kim JY, Hughes L, Lu L, Kim S, et al. Dual function of Yap in the regulation of lens progenitor cells and cellular polarity. Dev Biol 2014;386(2):281–90.

[53] Cho SW, Park JS, Heo HJ, Park SW, Song S, Kim I, et al. Dual modulation of the mitochondrial permeability transition pore and redox signaling synergistically promotes cardiomyocyte differentiation from pluripotent stem cells. J Am Heart Assoc 2014;3(2):e000693.

[54] Folmes CD, Dzeja PP, Nelson TJ, Terzic A. Metabolic plasticity in stem cell homeostasis and differentiation. Cell Stem Cell 2012;11(5):596–606.

[55] Han S, Auger C, Thomas SC, Beites CL, Appanna VD. Mitochondrial biogenesis and energy production in differentiating murine stem cells: a functional metabolic study. Cell Reprogram 2014;16(1):84–90.

[56] St John JC, Ramalho-Santos J, Gray HL, Petrosko P, Rawe VY, Navara CS, et al. The expression of mitochondrial DNA transcription factors during early cardiomyocyte in vitro differentiation from human embryonic stem cells. Cloning Stem Cells 2005;7(3):141–53.

[57] Prigione A, Lichtner B, Kuhl H, Struys EA, Wamelink M, Lehrach H, et al. Human induced pluripotent stem cells harbor homoplasmic and heteroplasmic mitochondrial DNA mutations while maintaining human embryonic stem cell-like metabolic reprogramming. Stem Cells 2011;29(9):1338–48.

[58] Birket MJ, Orr AL, Gerencser AA, Madden DT, Vitelli C, Swistowski A, et al. A reduction in ATP demand and mitochondrial activity with neural differentiation of human embryonic stem cells. J Cell Sci 2011;124(Pt 3):348–58.

[59] Okita K, Hong H, Takahashi K, Yamanaka S. Generation of mouse-induced pluripotent stem cells with plasmid vectors. Nat Protoc 2010;5(3):418–28.

[60] Folmes CD, Nelson TJ, Dzeja PP, Terzic A. Energy metabolism plasticity enables stemness programs. Ann N Y Acad Sci 2012;1254:82–9.

[61] Zhu S, Li W, Zhou H, Wei W, Ambasudhan R, Lin T, et al. Reprogramming of human primary somatic cells by OCT4 and chemical compounds. Cell Stem Cell 2010;7(6):651–5.

[62] Zaugg K, Yao Y, Reilly PT, Kannan K, Kiarash R, Mason J, et al. Carnitine palmitoyltransferase 1C promotes cell survival and tumor growth under conditions of metabolic stress. Genes Dev 2011;25(10):1041–51.

[63] Yanes O, Clark J, Wong DM, Patti GJ, Sanchez-Ruiz A, Benton HP, et al. Metabolic oxidation regulates embryonic stem cell differentiation. Nat Chem Biol 2010;6(6): 411–7.

[64] Wang J, Alexander P, Wu L, Hammer R, Cleaver O, McKnight SL. Dependence of mouse embryonic stem cells on threonine catabolism. Science 2009;325(5939):435–9.

[65] Ryu JM, Han HJ. L-threonine regulates G1/S phase transition of mouse embryonic stem cells via PI3K/Akt, MAPKs, and mTORC pathways. J Biol Chem 2011;286(27):23667–78.

[66] Burgess RJ, Agathocleous M, Morrison SJ. Metabolic regulation of stem cell function. J Intern Med 2014;276(1):12–24.

[67] Ito K, Suda T. Metabolic requirements for the maintenance of self-renewing stem cells. Nat Rev Mol Cell Biol 2014;15(4):243–56.

[68] Ochocki JD, Simon MC. Nutrient-sensing pathways and metabolic regulation in stem cells. J Cell Biol 2013;203(1):23–33.

[69] Greer EL, Brunet A. Signaling networks in aging. J Cell Sci 2008;121(Pt 4):407–12.

[70] Laplante M, Sabatini DM. mTOR signaling in growth control and disease. Cell 2012;149(2):274–93.

[71] Morita M, Gravel SP, Chenard V, Sikstrom K, Zheng L, Alain T, et al. mTORC1 controls mitochondrial activity and biogenesis through 4E-BP-dependent translational regulation. Cell Metab 2013;18(5):698–711.

[72] Efeyan A, Comb WC, Sabatini DM. Nutrient-sensing mechanisms and pathways. Nature 2015;517(7534):302–10.

[73] Inoki K, Li Y, Zhu T, Wu J, Guan KL. TSC2 is phosphorylated and inhibited by Akt and suppresses mTOR signalling. Nat Cell Biol 2002;4(9):648–57.

[74] Altomare DA, Khaled AR. Homeostasis and the importance for a balance between AKT/mTOR activity and intracellular signaling. Curr Med Chem 2012;19(22):3748–62.

[75] Lee JY, Kim YR, Park J, Kim S. Inositol polyphosphate multikinase signaling in the regulation of metabolism. Ann N Y Acad Sci 2012;1271:68–74.

[76] Schieke SM, Phillips D, McCoy Jr JP, Aponte AM, Shen RF, Balaban RS, et al. The mammalian target of rapamycin (mTOR) pathway regulates mitochondrial oxygen consumption and oxidative capacity. J Biol Chem 2006;281(37):27643–52.

[77] Zhou J, Su P, Wang L, Chen J, Zimmermann M, Genbacev O, et al. mTOR supports long-term self-renewal and suppresses mesoderm and endoderm activities of human embryonic stem cells. Proc Natl Acad Sci USA 2009;106(19):7840–5.

[78] Murakami M, Ichisaka T, Maeda M, Oshiro N, Hara K, Edenhofer F, et al. mTOR is essential for growth and proliferation in early mouse embryos and embryonic stem cells. Mol Cell Biol 2004;24(15):6710–8.

[79] Easley CA, Ben-Yehudah A, Redinger CJ, Oliver SL, Varum ST, Eisinger VM, et al. mTOR-mediated activation of p70 S6K induces differentiation of pluripotent human embryonic stem cells. Cell Reprogram 2010;12(3):263–73.

[80] Sampath P, Pritchard DK, Pabon L, Reinecke H, Schwartz SM, Morris DR, et al. A hierarchical network controls protein translation during murine embryonic stem cell self-renewal and differentiation. Cell Stem Cell 2008;2(5):448–60.

[81] Chen T, Shen L, Yu J, Wan H, Guo A, Chen J, et al. Rapamycin and other longevity-promoting compounds enhance the generation of mouse induced pluripotent stem cells. Aging Cell 2011;10(5):908–11.

[82] Winder WW, Hardie DG. AMP-activated protein kinase, a metabolic master switch: possible roles in type 2 diabetes. Am J Physiol 1999;277(1 Pt 1):E1–10.

[83] Hardie DG, Ross FA, Hawley SA. AMPK: a nutrient and energy sensor that maintains energy homeostasis. Nat Rev Mol Cell Biol 2012;13(4):251–62.

[84] Egan D, Kim J, Shaw RJ, Guan KL. The autophagy initiating kinase ULK1 is regulated via opposing phosphorylation by AMPK and mTOR. Autophagy 2011;7(6):643–4.
[85] Banko MR, Allen JJ, Schaffer BE, Wilker EW, Tsou P, White JL, et al. Chemical genetic screen for AMPKalpha2 substrates uncovers a network of proteins involved in mitosis. Mol Cell 2011;44(6):878–92.
[86] Mihaylova MM, Shaw RJ. The AMPK signalling pathway coordinates cell growth, autophagy and metabolism. Nat Cell Biol 2011;13(9):1016–23.
[87] Shi X, Wu Y, Ai Z, Liu X, Yang L, Du J, et al. AICAR sustains J1 mouse embryonic stem cell self-renewal and pluripotency by regulating transcription factor and epigenetic modulator expression. Cell Physiol Biochem 2013;32(2):459–75.
[88] Vazquez-Martin A, Vellon L, Quiros PM, Cufi S, Ruiz de Galarreta E, Oliveras-Ferraros C, et al. Activation of AMP-activated protein kinase (AMPK) provides a metabolic barrier to reprogramming somatic cells into stem cells. Cell Cycle 2012;11(5):974–89.
[89] Lai D, Chen Y, Wang F, Jiang L, Wei C. LKB1 controls the pluripotent state of human embryonic stem cells. Cell Reprogram 2012;14(2):164–70.
[90] Rafalski VA, Brunet A. Energy metabolism in adult neural stem cell fate. Prog Neurobiol 2011;93(2):182–203.
[91] Tothova Z, Gilliland DG. FoxO transcription factors and stem cell homeostasis: insights from the hematopoietic system. Cell Stem Cell 2007;1(2):140–52.
[92] Miyamoto K, Araki KY, Naka K, Arai F, Takubo K, Yamazaki S, et al. Foxo3a is essential for maintenance of the hematopoietic stem cell pool. Cell Stem Cell 2007;1(1):101–12.
[93] Tothova Z, Kollipara R, Huntly BJ, Lee BH, Castrillon DH, Cullen DE, et al. FoxOs are critical mediators of hematopoietic stem cell resistance to physiologic oxidative stress. Cell 2007;128(2):325–39.
[94] Webb AE, Brunet A. FOXO transcription factors: key regulators of cellular quality control. Trends Biochem Sci 2014;39(4):159–69.
[95] Lutzner N, De-Castro Arce J, Rosl F. Gene expression of the tumour suppressor LKB1 is mediated by Sp1, NF-Y and FOXO transcription factors. PloS One 2012;7(3):e32590.
[96] Zhang X, Yalcin S, Lee DF, Yeh TY, Lee SM, Su J, et al. FOXO1 is an essential regulator of pluripotency in human embryonic stem cells. Nat Cell Biol 2011;13(9):1092–9.
[97] Rimmele P, Zhang X, Ghaffari S. FoxO proteins in the control of stem cells. Med Sci 2012;28(3):250–4.
[98] Suda T, Takubo K, Semenza GL. Metabolic regulation of hematopoietic stem cells in the hypoxic niche. Cell Stem Cell 2011;9(4):298–310.
[99] Spencer JA, Ferraro F, Roussakis E, Klein A, Wu J, Runnels JM, et al. Direct measurement of local oxygen concentration in the bone marrow of live animals. Nature 2014;508(7495):269–73.
[100] Parmar K, Mauch P, Vergilio JA, Sackstein R, Down JD. Distribution of hematopoietic stem cells in the bone marrow according to regional hypoxia. Proc Natl Acad Sci USA 2007;104(13):5431–6.
[101] Simsek T, Kocabas F, Zheng J, Deberardinis RJ, Mahmoud AI, Olson EN, et al. The distinct metabolic profile of hematopoietic stem cells reflects their location in a hypoxic niche. Cell Stem Cell 2010;7(3):380–90.
[102] Unwin RD, Smith DL, Blinco D, Wilson CL, Miller CJ, Evans CA, et al. Quantitative proteomics reveals posttranslational control as a regulatory factor in primary hematopoietic stem cells. Blood 2006;107(12):4687–94.
[103] Takubo K, Nagamatsu G, Kobayashi CI, Nakamura-Ishizu A, Kobayashi H, Ikeda E, et al. Regulation of glycolysis by Pdk functions as a metabolic checkpoint for cell cycle quiescence in hematopoietic stem cells. Cell Stem Cell 2013;12(1):49–61.

[104] Klimmeck D, Hansson J, Raffel S, Vakhrushev SY, Trumpp A, Krijgsveld J. Proteomic cornerstones of hematopoietic stem cell differentiation: distinct signatures of multipotent progenitors and myeloid committed cells. Mol Cell Proteomics 2012;11(8):286–302.

[105] Wilson JE. Isozymes of mammalian hexokinase: structure, subcellular localization and metabolic function. J Exp Biol 2003;206(Pt 12):2049–57.

[106] Yu WM, Liu X, Shen J, Jovanovic O, Pohl EE, Gerson SL, et al. Metabolic regulation by the mitochondrial phosphatase PTPMT1 is required for hematopoietic stem cell differentiation. Cell Stem Cell 2013;12(1):62–74.

[107] Liu X, Zheng H, Yu WM, Cooper TM, Bunting KD, Qu CK. Maintenance of mouse hematopoietic stem cells ex vivo by reprogramming cellular metabolism. Blood 2015;125(10):1562–5.

[108] Ito K, Hirao A, Arai F, Matsuoka S, Takubo K, Hamaguchi I, et al. Regulation of oxidative stress by ATM is required for self-renewal of haematopoietic stem cells. Nature 2004;431(7011):997–1002.

[109] Ito K, Carracedo A, Weiss D, Arai F, Ala U, Avigan DE, et al. A PML-PPAR-delta pathway for fatty acid oxidation regulates hematopoietic stem cell maintenance. Nat Med 2012;18(9):1350–8.

[110] Chen C, Liu Y, Liu R, Ikenoue T, Guan KL, Liu Y, et al. TSC-mTOR maintains quiescence and function of hematopoietic stem cells by repressing mitochondrial biogenesis and reactive oxygen species. J Exp Med 2008;205(10):2397–408.

[111] Chen C, Liu Y, Liu Y, Zheng P. Mammalian target of rapamycin activation underlies HSC defects in autoimmune disease and inflammation in mice. J Clin Invest 2010;120(11):4091–101.

[112] Yilmaz OH, Valdez R, Theisen BK, Guo W, Ferguson DO, Wu H, et al. Pten dependence distinguishes haematopoietic stem cells from leukaemia-initiating cells. Nature 2006;441(7092):475–82.

[113] Zhang J, Grindley JC, Yin T, Jayasinghe S, He XC, Ross JT, et al. PTEN maintains haematopoietic stem cells and acts in lineage choice and leukaemia prevention. Nature 2006;441(7092):518–22.

[114] Lee JY, Nakada D, Yilmaz OH, Tothova Z, Joseph NM, Lim MS, et al. mTOR activation induces tumor suppressors that inhibit leukemogenesis and deplete hematopoietic stem cells after Pten deletion. Cell Stem Cell 2010;7(5):593–605.

[115] Kharas MG, Gritsman K. Akt: a double-edged sword for hematopoietic stem cells. Cell Cycle 2010;9(7):1223–4.

[116] Kalaitzidis D, Sykes SM, Wang Z, Punt N, Tang Y, Ragu C, et al. mTOR complex 1 plays critical roles in hematopoiesis and Pten-loss-evoked leukemogenesis. Cell Stem Cell 2012;11(3):429–39.

[117] Mortensen M, Soilleux EJ, Djordjevic G, Tripp R, Lutteropp M, Sadighi-Akha E, et al. The autophagy protein Atg7 is essential for hematopoietic stem cell maintenance. J Exp Med 2011;208(3):455–67.

[118] Nakada D, Saunders TL, Morrison SJ. Lkb1 regulates cell cycle and energy metabolism in haematopoietic stem cells. Nature 2010;468(7324):653–8.

[119] Gan B, Hu J, Jiang S, Liu Y, Sahin E, Zhuang L, et al. Lkb1 regulates quiescence and metabolic homeostasis of haematopoietic stem cells. Nature 2010;468(7324):701–4.

[120] Gurumurthy S, Xie SZ, Alagesan B, Kim J, Yusuf RZ, Saez B, et al. The Lkb1 metabolic sensor maintains haematopoietic stem cell survival. Nature 2010;468(7324):659–63.

[121] Warr MR, Binnewies M, Flach J, Reynaud D, Garg T, Malhotra R, et al. FoxO3a directs a protective autophagy program in haematopoietic stem cells. Nature 2013;494(7437):323–7.

[122] Yalcin S, Zhang X, Luciano JP, Mungamuri SK, Marinkovic D, Vercherat C, et al. Foxo3 is essential for the regulation of ataxia telangiectasia mutated and oxidative stress-mediated homeostasis of hematopoietic stem cells. J Biol Chem 2008;283(37): 25692–705.

[123] Mazumdar J, O'Brien WT, Johnson RS, LaManna JC, Chavez JC, Klein PS, et al. O_2 regulates stem cells through Wnt/beta-catenin signalling. Nat Cell Biol 2010;12(10):1007–13.

[124] Gershon TR, Crowther AJ, Tikunov A, Garcia I, Annis R, Yuan H, et al. Hexokinase-2-mediated aerobic glycolysis is integral to cerebellar neurogenesis and pathogenesis of medulloblastoma. Cancer Metab 2013;1(1):2.

[125] Knobloch M, Braun SM, Zurkirchen L, von Schoultz C, Zamboni N, Arauzo-Bravo MJ, et al. Metabolic control of adult neural stem cell activity by Fasn-dependent lipogenesis. Nature 2013;493(7431):226–30.

[126] Zhou J, Shrikhande G, Xu J, McKay RM, Burns DK, Johnson JE, et al. Tsc1 mutant neural stem/progenitor cells exhibit migration deficits and give rise to subependymal lesions in the lateral ventricle. Genes Dev 2011;25(15):1595–600.

[127] Dasgupta B, Milbrandt J. AMP-activated protein kinase phosphorylates retinoblastoma protein to control mammalian brain development. Dev Cell 2009;16(2):256–70.

[128] Paik JH, Ding Z, Narurkar R, Ramkissoon S, Muller F, Kamoun WS, et al. FoxOs cooperatively regulate diverse pathways governing neural stem cell homeostasis. Cell Stem Cell 2009;5(5):540–53.

[129] Renault VM, Rafalski VA, Morgan AA, Salih DA, Brett JO, Webb AE, et al. FoxO3 regulates neural stem cell homeostasis. Cell Stem Cell 2009;5(5):527–39.

[130] Mauro A. Satellite cell of skeletal muscle fibers. J Biophys Biochem Cytol 1961;9:493–5.

[131] Bentzinger CF, Wang YX, Dumont NA, Rudnicki MA. Cellular dynamics in the muscle satellite cell niche. EMBO Rep 2013;14(12):1062–72.

[132] Ryall JG. Metabolic reprogramming as a novel regulator of skeletal muscle development and regeneration. FEBS J 2013;280(17):4004–13.

[133] Putman CT, Dusterhoft S, Pette D. Changes in satellite cell content and myosin isoforms in low-frequency-stimulated fast muscle of hypothyroid rat. J Appl Physiol (1985) 1999;86(1):40–51.

[134] Cerletti M, Jang YC, Finley LW, Haigis MC, Wagers AJ. Short-term calorie restriction enhances skeletal muscle stem cell function. Cell Stem Cell 2012;10(5):515–9.

[135] Fulco M, Cen Y, Zhao P, Hoffman EP, McBurney MW, Sauve AA, et al. Glucose restriction inhibits skeletal myoblast differentiation by activating SIRT1 through AMPK-mediated regulation of Nampt. Dev Cell 2008;14(5):661–73.

[136] Rocheteau P, Gayraud-Morel B, Siegl-Cachedenier I, Blasco MA, Tajbakhsh S. A subpopulation of adult skeletal muscle stem cells retains all template DNA strands after cell division. Cell 2012;148(1–2):112–25.

[137] Rocheteau P, Vinet M, Chretien F. Dormancy and quiescence of skeletal muscle stem cells. Results Probl Cell Differ 2015;56:215–35.

[138] Christov C, Chretien F, Abou-Khalil R, Bassez G, Vallet G, Authier FJ, et al. Muscle satellite cells and endothelial cells: close neighbors and privileged partners. Mol Biol Cell 2007;18(4):1397–409.

[139] Manzano R, Toivonen JM, Calvo AC, Miana-Mena FJ, Zaragoza P, Munoz MJ, et al. Sex, fiber-type, and age dependent in vitro proliferation of mouse muscle satellite cells. J Cell Biochem 2011;112(10):2825–36.

[140] Fukada S, Uezumi A, Ikemoto M, Masuda S, Segawa M, Tanimura N, et al. Molecular signature of quiescent satellite cells in adult skeletal muscle. Stem Cells 2007;25(10):2448–59.

[141] Kraft CS, LeMoine CM, Lyons CN, Michaud D, Mueller CR, Moyes CD. Control of mitochondrial biogenesis during myogenesis. Am J Physiol Cell Physiol 2006;290(4):C1119–27.
[142] Leary SC, Battersby BJ, Hansford RG, Moyes CD. Interactions between bioenergetics and mitochondrial biogenesis. Biochim Biophys Acta 1998;1365(3):522–30.
[143] Rathbone CR, Booth FW, Lees SJ. FoxO3a preferentially induces p27Kip1 expression while impairing muscle precursor cell-cycle progression. Muscle Nerve 2008;37(1):84–9.
[144] Rennie MJ, Edwards RH, Halliday D, Matthews DE, Wolman SL, Millward DJ. Muscle protein synthesis measured by stable isotope techniques in man: the effects of feeding and fasting. Clin Sci 1982;63(6):519–23.
[145] Volpi E, Mittendorfer B, Wolf SE, Wolfe RR. Oral amino acids stimulate muscle protein anabolism in the elderly despite higher first-pass splanchnic extraction. Am J Physiol 1999;277(3 Pt 1):E513–20.
[146] Paddon-Jones D, Sheffield-Moore M, Zhang XJ, Volpi E, Wolf SE, Aarsland A, et al. Amino acid ingestion improves muscle protein synthesis in the young and elderly. Am J Physiol Endocrinol Metab 2004;286(3):E321–8.
[147] Peserico A, Chiacchiera F, Grossi V, Matrone A, Latorre D, Simonatto M, et al. A novel AMPK-dependent FoxO3A-SIRT3 intramitochondrial complex sensing glucose levels. Cell Mol Life Sci 2013;70(11):2015–29.
[148] Chen CT, Shih YR, Kuo TK, Lee OK, Wei YH. Coordinated changes of mitochondrial biogenesis and antioxidant enzymes during osteogenic differentiation of human mesenchymal stem cells. Stem Cells 2008;26(4):960–8.
[149] Estrada JC, Albo C, Benguria A, Dopazo A, Lopez-Romero P, Carrera-Quintanar L, et al. Culture of human mesenchymal stem cells at low oxygen tension improves growth and genetic stability by activating glycolysis. Cell Death Differ 2012;19(5):743–55.
[150] Sugrue T, Lowndes NF, Ceredig R. Hypoxia enhances the radioresistance of mouse mesenchymal stromal cells. Stem Cells 2014;32(8):2188–200.
[151] Pattappa G, Heywood HK, de Bruijn JD, Lee DA. The metabolism of human mesenchymal stem cells during proliferation and differentiation. J Cell Physiol 2011;226(10):2562–70.
[152] Ivanovic Z. Hypoxia or in situ normoxia: the stem cell paradigm. J Cell Physiol 2009;219(2):271–5.
[153] Nathan C, Cunningham-Bussel A. Beyond oxidative stress: an immunologist's guide to reactive oxygen species. Nat Rev Immunol 2013;13(5):349–61.
[154] Tahara EB, Navarete FD, Kowaltowski AJ. Tissue-, substrate-, and site-specific characteristics of mitochondrial reactive oxygen species generation. Free Radic Biol Med 2009;46(9):1283–97.
[155] Ferrer-Sueta G, Radi R. Chemical biology of peroxynitrite: kinetics, diffusion, and radicals. ACS Chem Biol 2009;4(3):161–77.
[156] Wardman P. Fluorescent and luminescent probes for measurement of oxidative and nitrosative species in cells and tissues: progress, pitfalls, and prospects. Free Radic Biol Med 2007;43(7):995–1022.
[157] Winterbourn CC. The challenges of using fluorescent probes to detect and quantify specific reactive oxygen species in living cells. Biochim Biophys Acta 2014;1840(2):730–8.
[158] D'Autreaux B, Toledano MB. ROS as signalling molecules: mechanisms that generate specificity in ROS homeostasis. Nat Rev Mol Cell Biol 2007;8(10):813–24.
[159] Bigarella CL, Liang R, Ghaffari S. Stem cells and the impact of ROS signaling. Development 2014;141(22):4206–18.
[160] Stamler JS, Singel DJ, Loscalzo J. Biochemistry of nitric oxide and its redox-activated forms. Science 1992;258(5090):1898–902.

[161] Reykdal S, Abboud C, Liesveld J. Effect of nitric oxide production and oxygen tension on progenitor preservation in ex vivo culture. Exp Hematol 1999;27(3):441–50.
[162] Michurina T, Krasnov P, Balazs A, Nakaya N, Vasilieva T, Kuzin B, et al. Nitric oxide is a regulator of hematopoietic stem cell activity. Mol Ther 2004;10(2):241–8.
[163] Epperly MW, Cao S, Zhang X, Franicola D, Shen H, Greenberger EE, et al. Increased longevity of hematopoiesis in continuous bone marrow cultures derived from NOS1 (nNOS, mtNOS) homozygous recombinant negative mice correlates with radioresistance of hematopoietic and marrow stromal cells. Exp Hematol 2007;35(1):137–45.
[164] Nogueira-Pedro A, Dias CC, Segreto HR, Addios PC, Lungato L, D'Almeida V, et al. Nitric oxide-induced murine hematopoietic stem cell fate involves multiple signaling proteins, gene expression and redox modulation. Stem Cells 2014;32(11):2949–60.
[165] Saretzki G, Armstrong L, Leake A, Lako M, von Zglinicki T. Stress defense in murine embryonic stem cells is superior to that of various differentiated murine cells. Stem Cells 2004;22(6):962–71.
[166] Gnaiger E. Bioenergetics at low oxygen: dependence of respiration and phosphorylation on oxygen and adenosine diphosphate supply. Respir Physiol 2001;128(3):277–97.
[167] Jang YY, Sharkis SJ. A low level of reactive oxygen species selects for primitive hematopoietic stem cells that may reside in the low-oxygenic niche. Blood 2007;110(8): 3056–63.
[168] Kunisaki Y, Bruns I, Scheiermann C, Ahmed J, Pinho S, Zhang D, et al. Arteriolar niches maintain haematopoietic stem cell quiescence. Nature 2013;502(7473):637–43.
[169] Hole PS, Pearn L, Tonks AJ, James PE, Burnett AK, Darley RL, et al. Ras-induced reactive oxygen species promote growth factor-independent proliferation in human CD34+ hematopoietic progenitor cells. Blood 2010;115(6):1238–46.
[170] Guzy RD, Schumacker PT. Oxygen sensing by mitochondria at complex III: the paradox of increased reactive oxygen species during hypoxia. Exp Physiol 2006;91(5):807–19.
[171] Sattler M, Winkler T, Verma S, Byrne CH, Shrikhande G, Salgia R, et al. Hematopoietic growth factors signal through the formation of reactive oxygen species. Blood 1999;93(9):2928–35.
[172] Hosokawa K, Arai F, Yoshihara H, Nakamura Y, Gomei Y, Iwasaki H, et al. Function of oxidative stress in the regulation of hematopoietic stem cell-niche interaction. Biochem Biophys Res Commun 2007;363(3):578–83.
[173] Tesio M, Golan K, Corso S, Giordano S, Schajnovitz A, Vagima Y, et al. Enhanced c-Met activity promotes G-CSF-induced mobilization of hematopoietic progenitor cells via ROS signaling. Blood 2011;117(2):419–28.
[174] Huang H, Kim HJ, Chang EJ, Lee ZH, Hwang SJ, Kim HM, et al. IL-17 stimulates the proliferation and differentiation of human mesenchymal stem cells: implications for bone remodeling. Cell Death Differ 2009;16(10):1332–43.
[175] Lan ML, Acharya MM, Tran KK, Bahari-Kashani J, Patel NH, Strnadel J, et al. Characterizing the radioresponse of pluripotent and multipotent human stem cells. PloS One 2012;7(12):e50048.
[176] Yahata T, Takanashi T, Muguruma Y, Ibrahim AA, Matsuzawa H, Uno T, et al. Accumulation of oxidative DNA damage restricts the self-renewal capacity of human hematopoietic stem cells. Blood 2011;118(11):2941–50.
[177] Kanda Y, Hinata T, Kang SW, Watanabe Y. Reactive oxygen species mediate adipocyte differentiation in mesenchymal stem cells. Life Sci 2011;89(7–8):250–8.
[178] Tormos KV, Anso E, Hamanaka RB, Eisenbart J, Joseph J, Kalyanaraman B, et al. Mitochondrial complex III ROS regulate adipocyte differentiation. Cell Metab 2011;14(4):537–44.

[179] Xiao Q, Luo Z, Pepe AE, Margariti A, Zeng L, Xu Q. Embryonic stem cell differentiation into smooth muscle cells is mediated by Nox4-produced H_2O_2. Am J Physiol Cell Physiol 2009;296(4):C711–23.

[180] Imhoff BR, Hansen JM. Differential redox potential profiles during adipogenesis and osteogenesis. Cell Mol Biol Lett 2011;16(1):149–61.

[181] Sart S, Tsai AC, Li Y, Ma T. Three-dimensional aggregates of mesenchymal stem cells: cellular mechanisms, biological properties, and applications. Tissue Eng Part B Rev 2014;20(5):365–80.

[182] Ji J, Sharma V, Qi S, Guarch ME, Zhao P, Luo Z, et al. Antioxidant supplementation reduces genomic aberrations in human induced pluripotent stem cells. Stem Cell Rep 2014;2(1):44–51.

[183] Naka K, Muraguchi T, Hoshii T, Hirao A. Regulation of reactive oxygen species and genomic stability in hematopoietic stem cells. Antioxid Redox Signal 2008;10(11):1883–94.

[184] Navarro-Yepes J, Burns M, Anandhan A, Khalimonchuk O, del Razo LM, Quintanilla-Vega B, et al. Oxidative stress, redox signaling, and autophagy: cell death versus survival. Antioxid Redox Signal 2014;21(1):66–85.

[185] Shao L, Li H, Pazhanisamy SK, Meng A, Wang Y, Zhou D. Reactive oxygen species and hematopoietic stem cell senescence. Int J Hematol 2011;94(1):24–32.

[186] Ito K, Hirao A, Arai F, Takubo K, Matsuoka S, Miyamoto K, et al. Reactive oxygen species act through p38 MAPK to limit the lifespan of hematopoietic stem cells. Nat Med 2006;12(4):446–51.

[187] Guo YL, Chakraborty S, Rajan SS, Wang R, Huang F. Effects of oxidative stress on mouse embryonic stem cell proliferation, apoptosis, senescence, and self-renewal. Stem Cells Dev 2010;19(9):1321–31.

[188] Anastasiou D, Poulogiannis G, Asara JM, Boxer MB, Jiang JK, Shen M, et al. Inhibition of pyruvate kinase M2 by reactive oxygen species contributes to cellular antioxidant responses. Science 2011;334(6060):1278–83.

[189] Sarbassov DD, Sabatini DM. Redox regulation of the nutrient-sensitive raptor-mTOR pathway and complex. J Biol Chem 2005;280(47):39505–9.

[190] Brunelle JK, Bell EL, Quesada NM, Vercauteren K, Tiranti V, Zeviani M, et al. Oxygen sensing requires mitochondrial ROS but not oxidative phosphorylation. Cell Metab 2005;1(6):409–14.

[191] Ditch S, Paull TT. The ATM protein kinase and cellular redox signaling: beyond the DNA damage response. Trends Biochem Sci 2012;37(1):15–22.

[192] Guo Z, Kozlov S, Lavin MF, Person MD, Paull TT. ATM activation by oxidative stress. Science 2010;330(6003):517–21.

[193] Maryanovich M, Oberkovitz G, Niv H, Vorobiyov L, Zaltsman Y, Brenner O, et al. The ATM-BID pathway regulates quiescence and survival of haematopoietic stem cells. Nat Cell Biol 2012;14(5):535–41.

[194] Kim J, Wong PK. Loss of ATM impairs proliferation of neural stem cells through oxidative stress-mediated p38 MAPK signaling. Stem Cells 2009;27(8):1987–98.

[195] Ding L, Liang XG, Hu Y, Zhu DY, Lou YJ. Involvement of p38MAPK and reactive oxygen species in icariin-induced cardiomyocyte differentiation of murine embryonic stem cells in vitro. Stem Cells Dev 2008;17(4):751–60.

[196] Bhandari DR, Seo KW, Roh KH, Jung JW, Kang SK, Kang KS. REX-1 expression and p38 MAPK activation status can determine proliferation/differentiation fates in human mesenchymal stem cells. PloS One 2010;5(5):e10493.

[197] Sarbassov DD, Ali SM, Sabatini DM. Growing roles for the mTOR pathway. Curr Opin Cell Biol 2005;17(6):596–603.

[198] Juntilla MM, Patil VD, Calamito M, Joshi RP, Birnbaum MJ, Koretzky GA. AKT1 and AKT2 maintain hematopoietic stem cell function by regulating reactive oxygen species. Blood 2010;115(20):4030–8.
[199] Le Belle JE, Orozco NM, Paucar AA, Saxe JP, Mottahedeh J, Pyle AD, et al. Proliferative neural stem cells have high endogenous ROS levels that regulate self-renewal and neurogenesis in a PI3K/Akt-dependant manner. Cell Stem Cell 2011;8(1):59–71.
[200] Abbas HA, Maccio DR, Coskun S, Jackson JG, Hazen AL, Sills TM, et al. Mdm2 is required for survival of hematopoietic stem cells/progenitors via dampening of ROS-induced p53 activity. Cell Stem Cell 2010;7(5):606–17.
[201] Takubo K, Goda N, Yamada W, Iriuchishima H, Ikeda E, Kubota Y, et al. Regulation of the HIF-1alpha level is essential for hematopoietic stem cells. Cell Stem Cell 2010;7(3):391–402.
[202] Piccoli C, D'Aprile A, Ripoli M, Scrima R, Boffoli D, Tabilio A, et al. The hypoxia-inducible factor is stabilized in circulating hematopoietic stem cells under normoxic conditions. FEBS Lett 2007;581(16):3111–9.
[203] Cam H, Easton JB, High A, Houghton PJ. mTORC1 signaling under hypoxic conditions is controlled by ATM-dependent phosphorylation of HIF-1alpha. Mol Cell 2010;40(4):509–20.
[204] Gordan JD, Bertout JA, Hu CJ, Diehl JA, Simon MC. HIF-2alpha promotes hypoxic cell proliferation by enhancing c-myc transcriptional activity. Cancer Cell 2007;11(4): 335–47.
[205] Bertout JA, Majmundar AJ, Gordan JD, Lam JC, Ditsworth D, Keith B, et al. HIF-2alpha inhibition promotes p53 pathway activity, tumor cell death, and radiation responses. Proc Natl Acad Sci USA 2009;106(34):14391–6.
[206] Rouault-Pierre K, Lopez-Onieva L, Foster K, Anjos-Afonso F, Lamrissi-Garcia I, Serrano-Sanchez M, et al. HIF-2alpha protects human hematopoietic stem/progenitors and acute myeloid leukemic cells from apoptosis induced by endoplasmic reticulum stress. Cell Stem Cell 2013;13(5):549–63.
[207] Kocabas F, Zheng J, Thet S, Copeland NG, Jenkins NA, DeBerardinis RJ, et al. Meis1 regulates the metabolic phenotype and oxidant defense of hematopoietic stem cells. Blood 2012;120(25):4963–72.
[208] Storz P. Forkhead homeobox type O transcription factors in the responses to oxidative stress. Antioxid Redox Signal 2011;14(4):593–605.
[209] Aruoma OI, Halliwell B, Hoey BM, Butler J. The antioxidant action of N-acetylcysteine: its reaction with hydrogen peroxide, hydroxyl radical, superoxide, and hypochlorous acid. Free Radic Biol Med 1989;6(6):593–7.
[210] Liu J, Cao L, Chen J, Song S, Lee IH, Quijano C, et al. Bmi1 regulates mitochondrial function and the DNA damage response pathway. Nature 2009;459(7245):387–92.
[211] Chuikov S, Levi BP, Smith ML, Morrison SJ. Prdm16 promotes stem cell maintenance in multiple tissues, partly by regulating oxidative stress. Nat Cell Biol 2010;12(10):999–1006.
[212] Alfadda AA, Sallam RM. Reactive oxygen species in health and disease. J Biomed Biotechnol 2012;2012:936486.
[213] Imai S, Guarente L. Ten years of NAD-dependent SIR2 family deacetylases: implications for metabolic diseases. Trends Pharmacol Sci 2010;31(5):212–20.
[214] Gomes AP, Price NL, Ling AJ, Moslehi JJ, Montgomery MK, Rajman L, et al. Declining NAD(+) induces a pseudohypoxic state disrupting nuclear-mitochondrial communication during aging. Cell 2013;155(7):1624–38.
[215] Haigis MC, Sinclair DA. Mammalian sirtuins: biological insights and disease relevance. Annu Rev Pathol 2010;5:253–95.

[216] Oberdoerffer P, Michan S, McVay M, Mostoslavsky R, Vann J, Park SK, et al. SIRT1 redistribution on chromatin promotes genomic stability but alters gene expression during aging. Cell 2008;135(5):907–18.

[217] Matsui K, Ezoe S, Oritani K, Shibata M, Tokunaga M, Fujita N, et al. NAD-dependent histone deacetylase, SIRT1, plays essential roles in the maintenance of hematopoietic stem cells. Biochem Biophys Res Commun 2012;418(4):811–7.

[218] Chae HD, Broxmeyer HE. SIRT1 deficiency downregulates PTEN/JNK/FOXO1 pathway to block reactive oxygen species-induced apoptosis in mouse embryonic stem cells. Stem Cells Dev 2011;20(7):1277–85.

[219] Han MK, Song EK, Guo Y, Ou X, Mantel C, Broxmeyer HE. SIRT1 regulates apoptosis and Nanog expression in mouse embryonic stem cells by controlling p53 subcellular localization. Cell Stem Cell 2008;2(3):241–51.

[220] Canto C, Auwerx J. PGC-1alpha, SIRT1 and AMPK, an energy sensing network that controls energy expenditure. Curr Opin Lipidol 2009;20(2):98–105.

[221] Price NL, Gomes AP, Ling AJ, Duarte FV, Martin-Montalvo A, North BJ, et al. SIRT1 is required for AMPK activation and the beneficial effects of resveratrol on mitochondrial function. Cell Metab 2012;15(5):675–90.

[222] Rimmele P, Bigarella CL, Liang R, Izac B, Dieguez-Gonzalez R, Barbet G, et al. Aging-like phenotype and defective lineage specification in SIRT1-deleted hematopoietic stem and progenitor cells. Stem Cell Rep 2014;3(1):44–59.

[223] Calvanese V, Lara E, Suarez-Alvarez B, Abu Dawud R, Vazquez-Chantada M, Martinez-Chantar ML, et al. Sirtuin 1 regulation of developmental genes during differentiation of stem cells. Proc Natl Acad Sci USA 2010;107(31):13736–41.

[224] McBurney MW, Yang X, Jardine K, Bieman M, Th'ng J, Lemieux M. The absence of SIR2alpha protein has no effect on global gene silencing in mouse embryonic stem cells. Mol Cancer Res 2003;1(5):402–9.

[225] Teodoro JS, Rolo AP, Palmeira CM. The NAD ratio redox paradox: why does too much reductive power cause oxidative stress? Toxicol Mech Methods 2013;23(5):297–302.

[226] Harman D. Aging: a theory based on free radical and radiation chemistry. J Gerontol 1956;11(3):298–300.

[227] Osawa M, Hanada K, Hamada H, Nakauchi H. Long-term lymphohematopoietic reconstitution by a single CD34-low/negative hematopoietic stem cell. Science 1996;273(5272):242–5.

[228] Bryder D, Rossi DJ, Weissman IL. Hematopoietic stem cells: the paradigmatic tissue-specific stem cell. Am J Pathol 2006;169(2):338–46.

Molecular Basis of "Hypoxic" Signaling, Quiescence, Self-Renewal, and Differentiation in Stem Cells

Chapter Outline

7.1 Stem Cell Signaling Transducing Pathways Triggered by Extrinsic Factors

Various molecules produced in the niche affect stem cell maintenance by triggering intracellular signaling pathways. Most of the studies are performed on the hematopoietic stem cells (HSCs) since these cells are the best characterized adult stem cells [1]. Here we present those identified as most critical for stem cell functioning.

Anaerobiosis and Stemness. http://dx.doi.org/10.1016/B978-0-12-800540-8.00007-7

7.1.1 Wnt Signaling Pathway

Canonical Wnt cascade is one of the signaling pathways that emerged as a critical regulator of stem cells. Signaling is initiated when secreted Wnt ligands get attached to their cognate receptor complex consisting of cell surface receptors belonging to LRP-5/6 and Frizzled families. This interaction enables stabilization of the central player of the Wnt cascade, β-catenin, that translocates to the nucleus where it interacts with the LEF/TCF family of transcriptional activators, to activate target genes [2]. In the absence of the Wnt ligands, β-catenin is attached to the destruction complex consisting of the tumor suppressors, adenomatous polyposis coli (APC) and axin. Also two kinases residing in the destruction complex, casein kinase I (CKI) and glycogen synthase kinase 3 (GSK3), phosphorylate specific residues in the β-catenin that tag it for the proteosomal degradation [3].

Wnt signaling cascade is involved in the control of stem cells self-renewal and cancer formation upon its dysregulation in various tissues [2,4]. Inhibition of the canonical Wnt signaling reduced quiescence and self-renewal and reconstitutive capacity of the HSCs [5]. Accordingly, another study showed that β-catenin activity promotes long-term maintenance of the HSCs [6]. Also, the Wnt pathway is involved in the maintenance of the pluripotent state in the mESCs and hESCs [7] and other adult stem cells (MSCs, NCS, intestinal and skin stem cells) [8,9]. In this, the Wnt cascade is partly regulated by hypoxia-inducible factors (HIFs) (see below). Thus, HIF-1α favors Wnt/β-catenin signaling in hypoxic embryonic stem cells (ESCs) by enhancing β-catenin activation and expression of downstream effectors, LEF-1 and TCF-1 and NSCs [4]. Also, Wnt/β-catenin activity is closely associated with low O_2 NSC niche in the subgranular zone of the hippocampus. But in colon carcinoma cells, low O_2 incubation inhibits of the Wnt/β-cateinin cascade that down-regulated expression of the Wnt/β-catenin target c-MYC, a potent cell cycle regulator. This is mediated by the interaction of the stabilized low oxygen of HIF-1α with β-catenin that results in reduced formation of β-catenin–TCF-4 complexes. In addition, β-catenin/HIF-1α interaction was found to increase HIF transcriptional activity [10]. Conversely, constitutively activated Wnt signaling led to the loss of HSCs self-renewal and multipotency [11]. Altogether, these data suggest that fine-tuning of the Wnt signaling is essential for the maintenance of HSCs [12].

7.1.2 Notch Signaling

Notch signaling is involved in tissue maintenance and control of the cell differentiation during the ontogenetic and adult stages of metazoa [13]. Activation of Notch signaling starts with a paracrine interaction between single-chain transmembrane receptors for Notch 1–4 in vertebrates with two specific ligands—Delta (DII 1, 3, 4) and Jagged/Serrate (Jag1, 2) [14]. Notch receptors and their ligands are expressed in a variety of tissues [15–22]. Signaling is initiated when the Jagged or Delta family of ligands binds a Notch receptor, triggering a two-step receptor proteolysis event. The second proteolytic cleavage, mediated by γ-secretase, results in the release of the Notch intracellular domain (ICD) from the plasma membrane and its transport to the nucleus where it reacts with the DNA-binding proteins, including RBP-Jk (also known as CBF1 or suppressor Hairless), MAML, CSL, and p300, and forms DNA-binding complex that activates target gene expression [23]. This complex stimulates transcription of target genes among which the most famous families are HEY (HEY-1, 2, and L) and HES (HES 1-5) [24,25].

It was evidenced that Notch signaling is critical in regulating stem cell quiescence and self-renewal. In muscle stem cells (MuSCs) and NSCs, Notch signaling promotes quiescence, and a disruption of Notch activity results in a depletion of the quiescent stem cell pool associated with spontaneous activation and premature differentiation of stem cells [26,27]. In contrast, Notch-deprived HPCs provided normal long-term reconstitution [28]. On the other hand, Notch signaling is required for expansion of the primitive hematopoietic cells [29]. Also, Notch is needed for the maintenance of undifferentiated stem and progenitor cell populations in the mammalian intestinal crypt [30], myogenic cell lines [31], and differentiation and proliferation of the erythroid progenitors [32]. Forced Notch activation in bone marrow or T cell progenitor cells inhibits differentiation and results in T cell acute lymphoblastic leukemia [33].

Notch signaling is stimulated in the low O_2 stem cell niche. Thus HIF-1α was demonstrated to potentiate Notch signaling by Notch ICD stabilization: HIF-1α physically interacts with Notch 1 ICD and accompanied it to Notch-responsive promoters to activate target genes. Also, low O_2 concentration stimulates the expression of the Delta4 ligand that stimulates Notch signaling [14,34].

7.1.3 Hedgehog Signaling

Binding of Hedgehog (Hh) ligand to patched receptor releases activity of the membrane protein, smoothened, which triggers signaling cascade [35]. In mutant mice, constitutive activation of Hh signaling increases cycling of the HSCs, which leads to their functional exhaustion and impairment of the reconstitutive hematopoietic capacity under homeostatic conditions and during stressed hematopoiesis. Also, in vivo inhibition of the Hh signaling preserves the functions of the HSCs [36]. Interestingly, there is a basal level of Hh signaling in undifferentiated cells that appears necessary for the maintenance, proliferation, and clonogenic capacity of adipose-derived multipotent mesenchymal cells, probably through regulation of retinoblastoma (RB) phosphorylation and cyclin A expression [37]. Hedgehog signaling is functional in anoxic/ischemic conditions as demonstrated in experiments in which it mediated the increase of neural stem cell proliferation upon treatment with a drug (resveratrol) [38]. Furthermore, under a low O_2 concentration, Hh and Notch signaling cooperate to maintain self-renewal of hESCs [39] and the hypoxia-induced Hh response is mediated by HIF-1α [40].

7.1.4 Other Signaling Pathways

c-Kit signaling: Interaction of the stem cell factor (membrane bound or soluble) with its receptor, c-kit, expressed on the HSCs triggered the signaling, which is essential for the HSCs' quiescence and function [41]. Impaired c-kit signaling led to the loss of the HSCs [42,43]. The relations between c-kit and hypoxic response are elaborated in Chapter 13.

Tie2/Ang-1 signaling: HSCs express the receptor tyrosine kinase Tie2 activated by the Ang-1 ligand presented in the niche. Tie2/Ang-1 signaling activates the expression of the adhesive molecules β1-integrines and *N*-cadherin that promote HSC interactions with the cellular components of the niche. This enhances the quiescence and survival of HSCs [44]. Also, Tie2/Ang-1 signaling activates the PI3-kinase pathway [45] that affected the stem cell fate by regulating cellular metabolism (see Chapter 6).

Trombopoietin (TPO)/myeloproliferative leukemia protein (MPL) signaling: Interaction of the myeloproliferative leukemia virus proto-oncogene (MPL) expressed on the HSCs and HPCs with the TPO produced in the niche that activate signaling cascade that modulates long-term HSCs (LT-HSCs) cell cycle progression and promotes retention of the quiescent LT-HSCs in the niche [46,47]. Engraftment potential of the HSCs and HPCs ($CD34^{+}CD38^{-}$ cells) correlate with the MPL expression [48]. The relations between MPL signaling and hypoxic response are elaborated in Chapter 13.

Transforming growth factor-β (TGF-β)—Bone morphogenetic proteins (BMPs)/ transforming growth factor-β receptor (TGF-βR): TGF-β signaling is evidenced to promote quiescence of the HSCs in the stress condition (myelosupressive chemotherapy) [49]. But the conditional TGF-βR knockout mutant does not show impaired HSCs self-renewal and reconstitutive capacity in the steady-state condition [50]. Also, an emerging role of the TGFβ pathway in self-renewal and differentiation of human adipose tissue-derived MSCs was described [51]. The TGF signaling is closely related to the low O_2 conditions [52].

CXCR4/CXCL12 signaling: In addition to being a major chemoattractant for HSCs, CXCR4/CXCL12 signaling is required for the control of the HSCs quiescence and maintenance. In vitro, addition of the CXCL12 inhibits HSCs' and HPCs' entry in the cell cycle, while inactivation of its receptor CXCR4 causes excessive HSC and HPC proliferation [53]. CXCR4 expression is modulated by the O_2 concentration in mesenchymal stromal cells (MStroC) [54] and a transient increase in HIF-1α is required for dmPGE2-enhanced CXCR4 up-regulation and enhanced migration and homing of HSC and HPC [55].

7.2 Intrinsic Factors Associated with Stem Cell Maintenance

7.2.1 p53 and RB Protein

p53, a transcription factor with tumor suppressor activity, is a master regulator of diverse cellular processes, especially those involved in the maintenance of genomic integrity. In the presence of genomic damage, p53 interrupts cycling to allow time for DNA repair. This is accomplished by p53 inhibition of phosphorylation of another tumor suppressor protein RB and consecutive cell cycle arrest [56]. E3 ubiquitin ligase, mouse double minute 2 homolog (MDM2), is a critical cellular inhibitor of p53 function: MDM2 targets p53 for degradation via ubiquitin proteosome pathway [57–59]. p53 is involved in the stem cell function control in adult and ESCs. For instance, p53 deficiency in HSCs promotes cell cycle entry with a reduction in the number of quiescent HSCs [60]. The mechanism by which p53 regulates HSCs self-renewal in $Lin^{-}Sca1^{+}Kit^{+}$ (LSK) population involve activation of the transcription of the target genes *Necdin* and *Gfi1*, which both restrict cell proliferation [60]. In the absence of MDM2, elevated reactive oxygen species (ROS) stabilized p53 activity, which induces cell cycle arrest, senescence, and depletion the HSCs in LSK population [61]. Also, activation of the p53 induces differentiation of ESCs and repression of the pluripotent genes expression (see below) [62].

An et al. demonstrated that HIF-1α can stabilize wild-type p53 [63], and HIF-1 binds p53 via its oxygen-dependent degradation (ODD) domain [64]. In turn, p53 can also promote the degradation of HIF-1α via MDM2-mediated activity [65]. Furthermore, while HIF-1α promotes p53 activity, HIF-2α actually inhibits it [66].

In addition to p53, RB protein is the crucial regulator of the cell cycle, and regulator of stem cell quiescence [67]. The activated cyclin-dependent kinase (CDK) phosphorylates and inactivates RB protein. In turn, it enables passage through a restriction point in the latter third of G1 and progression of the cell cycle. In NSCs, genetic ablation of RB together with p53 triggers NSC overproliferation and tumor formation [68]. In ESCs, ablation of all three RB family members (RB, p107, and p130) results in impaired differentiation and an increase in cell turnover under growth arrest conditions [69]. Genetic ablation of RB induces loss of the quiescent MuSCs and HSCs pool and impairment of the hematopoietic reconstitution, self-renewal, and expansion of the muscle and HPCs [70,71].

7.2.2 Genes Involved in the Cell Cycle Progression

Cell cycle is regulated by special protein kinases—CDKs, which complex with corresponding, regulatory units, cyclins. Formation of the cyclin/CDK complexes results in phosphorylation and activation of the CDKs. Activated cyclin/CDK complexes phosphorylate specific protein substrates that enable cell cycle progression. Scrupulous regulation of the cell cycle phase transitions is accomplished by fine-tuned activation and inhibition of specific cyclin/CDK complexes [56]. Exit from the G0 phase and entry into the cell cycle are regulated by cyclin D/cdk-4/6 and cyclin E/cdk-2 complexes, respectively. In turn, cyclin-dependent kinase inhibitors (CKIs) of the INK (p15, p16, p18, p19) and CIP/KIP (p21, p27, p57) families control the activity of these cyclin/CDK complexes [72].

Not surprisingly, implication of these factors is shown to be crucial in the control of quiescence/proliferation balance of the stem cells. For instance, in HSCs and NSCs, inhibition of p21 results in a decrease in the quiescent stem cell population and impaired self-renewal capacity of p21-deficient cells, which ultimately results in an exhaustion of the stem cell pool [73,74]. In double knockout mice lacking both p57 and p27, HSCs' quiescence and self-renewal are severely impaired [75]. Conversely, the loss of p18 increases self-renewing, cell divisions, and long-term engraftment in the HSCs/HPCs population (CD34-LSK population) [76]. Quiescent state is characterized by the down-regulation of genes encoding cyclin A2, cyclin B1, cyclin E2, and survivin, which control various aspects of cell cycle progression and genes correlated with the proliferation status, such as proliferating cell nuclear antigen (PCNA) in populations enriched in the stem cells (HSCs, MuSCs, and hair follicle stem cells) [77–79].

7.2.3 Transcriptional Regulators Associated with Stem Cell Maintenance

Numerous transcription factors coordinate stem cell functions. Using "gain of function" and "loss of function" approaches enables identification of the different transcription factors implicated in stem cell maintenance (see review [72]). The conditional deletion of the pre-B cell leukemia transcription factor 1 (PBX1) [80] or homozygous deletion of the forkhead transcription factors FOXO1, FOXO2, and FOXO3 [81] provoke in the reduction of the LSK cells, defective HSCs self-renewal and defective long-term repopulating capacity. Stem cell leukemia/T cell acute lymphoblastic leukemia 1 (SCL/TAL1) is highly expressed in the LT-HSCs and regulates HSCs quiescence by promoting expression of cell cycle regulators Id1 and p21 [72]. Inhibition of the expression of SCL/TAL1 impaired hematological reconstitution after transplantation [82]. Stimulated expression of the orphan receptor Nurrl (Nr4a2) increased the percentage of the quiescent cells in population, highly enriched in the HSCs (Hoechst 33,342 side population: LSK) by up-regulating the cell cycle inhibitor p18Ink4c [83]. The transcriptional repressor thioredoxin interacting protein (Txnip) [84] acts in the same manner. Txnip deletion causes defect in self-renewal repopulating capacity of LT-HSCs and exhaustion of HSCs associated with their lower retention in bone marrow. Knockout model for Gfi1 displayed increased cycling within HSCs' compartment and impaired reconstitutive capacity [85]. Also, quiescence promoting effects in stem cells are associated with expression of the G0S2 [86], AML1 (also known as Runx1), a binding partner of ELF4, early growth response 1 (Egr1), and STAT5 (signal transducer and activator of transcription 5) [87–89]. STAT5-induced HSCs self-renewal is mediated at least in part via the up-regulation of HIF-2α [88]. Also, Wang et al. evidenced that c-MPL effects in the HSCs maintaining are mediated via STAT5 in the steady-state hematopoiesis [88]. Enforced expression of the GATA2 in HSCs and HPCs increased quiescence but reduced reconstitutive capacity [90]. In contrast, c-MYC, proto-oncogen, and regulators of cell proliferation differentiation and apoptosis control stem cells self-renewal/differentiation balance [91,92]. Conditional inactivation of the c-MYC promotes HSCs quiescence and retention in the niche [92]. Similarly, negative regulation of the quiescence is associated with MEF/ELF4 transcription factors, which inhibition stimulates proliferation of the HSCs and HPCs [93]. A study using a culture model of mouse NSCs through motif analysis and Chip-seq identified important transcription factors for the maintenance of the NSCs that involve nuclear factor I, Sox and Fox family members, and OLIGO2 as multifunctional regulators of NSCs self-renewal [94].

7.2.3.1 Pluripotency Genes

OCT4, SOX2, and *NANOG* constitute a core transcriptional regulatory network that maintains the ESCs (mESCs and hESCs) in the pluripotent state [95,96]. Interestingly, adult stem cells such as MSCs also express embryonic genes *OCT4, SOX2*, and

REX1 [97]. OCT-4 interacts with SOX2, forming the complex that regulates expression of the downstream targets including the OCT-4, SOX2, and NANOG [98,99]. Using chromatin immunoprecipitation (ChIP)-based assay, additional pluripotency-associated transcriptional network has been identified in mESC. Thus, the OCT-4 centric module in addition to OCT-4, SOX2, and NANOG include Smad1, STAT3, Tcf3, which are downstream effectors for BMP, leukemia inhibitory factor (LIF), and Wnt signaling pathway, respectively. It was proposed that OCT-4 interacts with the factors as an anchor point for their assembly and functioning [96,100]. Another binding site module, MYC-centric module, includes c-MYC, n-MYC ELF1, ZFX, REX1, and RONIN transcription factors [101–103]. Many of the OCT-4 centric module target genes are transcription factors but also are noncoding RNAs (e.g., mir-302 and mir-290 clusters) [96,104]. In contrast, MYC-centric module regulates transcription of the genes associated with protein metabolism [101,103]. Also, maintenance of the pluripotent state could be regulated by the repression of the transcription of the genes involved in the differentiation such as *GATA4* and *GATA6* [105,106]. In addition, INO80 and TAF (TFIID-associated factor transcription complex) and PRDM14 were found to be important for the maintenance of the hESC [96].

Beyond the identification of the different transcriptional regulators, genetic studies revealed also that ESCs persist in the interconvertible, metastable pluripotent state [96]. It represents a heterogeneous population in dynamic equilibrium in which subpopulations are interconvertible in terms of expression of pluripotent markers. Thus, ectopic expression of the transcription factors (KLF4, KLF2, NR5A2, NRA1, NANOG, or STAT3) or culture in the medium containing LIF can convert Epics into mESCs [107,108]. On the other hand, mESCs could be converted to Epics in the presence of bFGF and Activin [109,110]. Also, hESCs can be converted to mESCs upon transient transfection of hESCs with *OCT4, KLF4,* or *KLF4* and *KLF2* [111] or eventually cultivated hESCs in the mESCs condition in the presence of LIF [112]. Finally, ectopic expression of the pluripotent master genes *OCT4, SOX2, KLF4,* and *c-MYC,* or *OCT4, SOX2, LIN28,* and *NANOG* can induce reprogramming of the somatic cells into induced pluripotent stem cells (iPSCs) [113,114].

7.2.3.2 HOX Genes

HOX genes encode for highly conserved transcription factors that belong to the Homeobox family, which influence spatiotemporal developmental and differentiation fate of cells. Expression pattern of the *HOX* genes, HOX code, is specific for the various stem cells and governs their lineage differentiation in vivo and in vitro. It was suggested that this code may contribute to stem cell maintenance. Hox gene expression during development is tightly regulated, by BMP and Wnt or retinoic signaling pathways or by chromatin structure and epigenetic modifications [1,115]. In ESCs cells, *HOX* genes expression is globally repressed when the differentiation takes place [116]. It was shown that a number of *HOXA* and *HOXB* genes are expressed in the human $CD34^+$ population with the expression of the *HOXB3* and *HOXB4* mainly restricted to the long-term culture initiating cells (LTC-IC) [117]. Overexpression of the *HOXB4* in murine bone marrow $CD34^+$ cell augments their long-term reconstitutive potential, amplification for the progenitors CFU-S and CFCs [118]. Ectopic expression of the *HOXB4* in the primitive progenitors from yolk sac or ESCs induces a swift to definitive HSCs enabled to reconstitute hematopoiesis in irradiated adult mice [119]. It suggests that primitive HSCs are poised to become definitive HSCs and that this transition can be promoted by *HOXB4* expression. Several Hox genes are expressed in the NSCs [120] and have been identified to be involved in the neurogenesis [121]. In MSCs, specific HOX code is associated with the different source of the MSCs, determines differentiation potential, and may regulate their self-renewal [1,122].

7.3 "Hypoxic Signaling" in Stem Cell Maintenance

Stem cells functions are related in vivo and in vitro to low O_2 concentration (see Chapters 3 and 4). This implicates "hypoxic signaling" as a critical regulator of the stem cell functions. HIFs, predominately HIF-1 and HIF-2, transcription factors are the key mediators mediating cellular response in a low O_2 environment [123–125]. They are heterodimeric proteins that comprise two subunits: α subunit, whose expression is stabilized when the O_2 concentration is <5% O_2, and β subunit, which is constitutively expressed [126–128]. Whereas HIF-1α is the master regulator of O_2 homeostasis in most mammalian cells, two *HIF1α* paralogues exist: *HIF2α* (also known as EPAS1) and *HIF3α* [128]. HIF-1α and HIF-2α are closely related and share a conserved basic helix–loop–helix (bHLH) domain. HIF-1α, -2α, and -3α each heterodimerize with one member of the β subunits class: aryl hydrocarbon receptor nuclear translocator (ARNT, HIF-1β), ARNT2, or ARNT3 [128].

Once stabilized, HIF-1α and HIF-2α are translocated in the nucleus where they are associated with β subunits in order to form functional HIF transcription factors: HIF-1 (HIF-1α/HIF-1β) and HIF-2 (HIF-2α/HIF-1β). They facilitate transcription of the overlapping and also distinct target genes associated with various cell functions (energetic metabolism, erythropoiesis, Fe metabolism, apoptosis, vasomotor functions, angiogenesis, cell migration, and so on) [129]. HIF-1 in the nucleus forms the transcription complex with the coactivators cp300 and CPB, which binds to the regulatory sequence hypoxia response element (HER) in the target gene promoters and activates their transcription [127].

By contrast, when the intracellular concentration of O_2 is higher than 5%, HIF-1α subunit levels decrease through oxygen-mediated degradation. In this situation, three prolyl hydroxylases, PHD1, PHD2, and PHD3 (which belong to the family of iron- and 2-oxoglutamate-dependent dioxygenases), are activated by oxygen, and they hydroxylate two proline (Pro) residues in the ODD domain of the α subunit (mainly PHD2 for HIF-1α); this primes the protein for degradation by the von Hippel–Lindau (VHL) ubiquitin ligase complex [129,130]. PHDs belong to the family of the α-ketoglutarate and Fe(II)-dependant dioxygenase, which catalyzes Pro hydroxylation of HIF-1α in the presence of Fe^{2+}, α-ketoglutarate and O_2, which generates succinate and CO_2. Another prolyl hydroxylase is factor inhibiting HIF (FIH) [131,132]. In contrast to PHDs, FIH catalyzes hydroxylation of the asparagine in Carboxy Terminal Activation Domain (C-TAD) of α subunit, which blocks the formation of the transcriptional activation complex of HIF-1 and coactivators CPB/p300 [129].

High retention for the pimonidazole suggested a "hypoxic" profile of stem cells. This is associated with high expression of both HIF-1α and HIF-2α transcripts, in LT-HSCs ($CD34^-LSK$) compared to primitive progenitor cell populations ($CD34^+LSK$ cells) [133]. Also, in situ tissue imaging evidenced high expression of HIF-1α subunit protein in LT-HSCs [134].

Genetic approaches proved that high expression HIFs has a functional importance for stem cell maintenance. Thus, ARNT-deficient mouse embryos or ARNT-targeted mutations that eliminate expression of the both HIF-1 and HIF-2 result in the decrease of the number of hematopoietic progenitors of all hematopoetic lineages in embryonic yolk sac or embryonic body assays and the death of mice embryos at day 10 of gestation [135,136]. In the case where physiological response to hypoxia was inhibited due to the absence of HIF-1α, the mice embryos develop disorders in the embryonic hematopoiesis [137].

Conditional depletion of *Hif1α* gene in mice results in the decrease of glycolysis and increased mitochondrial metabolism, which leads to disruption of the LT-HSCs ($CD34^-Flt3^-LSK$ cells), cell cycle quiescence, and long-term reconstitutive capacity in the secondary recipient (serial transplantation), while steady-state hematopoiesis and hematopoietic reconstitution in the primary recipient is preserved [133,138]. This suggests that HIF-1α is associated with long-term stem cell self-renewal. In addition, the same authors evidenced that this regulatory network implicates direct downstream HIF-1 target gene pyruvate dehydrogenase kinase 2 and 4, which prevent pyruvate from entering the tricarboxylic acid (TCA) cycle, *thus blocking mitochondrial respiration.* Overexpression of the PDK2 and PDK4 restores the cell cycle quiescence and reconstitutive capacity in *HIF1α*-deficient LSK cells [138]. In contrast, it was evidenced that mice lacking *HIF1α* only in the hematopoietic system, mature to adulthood, have normal adult hematopoiesis, and LSK cells from these mice have shown efficient long-term hematopoietic reconstitution [139]. The same observation was made with *HIF1α* gene knockdown in HSCs' and HPCs' cord blood-derived cells ($CD34^+$ cells) [140].

Considering the other family member, it was shown that HIF-2α maintains the survival and protects from endoplasmic reticulum (ER) stress-induced apoptosis the HSC's and HPC's cord blood-derived and acute myeloid leukemia cells [140]. Also in mutated mice modified to reach adult age, the loss of HIF-2α leads to pancytopenia and metabolic disorders [141]. However, inducible or constitutive conditional deletion of *Hif2α* specifically within the hematopoietic system had no impact on HSC (LSK-$CD48^-CD150^+$) survival, self-renewal, and long-term hematopoietic reconstitution under steady-state conditions or following severe stress such as hematopoietic injury (i.e., 5-fluorouracil [5-FU] treatment) or serial transplantation [142]. Also, it was suggested that HIF-1 and HIF-2 exercise a compensatory role in HSCs' maintenance since deletion of *Hif1α* in HSCs resulted in a compensatory increase in *Hif2α* mRNA levels (and no change in *Hif3α* mRNA levels) [143].

Conversely, conditional knockout mice that lack both *Hif1α* and *Hif2α* specifically within the hematopoietic system matured to adulthood without any obvious defects, displayed normal steady-state multilineage hematopoiesis, and have maintained long-term reconstitutive capacity in primary recipient mice [146].

Altogether these data suggest that the role of the HIF-1 and HIF-2 in the self-renewal of the HSCs should be better understood and defined [129].

Considering the role of HIF-3α, one of its splice variants (inhibitory PAS, iPAS) functions as a dominant-negative regulator of HIF-1α function. iPAS dimerizes with HIF-α protein and thereby impairs productive interaction between HIF-1α and HER of target genes [144]. Also, its role in the cellular response to a low O_2 environment is supposed by the fact that HIF-3α itself is a direct transcriptional target of HIF-1α [145]. Bearing in mind that HIF-3α is highly expressed in LT-HSCs ($CD34^-$ LSK) compared to primitive progenitor cell populations ($CD34^+$ LSK cells) [133], it will be of major interest to investigate its role in HSC maintenance.

7.3.1 HIF-Mediated Low Oxygen Metabolic Response

HIF deficiency that leads to stem cell functional disruption is associated with the metabolic/energetic changes, suggesting that HIF-mediated "hypoxia adaptive cellular response" underlines stem cells maintenance. Metabolic functions of HIF-1α are mostly associated with glycolysis. Thus, HIF-1α promotes glycolytic metabolic shift by stimulating the expression of glucose transporters, glycolytic enzymes, and lactate dehydrogenase, which replenishes NAD^+ for further glycolysis [146,147]. Also, HIF-1α target genes (*PDK1*, *PDK2*, *PDK4*) repress the flux of pyruvate into acetyl-CoA, diverting carbon away from mitochondria and suppressing O_2 consumption [149,146,148]. In addition, HIF-1α activity was shown to influence the pentose phosphate pathway [150], suggesting stimulation of the anabolic pathway generating metabolites for the nucleotide biosynthesis enabling cellular proliferation and antioxidative cellular protector NADPH (see Chapter 6). Also, very important for stem cell maintenance is the HIF-1 role in defense of oxidative stress and redox homeostasis (see Chapter 6). Semenza et al. demonstrated that both HIF-1α and HIF-2α can modulate the expression of cytochrome c oxidase isoforms in order to maximize the efficiency of the electron transport chain [146,151]. In addition, HIF-1 controls the expression of several genes essential for HSCs functions, including VEGF, FOXOa and CXCR4 [125]. FOXO3a is activated downstream of HIF-1 and mediates repression of a set of nuclear-encoded mitochondrial genes that inhibit mitochondrial O_2 consumption [152].

Another family member, HIF-2α, could contribute to the glycolytic metabolic shift indirectly by regulated PPARAα that may regulate PDK4 [153]. Also, HIF-2α represses lipid catabolism and oxidation and stimulates lipid storage (biosynthesis) [154].

Stimulation of the glycolysis by HIF-1 and lipid biosynthesis by HIF-2 both suggest that metabolic adaptation induced by HIFs are dedicated to decreased production of ATP through oxidative phosphorylation and less ROS as a byproduct [155].

It should be noted that a decrease in the O_2 concentration in situ could also initiate metabolic response independently of HIFs. For instance, mTOR signaling and the UPR play critical roles in hypoxic adaptations by modulating protein translation and cell metabolism [156].

Transcriptional regulators, such as PGC1a, can complement the HIF response in ischemic conditions by promoting production of the VEGF and neoangiogenesis [157]. In addition, PHDs and FIH have non-HIFα substrates, which may underlie some of their biological functions [158].

7.3.2 HIF Interactions with Metabolic Regulators

Several other potential upstream regulators and downstream targets of HIF-1 and HIF-2 affect HIFs-mediated adaptive metabolic response and stem cell activity. For instance, MEIS1, an evolutionarily conserved Hox family transcription factor [159], regulates both glycolysis and the oxidative stress response by activation of *Hif1α* and *HIf2α* transcription in HSCs. Deletion of *Meis1* (Meis1 knockout mice) leads to the loss of the long-term reconstitutive capacity associated with the decreased expression of the HIF-1α and HIF-2α and loss of the survival and self-renewal of LT-HSCs ($Lin^-Sca1^+Kit^+Flk2^-CD34^-$ cells) [143].

Moreover, the homeobox protein CBP/p300-interacting transactivator 2 (CITED2) is a transcriptional regulator induced by HIF-1. It competes with HIF-1 for CBP/p300 binding and thus provides a negative feedback loop that fine-tunes cellular responses to low oxygen availability. The conditional deletion of Cited2 leads to loss of LT-HSCs ($CD34^-Flt3^-LSK$ cells), leading to bone marrow failure, which can be rescued by the deletion of *p16, p19, p53*, or *Hif1* deletion [160].

Researchers have also identified extracellular factors that regulate HIF-1 activity in LT-HSC ($CD34^-Flt3^-LSK$ cells). The soluble protein CRIPTO (also known as TDGF1), and its cell surface receptor 78 kDa glucose-regulated protein (GRP78), has crucial roles in promoting quiescence and in the maintenance of HSCs. Under hypoxic conditions, HIF-1 binds to HREs in the promoter region of CRIPTO and activates its expression. CRIPTO then binds to GRP78 and stimulates glycolytic metabolism-related proteins [161,162].

Sirtuins, cellular sensors for the cellular redox state, have been shown to modulate HIFs' activity, enabling adaption of the HIF-mediated cellular response to stress condition [163,164]. Sir1 forms a complex with HIF-2α and deacetylates HIF-2α protein, which enhances transcriptional activity of the HIF-2 [163]. By contrast, deacetylation of HIF-1α protein blocks its interaction with p300 coactivator, which in turn inhibits HIF-1 transcriptional activity. Thus, in the situation when NAD^+ content is low in the cells (a state associated with low energy as well as oxidative stress), expression of the Sirt1 is decreased, but HIF-1 activity is increased, which augments the expression of the HIF-1 targets genes [165]. Also, another family member Sir6 deficiency increases HIF-1 protein activity and consecutive stimulation of the glycolytic machinery [164].

In addition, noncoding RNAs that control the expression of the gene post-transcriptionally (microRNAs [miRNAs], miR-107 and miR-17-92) decrease the expression of the ARNT and HIF-1α, respectively [86,166].

Also, in response to the growth factors activated, PI3K/Akt pathway inhibits TSC1/2 and activates mTOR signaling cascade (see Chapter 6). PI3K/Akt activation also increases HIF-1α protein stabilization [167,168], and mTORC1 increases HIF-1 mRNA translation [169,170]. Based on this, these data suggest that mTOR stimulates HIF activity.

Activated forms of Ras oncogene have also been correlated with the production of high amounts of ROS [171]. In this situation, Ras-mediated ROS generation reduces activity of the PHD that in turn decreases HIF-1α degradation [172], which then enables establishment of the metabolic antioxidative response.

7.4 Epigenetic Regulation of the Stem Cell Fate

Gene expression pattern in the stem cells is critically regulated by epigenetic modification including alternation in the DNA methylation, covalent histone modifications, polycomb/trithorax protein, and miRNA expression.

Posttranslational modification of the histones, major chromatin proteins, represent a powerful way for reversible control of gene expression [173]. Trimethylation of histone H3 at Lys4 (H3K4me3) and at Lys27 (H3K27me3) is of particular interest because of their roles in the positive and negative regulation of transcription, respectively [173]. Not surprisingly, important roles of H3K27 methyltransferases in regulating stem cell quiescence was also shown.

7.4.1 Histone Modifications and DNA Methylation

Histone modifications in stem cells are different than in differentiated cells. For instance, ESCs' chromatin is mainly open euchromatin, accessible for the nucleases, and with abundantly acetylated histones [174,175]. During the in vitro ESCs' differentiation, increase of the hetochromatin marker for gene silencing (H3-triMeK9) and decrease of the histone H3 and H4 level acetylation were observed [175,176]. Intriguingly, lineage-specific genes in ESC have a high level of the acetylated H3K9 and methylated H3K4 that are euchromatin marks for the genes ready to be transcribed [177,178]. But their expression is prevented by the H3K27 methylation. This suggests that the lineage-specific genes are primed for expression in ESCs but are held in check by opposing chromatin modifications [177].

DNA methylation at CpG islands is another way for gene silencing, and it is required for the regulation of the self-renewal differentiation balance in stem cells. For instance, inhibition of the DNA methyltransferase (Dnmt1, 3a, and 3b) and deficiency in the CpG island-binding protein CGBP induce severe DNA hypometilation and blocking of the differentiation [179,180]. Also, hypermethylation of the *OCT4* promoter in differentiated cells correlates with gene silencing, while hypomethylation of *OCT4* enables its expression in pluripotent state [95].

7.4.2 Polycomb and Trithorax Proteins

Evolutionary highly conserved Polycomb (PcG) and Trithorax (TrxG) proteins antagonistically regulate expression of the various genes [181]. As a part of multiprotein complexes that interact with chromatin and catalyze or recognize histone methylation, ubiquitination, and acetylation marks, PcG and TrxG function as repressor or activators, respectively, of the same target gene [182]. PcG complexes catalyze trymethylation of histone 3 lysine 27 (H3K27me3), which is recognized by other PcG complexes that bind to H3K27me3 and induce gene silencing by interfering with transcription initiation [183] by recruitment of additional histone modifying complexes [184,185] or by compacting chromatin [186]. In contrast, TrxG activate gene expression by catalyzing histone modifications that are a general mark of actively transcribed genes trimethylation of the histone 3 lysine 4 (H3K4me3), dimethylation of the histone 3 lysine 26 (H3K26me3) [187], or by working as ATP–dependent chromosome remodelers [188].

Expectedly, PcG and Trx G regulation play an important role in self-renewal and pluripotency. It was evidenced that PcG proteins in ESCs repress in the reversible manner of expression of the genes implicated in the differentiation that would be otherwise derepressed in the absence of the PcG [173,174,189]. However, in response to different cellular signals, the complex containing TrxG mediates H3K4 trimethylation of the PcG target genes [190] that activate their transcription. Also, PcG and TrxG proteins influence transcriptional or posttranscriptional expression of the cell cycle-involved genes [191,192] that implicate their role in the quiescence.

Also, some PcG and TrxG proteins function as "erasers," working as demethylases, deubiquitinases, or deacetylases that prime histones for addition of the activating or silencing marks. This facilitates immediate transition from inactive to active chromatin state and vice versa [184].

In order to couple acute or long-term epigenetic modifications to cellular states and overall homeostasis, PcG and TrxG are regulated in a sophisticated manner. Cellular kinase network regulates PcG and TrxG activity either by direct phosphorylation or by phosphorylation of the histones tails in response to the various intra- and extracellular signals. For instance, cell cycle-regulated phosphorylation of the PcG protein EZH2 by CDK1 and CDK2 or by AKT in mammals induces reduction in H3K27me3 levels, increased degradation of the EZH2, and interference with EZH2-binding to the miRNAs, which results in the overall increased expression of the PcG-silenced target genes [193–195]. Also, phosphorylation of the PcG complex protein Bmi-1 enhances ubiquitylation of histone H2A, resulting in tumor suppressor gene silencing [196]. In addition, in response to oxidative or other cellular stress, activated MAPK phosphorylates Bmi-1, leading to its expulsion from chromatin [197,198]. This might enable development of antistress, protective cellular response [197]. Also, phosphorylation of the other member of the PcG proteins by MK2 (mitogen-activated protein kinase-activated protein kinase 2) is required for the silencing of the differentiation genes and has been found to be critical for HSC maintenance [199].

7.4.3 miRNAs

miRNAs control the expression of the gene posttranscriptionally [200,201]. miRNAs bind to the 3′ untranslated region of the target mRNAs, resulting in their cleavage or translational repression [202]. miRNA profiling revealed the function of various miRNAs in stem cell maintenance regulation [203,204]. In HSCs, miR-126 controls stem cell quiescence by attenuating PI3K–AKT signaling pathway [205]. In MuSCs, miR-489 is an important regulator of the quiescent state by suppressing the oncogene *DEK* [203]. Interestingly, another study showed that the myogenic factor 5 mRNA (key regulator of myogenesis) and its regulatory miRNA miR-31 were sequestered in mRNA ribonucleoprotein particle granules in quiescent MuSCs [206]. That suggests that quiescent stem cells are primed for differentiation, as the storage of mRNAs makes them readily available for the activation of differentiation program upon stimulation [67]. In MSCs, miR-143 is involved in the control of differentiation partly by regulating expression of target gene ERK5 [207]. Specific miRNAs are important for the self-renewal and differentiation of ESCs and NSCs [104,208]. Interestingly, using miR-302 it was shown that human skin cancer cells could be reprogrammed to iPSCs. In this situation, miR-302 was sufficient to maintain iPSCs even in the feeder-free culture (condition that is otherwise completely necessary for the maintenance of the hESCs in the standard culture) [209].

7.4.4 Metabolic-Based Epigenetic Control of the Transcription Program in Stem Cells

Reprogramming somatic cells into iPSCs has shown that metabolic changes precede pluripotent marker expression [210]. This pointed to the epigenetic regulation of gene expression by the metabolic intermediaries. Several examples are evidenced to prove this hypothesis.

Metabolic intermediates and ATP can be used as cofactors by histone and DNA-modifying enzymes that by chromatin modification change target gene expression. In that way, epigenetic transcriptional regulation couples cellular energetic and nutrient status with the gene expression.

Histone methylation is catalyzed by lysine methyltransferases, requiring *S*-adenosylmethionine (SAM), a main physiological methyl group donor in cellular methyltransferase reactions. Recent observations relay physiological SAM levels and histone methylation potential of PcG/TrXG methytransferase and mammalian DNA methyltransferase [211,212].

Furthermore, epigenetic regulation is critically dependent on the activity of the histone demethylases that utilize FAD and α-ketoglutarate, Fe(II), and O_2 as cofactors. Ten-eleven translocation (TET) family of DNA hydroxylases, which catalyzes the conversion of 5-methylcytosine to 5-hydroxymethylcytosine on DNA [213]. Not surprisingly, activity of these enzymes is highly dependent on the cellular metabolism and sensitive to ROS [214,215].

O-GlcNAc is the end product of the hexosamine pathway depending critically on glucose availability. O-GlcNAcylation of histones (H2A; H2B, H3, and H4) influences transcriptional activity, gene silencing, mitotic chromatin regulation, and the modulation of other histone modifications [216]. It was shown that this posttranslation modification regulates expression of the pluripotent genes, *OCT4* and *SOX2* [217]. O-GlcNAcylation of both members of the PcGs and TrxG components suggest that this posttranslation modification is critical for controlling activating and repressive functions of the PcG and TrxG complexes.

Acetyl-CoA derived from citrate (ATP citrate liase) or by ligation of acetyl and CoA (acetyl-CoA synthetase) is a substrate for the acetylation of the proteins involved in energy metabolism, nuclear import, and gene expression [218]. Histone acetylation is catalyzed by lysine acetyltransferases in a reaction dependent on acetyl-CoA [219]. The physiological role of NAD^+-dependent histone deacetylases (sirtuins) is exhibited by removing acetyl marks from lysine and cytosine residues on many proteins and of DNA, altering chromatin structure and gene expression [220].

7.5 Conclusion

Altogether, presented data show complex and multilevel (transcriptional, metabolic, epigenetic) control of the stem cell functions. New insights in the field imply that instead of considering the quiescent state as a dormant state, it should be regarded as an actively regulated primed state in which cells can sense environmental changes and respond by reentering the cell cycle for proliferation. For instance, it was evidenced that numerous genes, including lineage-specific ones, are in the permissive state for the transcription, albeit quiescent cells have a low transcription rate. This situation, however, allows for the rapid activation upon stimulation [67]. In this light as it was suggested by Cheung et al. [203], the quiescent state could be considered as a poised state in which activation is put on pause by one of the epigenetic mechanisms. Also, we could consider that by expressing the genes needed for quiescence and self-renewal in a low energetic metabolic state, quiescence is maintained along with silencing the differentiation genes epigenetically, which are primed to be rapidly transcribed upon activation.

From the data reviewed in this chapter, it is evident that the complete “machinery” employed in the maintenance of stemness (particularly in the stem cell self-renewal) is set up for an anaerobic/microaerophilic metabolic environment. This point should be remembered for further consideration (see Chapters 8–14).

References

[1] Seifert A, Werheid DF, Knapp SM, Tobiasch E. Role of Hox genes in stem cell differentiation. World J Stem Cells 2015;7(3):583–95.

[2] Reya T, Clevers H. Wnt signalling in stem cells and cancer. Nature 2005;434(7035):843–50.

[3] Verheyen EM, Gottardi CJ. Regulation of Wnt/beta-catenin signaling by protein kinases. Dev Dyn 2010;239(1):34–44.

[4] Mazumdar J, Dondeti V, Simon MC. Hypoxia-inducible factors in stem cells and cancer. J Cell Mol Med 2009;13(11–12):4319–28.

[5] Fleming HE, Janzen V, Lo Celso C, Guo J, Leahy KM, Kronenberg HM, et al. Wnt signaling in the niche enforces hematopoietic stem cell quiescence and is necessary to preserve self-renewal in vivo. Cell Stem Cell 2008;2(3):274–83.

[6] Nemeth MJ, Mak KK, Yang Y, Bodine DM. beta-Catenin expression in the bone marrow microenvironment is required for long-term maintenance of primitive hematopoietic cells. Stem Cells 2009;27(5):1109–19.
[7] Sato N, Meijer L, Skaltsounis L, Greengard P, Brivanlou AH. Maintenance of pluripotency in human and mouse embryonic stem cells through activation of Wnt signaling by a pharmacological GSK-3-specific inhibitor. Nat Med 2004;10(1):55–63.
[8] Boland GM, Perkins G, Hall DJ, Tuan RS. Wnt 3a promotes proliferation and suppresses osteogenic differentiation of adult human mesenchymal stem cells. J Cell Biochem 2004;93(6):1210–30.
[9] Kleber M, Sommer L. Wnt signaling and the regulation of stem cell function. Curr Opin Cell Biol 2004;16(6):681–7.
[10] Kaidi A, Williams AC, Paraskeva C. Interaction between beta-catenin and HIF-1 promotes cellular adaptation to hypoxia. Nat Cell Biol 2007;9(2):210–7.
[11] Kirstetter P, Anderson K, Porse BT, Jacobsen SE, Nerlov C. Activation of the canonical Wnt pathway leads to loss of hematopoietic stem cell repopulation and multilineage differentiation block. Nat Immunol 2006;7(10):1048–56.
[12] Li L, Bhatia R. Stem cell quiescence. Clin Cancer Res 2011;17(15):4936–41.
[13] Kojika S, Griffin JD. Notch receptors and hematopoiesis. Exp Hematol 2001;29(9): 1041–52.
[14] Diez H, Fischer A, Winkler A, Hu CJ, Hatzopoulos AK, Breier G, et al. Hypoxia-mediated activation of Dll4-Notch-Hey2 signaling in endothelial progenitor cells and adoption of arterial cell fate. Exp Cell Res 2007;313(1):1–9.
[15] Weinmaster G, Roberts VJ, Lemke G. Notch2: a second mammalian Notch gene. Development 1992;116(4):931–41.
[16] Mitsiadis TA, Lardelli M, Lendahl U, Thesleff I. Expression of Notch 1, 2 and 3 is regulated by epithelial-mesenchymal interactions and retinoic acid in the developing mouse tooth and associated with determination of ameloblast cell fate. J Cell Biol 1995;130(2):407–18.
[17] Lewis J. Notch signalling. A short cut to the nucleus. Nature 1998;393(6683):304–5.
[18] Powell BC, Passmore EA, Nesci A, Dunn SM. The Notch signalling pathway in hair growth. Mech Dev 1998;78(1–2):189–92.
[19] Artavanis-Tsakonas S, Rand MD, Lake RJ. Notch signaling: cell fate control and signal integration in development. Science 1999;284(5415):770–6.
[20] Hoyne GF, Dallman MJ, Champion BR, Lamb JR. Notch signalling in the regulation of peripheral immunity. Immunol Rev 2001;182:215–27.
[21] Varnum-Finney B, Purton LE, Yu M, Brashem-Stein C, Flowers D, Staats S, et al. The Notch ligand, Jagged-1, influences the development of primitive hematopoietic precursor cells. Blood 1998;91(11):4084–91.
[22] Sugimoto A, Yamamoto M, Suzuki M, Inoue T, Nakamura S, Motoda R, et al. Delta-4 Notch ligand promotes erythroid differentiation of human umbilical cord blood $CD34^+$ cells. Exp Hematol 2006;34(4):424–32.
[23] Bray SJ. Notch signalling: a simple pathway becomes complex. Nat Rev Mol Cell Biol 2006;7(9):678–89.
[24] Kadesch T. Notch signaling: the demise of elegant simplicity. Curr Opin Genet Dev 2004;14(5):506–12.
[25] Ohtsuka T, Ishibashi M, Gradwohl G, Nakanishi S, Guillemot F, Kageyama R. Hes1 and Hes5 as notch effectors in mammalian neuronal differentiation. EMBO J 1999;18(8):2196–207.
[26] Mourikis P, Sambasivan R, Castel D, Rocheteau P, Bizzarro V, Tajbakhsh S. A critical requirement for notch signaling in maintenance of the quiescent skeletal muscle stem cell state. Stem Cells 2012;30(2):243–52.

[27] Chapouton P, Skupien P, Hesl B, Coolen M, Moore JC, Madelaine R, et al. Notch activity levels control the balance between quiescence and recruitment of adult neural stem cells. J Neurosci 2010;30(23):7961–74.
[28] Maillard I, Koch U, Dumortier A, Shestova O, Xu L, Sai H, et al. Canonical notch signaling is dispensable for the maintenance of adult hematopoietic stem cells. Cell Stem Cell 2008;2(4):356–66.
[29] Delaney C, Heimfeld S, Brashem-Stein C, Voorhies H, Manger RL, Bernstein ID. Notch-mediated expansion of human cord blood progenitor cells capable of rapid myeloid reconstitution. Nat Med 2010;16(2):232–6.
[30] VanDussen KL, Carulli AJ, Keeley TM, Patel SR, Puthoff BJ, Magness ST, et al. Notch signaling modulates proliferation and differentiation of intestinal crypt base columnar stem cells. Development 2012;139(3):488–97.
[31] Gustafsson MV, Zheng X, Pereira T, Gradin K, Jin S, Lundkvist J, et al. Hypoxia requires notch signaling to maintain the undifferentiated cell state. Dev Cell 2005;9(5):617–28.
[32] Tachikawa Y, Matsushima T, Abe Y, Sakano S, Yamamoto M, Nishimura J, et al. Pivotal role of Notch signaling in regulation of erythroid maturation and proliferation. Eur J Haematol 2006;77(4):273–81.
[33] Pear WS, Aster JC. T cell acute lymphoblastic leukemia/lymphoma: a human cancer commonly associated with aberrant NOTCH1 signaling. Curr Opin Hematol 2004;11(6):426–33.
[34] Patel NS, Li JL, Generali D, Poulsom R, Cranston DW, Harris AL. Up-regulation of delta-like 4 ligand in human tumor vasculature and the role of basal expression in endothelial cell function. Cancer Res 2005;65(19):8690–7.
[35] Ingham PW, Nakano Y, Seger C. Mechanisms and functions of Hedgehog signalling across the metazoa. Nat Rev Genet 2011;12(6):393–406.
[36] Trowbridge JJ, Scott MP, Bhatia M. Hedgehog modulates cell cycle regulators in stem cells to control hematopoietic regeneration. Proc Natl Acad Sci USA 2006;103(38):14134–9.
[37] Plaisant M, Giorgetti-Peraldi S, Gabrielson M, Loubat A, Dani C, Peraldi P. Inhibition of hedgehog signaling decreases proliferation and clonogenicity of human mesenchymal stem cells. PloS One 2011;6(2):e16798.
[38] Cheng W, Yu P, Wang L, Shen C, Song X, Chen J, et al. Sonic hedgehog signaling mediates resveratrol to increase proliferation of neural stem cells after oxygen-glucose deprivation/reoxygenation injury in vitro. Cell Physiol Biochem 2015;35(5):2019–32.
[39] Weli SC, Fink T, Cetinkaya C, Prasad MS, Pennisi CP, Zachar V. Notch and hedgehog signaling cooperate to maintain self-renewal of human embryonic stem cells exposed to low oxygen concentration. Int J Stem Cells 2010;3(2):129–37.
[40] Bijlsma MF, Groot AP, Oduro JP, Franken RJ, Schoenmakers SH, Peppelenbosch MP, et al. Hypoxia induces a hedgehog response mediated by HIF-1alpha. J Cell Mol Med 2009;13(8B):2053–60.
[41] Kiel MJ, Morrison SJ. Maintaining hematopoietic stem cells in the vascular niche. Immunity 2006;25(6):862–4.
[42] Barker JE. Sl/Sld hematopoietic progenitors are deficient in situ. Exp Hematol 1994;22(2):174–7.
[43] Nilsson SK, Johnston HM, Coverdale JA. Spatial localization of transplanted hemopoietic stem cells: inferences for the localization of stem cell niches. Blood 2001;97(8):2293–9.
[44] Arai F, Suda T. Maintenance of quiescent hematopoietic stem cells in the osteoblastic niche. Ann N Y Acad Sci 2007;1106:41–53.

[45] Visnjic D, Kalajzic Z, Rowe DW, Katavic V, Lorenzo J, Aguila HL. Hematopoiesis is severely altered in mice with an induced osteoblast deficiency. Blood 2004;103(9):3258–64.
[46] Qian H, Buza-Vidas N, Hyland CD, Jensen CT, Antonchuk J, Mansson R, et al. Critical role of thrombopoietin in maintaining adult quiescent hematopoietic stem cells. Cell Stem Cell 2007;1(6):671–84.
[47] Yoshihara H, Arai F, Hosokawa K, Hagiwara T, Takubo K, Nakamura Y, et al. Thrombopoietin/MPL signaling regulates hematopoietic stem cell quiescence and interaction with the osteoblastic niche. Cell Stem Cell 2007;1(6):685–97.
[48] Solar GP, Kerr WG, Zeigler FC, Hess D, Donahue C, de Sauvage FJ, et al. Role of c-mpl in early hematopoiesis. Blood 1998;92(1):4–10.
[49] Brenet F, Kermani P, Spektor R, Rafii S, Scandura JM. TGFbeta restores hematopoietic homeostasis after myelosuppressive chemotherapy. J Exp Med 2013;210(3):623–39.
[50] Larsson J, Blank U, Helgadottir H, Bjornsson JM, Ehinger M, Goumans MJ, et al. TGF-beta signaling-deficient hematopoietic stem cells have normal self-renewal and regenerative ability in vivo despite increased proliferative capacity in vitro. Blood 2003;102(9):3129–35.
[51] Zamani N, Brown CW. Emerging roles for the transforming growth factor-{beta} superfamily in regulating adiposity and energy expenditure. Endocr Rev 2011;32(3):387–403.
[52] Bogaerts E, Heindryckx F, Vandewynckel YP, Van Grunsven LA, Van Vlierberghe H. The roles of transforming growth factor-β, Wnt, Notch and hypoxia on liver progenitor cells in primary liver tumours (Review). Int J Oncol 2014;44(4):1015–22.
[53] Nie Y, Han YC, Zou YR. CXCR4 is required for the quiescence of primitive hematopoietic cells. J Exp Med 2008;205(4):777–83.
[54] Kadivar M, Alijani N, Farahmandfar M, Rahmati S, Ghahhari NM, Mahdian R. Effect of acute hypoxia on CXCR4 gene expression in C57BL/6 mouse bone marrow-derived mesenchymal stem cells. Adv Biomed Res 2014;3:222.
[55] Speth JM, Hoggatt J, Singh P, Pelus LM. Pharmacologic increase in HIF1alpha enhances hematopoietic stem and progenitor homing and engraftment. Blood 2014;123(2):203–7.
[56] Israels ED, Israels LG. The cell cycle. Oncologist 2000;5(6):510–3.
[57] Bottger A, Bottger V, Sparks A, Liu WL, Howard SF, Lane DP. Design of a synthetic Mdm2-binding mini protein that activates the p53 response in vivo. Curr Biol 1997;7(11):860–9.
[58] Haupt Y, Maya R, Kazaz A, Oren M. Mdm2 promotes the rapid degradation of p53. Nature 1997;387(6630):296–9.
[59] Kubbutat MH, Jones SN, Vousden KH. Regulation of p53 stability by Mdm2. Nature 1997;387(6630):299–303.
[60] Liu Y, Elf SE, Miyata Y, Sashida G, Liu Y, Huang G, et al. p53 regulates hematopoietic stem cell quiescence. Cell Stem Cell 2009;4(1):37–48.
[61] Abbas HA, Maccio DR, Coskun S, Jackson JG, Hazen AL, Sills TM, et al. Mdm2 is required for survival of hematopoietic stem cells/progenitors via dampening of ROS-induced p53 activity. Cell Stem Cell 2010;7(5):606–17.
[62] Lin T, Chao C, Saito S, Mazur SJ, Murphy ME, Appella E, et al. p53 induces differentiation of mouse embryonic stem cells by suppressing Nanog expression. Nat Cell Biol 2005;7(2):165–71.
[63] An WG, Kanekal M, Simon MC, Maltepe E, Blagosklonny MV, Neckers LM. Stabilization of wild-type p53 by hypoxia-inducible factor 1alpha. Nature 1998;392(6674):405–8.
[64] Sanchez-Puig N, Veprintsev DB, Fersht AR. Binding of natively unfolded HIF-1alpha ODD domain to p53. Mol Cell 2005;17(1):11–21.

[65] Ravi R, Mookerjee B, Bhujwalla ZM, Sutter CH, Artemov D, Zeng Q, et al. Regulation of tumor angiogenesis by p53-induced degradation of hypoxia-inducible factor 1alpha. Genes Dev 2000;14(1):34–44.
[66] Bertout JA, Patel SA, Fryer BH, Durham AC, Covello KL, Olive KP, et al. Heterozygosity for hypoxia inducible factor 1alpha decreases the incidence of thymic lymphomas in a p53 mutant mouse model. Cancer Res 2009;69(7):3213–20.
[67] Cheung TH, Rando TA. Molecular regulation of stem cell quiescence. Nat Rev Mol Cell Biol 2013;14(6):329–40.
[68] Jacques TS, Swales A, Brzozowski MJ, Henriquez NV, Linehan JM, Mirzadeh Z, et al. Combinations of genetic mutations in the adult neural stem cell compartment determine brain tumour phenotypes. EMBO J 2010;29(1):222–35.
[69] Dannenberg JH, van Rossum A, Schuijff L, te Riele H. Ablation of the retinoblastoma gene family deregulates G(1) control causing immortalization and increased cell turnover under growth-restricting conditions. Genes Dev 2000;14(23):3051–64.
[70] Hosoyama T, Nishijo K, Prajapati SI, Li G, Keller C. Rb1 gene inactivation expands satellite cell and postnatal myoblast pools. J Biol Chem 2011;286(22):19556–64.
[71] Viatour P, Somervaille TC, Venkatasubrahmanyam S, Kogan S, McLaughlin ME, Weissman IL, et al. Hematopoietic stem cell quiescence is maintained by compound contributions of the retinoblastoma gene family. Cell Stem Cell 2008;3(4):416–28.
[72] Yamada T, Park CS, Lacorazza HD. Genetic control of quiescence in hematopoietic stem cells. Cell Cycle 2013;12(15):2376–83.
[73] Cheng T, Rodrigues N, Shen H, Yang Y, Dombkowski D, Sykes M, et al. Hematopoietic stem cell quiescence maintained by p21cip1/waf1. Science 2000;287(5459):1804–8.
[74] Kippin TE, Martens DJ, van der Kooy D. p21 loss compromises the relative quiescence of forebrain stem cell proliferation leading to exhaustion of their proliferation capacity. Genes Dev 2005;19(6):756–67.
[75] Zou P, Yoshihara H, Hosokawa K, Tai I, Shinmyozu K, Tsukahara F, et al. p57(Kip2) and p27(Kip1) cooperate to maintain hematopoietic stem cell quiescence through interactions with Hsc70. Cell Stem Cell 2011;9(3):247–61.
[76] Yuan Y, Shen H, Franklin DS, Scadden DT, Cheng T. In vivo self-renewing divisions of haematopoietic stem cells are increased in the absence of the early G1-phase inhibitor, p18INK4C. Nat Cell Biol 2004;6(5):436–42.
[77] Blanpain C, Lowry WE, Geoghegan A, Polak L, Fuchs E. Self-renewal, multipotency, and the existence of two cell populations within an epithelial stem cell niche. Cell 2004;118(5):635–48.
[78] Fukada S, Uezumi A, Ikemoto M, Masuda S, Segawa M, Tanimura N, et al. Molecular signature of quiescent satellite cells in adult skeletal muscle. Stem Cells 2007;25(10): 2448–59.
[79] Forsberg EC, Passegue E, Prohaska SS, Wagers AJ, Koeva M, Stuart JM, et al. Molecular signatures of quiescent, mobilized and leukemia-initiating hematopoietic stem cells. PloS One 2010;5(1):e8785.
[80] Ficara F, Murphy MJ, Lin M, Cleary ML. Pbx1 regulates self-renewal of long-term hematopoietic stem cells by maintaining their quiescence. Cell Stem Cell 2008;2(5):484–96.
[81] Tothova Z, Kollipara R, Huntly BJ, Lee BH, Castrillon DH, Cullen DE, et al. FoxOs are critical mediators of hematopoietic stem cell resistance to physiologic oxidative stress. Cell 2007;128(2):325–39.
[82] Lacombe J, Herblot S, Rojas-Sutterlin S, Haman A, Barakat S, Iscove NN, et al. Scl regulates the quiescence and the long-term competence of hematopoietic stem cells. Blood 2010;115(4):792–803.

[83] Sirin O, Lukov GL, Mao R, Conneely OM, Goodell MA. The orphan nuclear receptor Nurr1 restricts the proliferation of haematopoietic stem cells. Nat Cell Biol 2010;12(12):1213–9.
[84] Jeong M, Piao ZH, Kim MS, Lee SH, Yun S, Sun HN, et al. Thioredoxin-interacting protein regulates hematopoietic stem cell quiescence and mobilization under stress conditions. J Immunol 2009;183(4):2495–505.
[85] Hock H, Hamblen MJ, Rooke HM, Schindler JW, Saleque S, Fujiwara Y, et al. Gfi-1 restricts proliferation and preserves functional integrity of haematopoietic stem cells. Nature 2004;431(7011):1002–7.
[86] Yamada T, Park CS, Burns A, Nakada D, Lacorazza HD. The cytosolic protein G0S2 maintains quiescence in hematopoietic stem cells. PloS One 2012;7(5):e38280.
[87] Ichikawa M, Goyama S, Asai T, Kawazu M, Nakagawa M, Takeshita M, et al. AML1/Runx1 negatively regulates quiescent hematopoietic stem cells in adult hematopoiesis. J Immunol 2008;180(7):4402–8.
[88] Wang Z, Li G, Tse W, Bunting KD. Conditional deletion of STAT5 in adult mouse hematopoietic stem cells causes loss of quiescence and permits efficient nonablative stem cell replacement. Blood 2009;113(20):4856–65.
[89] Min IM, Pietramaggiori G, Kim FS, Passegue E, Stevenson KE, Wagers AJ. The transcription factor EGR1 controls both the proliferation and localization of hematopoietic stem cells. Cell Stem Cell 2008;2(4):380–91.
[90] Tipping AJ, Pina C, Castor A, Hong D, Rodrigues NP, Lazzari L, et al. High GATA-2 expression inhibits human hematopoietic stem and progenitor cell function by effects on cell cycle. Blood 2009;113(12):2661–72.
[91] Hoffman B, Liebermann DA. Apoptotic signaling by c-MYC. Oncogene 2008;27(50): 6462–72.
[92] Wilson A, Murphy MJ, Oskarsson T, Kaloulis K, Bettess MD, Oser GM, et al. c-Myc controls the balance between hematopoietic stem cell self-renewal and differentiation. Genes Dev 2004;18(22):2747–63.
[93] Lacorazza HD, Miyazaki Y, Di Cristofano A, Deblasio A, Hedvat C, Zhang J, et al. The ETS protein MEF plays a critical role in perforin gene expression and the development of natural killer and NK-T cells. Immunity 2002;17(4):437–49.
[94] Mateo JL, van den Berg DL, Haeussler M, Drechsel D, Gaber ZB, Castro DS, et al. Characterization of the neural stem cell gene regulatory network identifies OLIG2 as a multifunctional regulator of self-renewal. Genome Res 2015;25(1):41–56.
[95] Liu N, Lu M, Tian X, Han Z. Molecular mechanisms involved in self-renewal and pluripotency of embryonic stem cells. J Cell Physiol 2007;211(2):279–86.
[96] Ng HH, Surani MA. The transcriptional and signalling networks of pluripotency. Nat Cell Biol 2011;13(5):490–6.
[97] Izadpanah R, Trygg C, Patel B, Kriedt C, Dufour J, Gimble JM, et al. Biologic properties of mesenchymal stem cells derived from bone marrow and adipose tissue. J Cell Biochem 2006;99(5):1285–97.
[98] Kuroda T, Tada M, Kubota H, Kimura H, Hatano SY, Suemori H, et al. Octamer and Sox elements are required for transcriptional cis regulation of Nanog gene expression. Mol Cell Biol 2005;25(6):2475–85.
[99] Rodda DJ, Chew JL, Lim LH, Loh YH, Wang B, Ng HH, et al. Transcriptional regulation of nanog by OCT4 and SOX2. J Biol Chem 2005;280(26):24731–7.
[100] van den Berg DL, Snoek T, Mullin NP, Yates A, Bezstarosti K, Demmers J, et al. An Oct4-centered protein interaction network in embryonic stem cells. Cell Stem Cell 2010;6(4):369–81.

[101] Dejosez M, Levine SS, Frampton GM, Whyte WA, Stratton SA, Barton MC, et al. Ronin/Hcf-1 binds to a hyperconserved enhancer element and regulates genes involved in the growth of embryonic stem cells. Genes Dev 2010;24(14):1479–84.
[102] Chen X, Xu H, Yuan P, Fang F, Huss M, Vega VB, et al. Integration of external signaling pathways with the core transcriptional network in embryonic stem cells. Cell 2008;133(6):1106–17.
[103] Kim J, Chu J, Shen X, Wang J, Orkin SH. An extended transcriptional network for pluripotency of embryonic stem cells. Cell 2008;132(6):1049–61.
[104] Marson A, Levine SS, Cole MF, Frampton GM, Brambrink T, Johnstone S, et al. Connecting microRNA genes to the core transcriptional regulatory circuitry of embryonic stem cells. Cell 2008;134(3):521–33.
[105] Mitsui K, Tokuzawa Y, Itoh H, Segawa K, Murakami M, Takahashi K, et al. The homeoprotein Nanog is required for maintenance of pluripotency in mouse epiblast and ES cells. Cell 2003;113(5):631–42.
[106] Chambers I, Colby D, Robertson M, Nichols J, Lee S, Tweedie S, Smith A. Functional expression cloning of Nanog, a pluripotency sustaining factor in embryonic stem cells. Cell 2003;113(5):643–55.
[107] Guo G, Smith A. A genome-wide screen in EpiSCs identifies Nr5a nuclear receptors as potent inducers of ground state pluripotency. Development 2010;137(19):3185–92.
[108] Silva J, Nichols J, Theunissen TW, Guo G, van Oosten AL, Barrandon O, et al. Nanog is the gateway to the pluripotent ground state. Cell 2009;138(4):722–37.
[109] Greber B, Wu G, Bernemann C, Joo JY, Han DW, Ko K, et al. Conserved and divergent roles of FGF signaling in mouse epiblast stem cells and human embryonic stem cells. Cell Stem Cell 2010;6(3):215–26.
[110] Guo G, Yang J, Nichols J, Hall JS, Eyres I, Mansfield W, et al. Klf4 reverts developmentally programmed restriction of ground state pluripotency. Development 2009;136(7):1063–9.
[111] Hanna J, Cheng AW, Saha K, Kim J, Lengner CJ, Soldner F, et al. Human embryonic stem cells with biological and epigenetic characteristics similar to those of mouse ESCs. Proc Natl Acad Sci USA 2010;107(20):9222–7.
[112] Xu Y, Zhu X, Hahm HS, Wei W, Hao E, Hayek A, et al. Revealing a core signaling regulatory mechanism for pluripotent stem cell survival and self-renewal by small molecules. Proc Natl Acad Sci USA 2010;107(18):8129–34.
[113] Takahashi K, Tanabe K, Ohnuki M, Narita M, Ichisaka T, Tomoda K, et al. Induction of pluripotent stem cells from adult human fibroblasts by defined factors. Cell 2007;131(5):861–72.
[114] Yu J, Vodyanik MA, Smuga-Otto K, Antosiewicz-Bourget J, Frane JL, Tian S, et al. Induced pluripotent stem cell lines derived from human somatic cells. Science 2007;318(5858):1917–20.
[115] Barber BA, Rastegar M. Epigenetic control of Hox genes during neurogenesis, development, and disease. Ann Anat 2010;192(5):261–74.
[116] Bahrami SB, Veiseh M, Dunn AA, Boudreau NJ. Temporal changes in Hox gene expression accompany endothelial cell differentiation of embryonic stem cells. Cell Adh Migr 2011;5(2):133–41.
[117] Sauvageau G, Lansdorp PM, Eaves CJ, Hogge DE, Dragowska WH, Reid DS, et al. Differential expression of homeobox genes in functionally distinct $CD34^+$ subpopulations of human bone marrow cells. Proc Natl Acad Sci USA 1994;91(25):12223–7.
[118] Sauvageau G, Thorsteinsdottir U, Eaves CJ, Lawrence HJ, Largman C, Lansdorp PM, et al. Overexpression of HOXB4 in hematopoietic cells causes the selective expansion of more primitive populations in vitro and in vivo. Genes Dev 1995;9(14):1753–65.

[119] Kyba M, Perlingeiro RC, Daley GQ. HoxB4 confers definitive lymphoid-myeloid engraftment potential on embryonic stem cell and yolk sac hematopoietic progenitors. Cell 2002;109(1):29–37.

[120] Kelly JJ, Stechishin O, Chojnacki A, Lun X, Sun B, Senger DL, et al. Proliferation of human glioblastoma stem cells occurs independently of exogenous mitogens. Stem Cells 2009;27(8):1722–33.

[121] Bami M, Episkopou V, Gavalas A, Gouti M. Directed neural differentiation of mouse embryonic stem cells is a sensitive system for the identification of novel Hox gene effectors. PloS One 2011;6(5):e20197.

[122] Phinney DG, Gray AJ, Hill K, Pandey A. Murine mesenchymal and embryonic stem cells express a similar Hox gene profile. Biochem Biophys Res Commun 2005; 338(4):1759–65.

[123] Semenza GL. Hypoxia-inducible factor 1 (HIF-1) pathway. Sci STKE 2007;2007(407):cm8.

[124] Semenza GL. Life with oxygen. Science 2007;318(5847):62–4.

[125] Gezer D, Vukovic M, Soga T, Pollard PJ, Kranc KR. Concise review: genetic dissection of hypoxia signaling pathways in normal and leukemic stem cells. Stem Cells 2014;32(6):1390–7.

[126] Jiang BH, Semenza GL, Bauer C, Marti HH. Hypoxia-inducible factor 1 levels vary exponentially over a physiologically relevant range of O_2 tension. Am J Physiol 1996;271(4 Pt 1): C1172–80.

[127] Semenza GL. Expression of hypoxia-inducible factor 1: mechanisms and consequences. Biochem Pharmacol 2000;59(1):47–53.

[128] Semenza GL. HIF-1: mediator of physiological and pathophysiological responses to hypoxia. J Appl Physiol 2000;88(4):1474–80.

[129] Schofield CJ, Ratcliffe PJ. Oxygen sensing by HIF hydroxylases. Nat Rev Mol Cell Biol 2004;5(5):343–54.

[130] Ivan M, Kondo K, Yang H, Kim W, Valiando J, Ohh M, et al. HIFalpha targeted for VHL-mediated destruction by proline hydroxylation: implications for O_2 sensing. Science 2001;292(5516):464–8.

[131] Metzen E, Berchner-Pfannschmidt U, Stengel P, Marxsen JH, Stolze I, Klinger M, et al. Intracellular localisation of human HIF-1 alpha hydroxylases: implications for oxygen sensing. J Cell Sci 2003;116(Pt 7):1319–26.

[132] Soilleux EJ, Turley H, Tian YM, Pugh CW, Gatter KC, Harris AL. Use of novel monoclonal antibodies to determine the expression and distribution of the hypoxia regulatory factors PHD-1, PHD-2, PHD-3 and FIH in normal and neoplastic human tissues. Histopathology 2005;47(6):602–10.

[133] Takubo K, Goda N, Yamada W, Iriuchishima H, Ikeda E, Kubota Y, et al. Regulation of the HIF-1alpha level is essential for hematopoietic stem cells. Cell Stem Cell 2010;7(3):391–402.

[134] Nombela-Arrieta C, Pivarnik G, Winkel B, Canty KJ, Harley B, Mahoney JE, et al. Quantitative imaging of haematopoietic stem and progenitor cell localization and hypoxic status in the bone marrow microenvironment. Nat Cell Biol 2013;15(5):533–43.

[135] Adelman DM, Maltepe E, Simon MC. Multilineage embryonic hematopoiesis requires hypoxic ARNT activity. Genes Dev 1999;13(19):2478–83.

[136] Ramirez-Bergeron DL, Simon MC. Hypoxia-inducible factor and the development of stem cells of the cardiovascular system. Stem Cells 2001;19(4):279–86.

[137] Yoon D, Pastore YD, Divoky V, Liu E, Mlodnicka AE, Rainey K, et al. Hypoxia-inducible factor-1 deficiency results in dysregulated erythropoiesis signaling and iron homeostasis in mouse development. J Biol Chem 2006;281(35):25703–11.

[138] Takubo K, Nagamatsu G, Kobayashi CI, Nakamura-Ishizu A, Kobayashi H, Ikeda E, et al. Regulation of glycolysis by Pdk functions as a metabolic checkpoint for cell cycle quiescence in hematopoietic stem cells. Cell Stem Cell 2013;12(1):49–61.

[139] Imanirad P, Dzierzak E. Hypoxia and HIFs in regulating the development of the hematopoietic system. Blood Cells Mol Dis 2013;51(4):256–63.

[140] Rouault-Pierre K, Lopez-Onieva L, Foster K, Anjos-Afonso F, Lamrissi-Garcia I, Serrano-Sanchez M, et al. HIF-2alpha protects human hematopoietic stem/progenitors and acute myeloid leukemic cells from apoptosis induced by endoplasmic reticulum stress. Cell Stem Cell 2013;13(5):549–63.

[141] Scortegagna M, Morris MA, Oktay Y, Bennett M, Garcia JA. The HIF family member EPAS1/HIF-2alpha is required for normal hematopoiesis in mice. Blood 2003;102(5):1634–40.

[142] Guitart AV, Subramani C, Armesilla-Diaz A, Smith G, Sepulveda C, Gezer D, et al. Hif-2alpha is not essential for cell-autonomous hematopoietic stem cell maintenance. Blood 2013;122(10):1741–5.

[143] Kocabas F, Zheng J, Thet S, Copeland NG, Jenkins NA, DeBerardinis RJ, et al. Meis1 regulates the metabolic phenotype and oxidant defense of hematopoietic stem cells. Blood 2012;120(25):4963–72.

[144] Makino Y, Cao R, Svensson K, Bertilsson G, Asman M, Tanaka H, et al. Inhibitory PAS domain protein is a negative regulator of hypoxia-inducible gene expression. Nature 2001;414(6863):550–4.

[145] Tanaka T, Wiesener M, Bernhardt W, Eckardt KU, Warnecke C. The human HIF (hypoxia-inducible factor)-3alpha gene is a HIF-1 target gene and may modulate hypoxic gene induction. Biochem J 2009;424(1):143–51.

[146] Gordan JD, Simon MC. Hypoxia-inducible factors: central regulators of the tumor phenotype. Curr Opin Genet Dev 2007;17(1):71–7.

[147] Suda T, Takubo K, Semenza GL. Metabolic regulation of hematopoietic stem cells in the hypoxic niche. Cell Stem Cell 2011;9(4):298–310.

[148] Aad G, Abajyan T, Abbott B, Abdallah J, Abdel Khalek S, Abdelalim AA, et al. Search for dark matter candidates and large extra dimensions in events with a photon and missing transverse momentum in pp collision data at sqrt[s]=7 TeV with the ATLAS detector. Phys Rev Lett 2013;110(1):011802.

[149] Simon MC. Coming up for air: HIF-1 and mitochondrial oxygen consumption. Cell Metab 2006;3(3):150–1.

[150] Zhao F, Mancuso A, Bui TV, Tong X, Gruber JJ, Swider CR, et al. Imatinib resistance associated with BCR-ABL upregulation is dependent on HIF-1alpha-induced metabolic reprograming. Oncogene 2010;29(20):2962–72.

[151] Semenza GL. Hypoxia-inducible factor 1: regulator of mitochondrial metabolism and mediator of ischemic preconditioning. Biochim Biophys Acta 2011;1813(7):1263–8.

[152] Jensen KS, Binderup T, Jensen KT, Therkelsen I, Borup R, Nilsson E, et al. FoxO3A promotes metabolic adaptation to hypoxia by antagonizing Myc function. EMBO J 2011;30(22):4554–70.

[153] Aragones J, Schneider M, Van Geyte K, Fraisl P, Dresselaers T, Mazzone M, et al. Deficiency or inhibition of oxygen sensor Phd1 induces hypoxia tolerance by reprogramming basal metabolism. Nat Genet 2008;40(2):170–80.

[154] Rankin EB, Rha J, Selak MA, Unger TL, Keith B, Liu Q, et al. Hypoxia-inducible factor 2 regulates hepatic lipid metabolism. Mol Cell Biol 2009;29(16):4527–38.

[155] Majmundar AJ, Wong WJ, Simon MC. Hypoxia-inducible factors and the response to hypoxic stress. Mol Cell 2010;40(2):294–309.

[156] Wouters BG, Koritzinsky M. Hypoxia signalling through mTOR and the unfolded protein response in cancer. Nat Rev Cancer 2008;8(11):851–64.
[157] Arany Z, Foo SY, Ma Y, Ruas JL, Bommi-Reddy A, Girnun G, et al. HIF-independent regulation of VEGF and angiogenesis by the transcriptional coactivator PGC-1alpha. Nature 2008;451(7181):1008–12.
[158] Webb JD, Coleman ML, Pugh CW. Hypoxia, hypoxia-inducible factors (HIF), HIF hydroxylases and oxygen sensing. Cell Mol Life Sci 2009;66(22):3539–54.
[159] Krumlauf R. Hox genes in vertebrate development. Cell 1994;78(2):191–201.
[160] Kranc KR, Schepers H, Rodrigues NP, Bamforth S, Villadsen E, Ferry H, et al. Cited2 is an essential regulator of adult hematopoietic stem cells. Cell Stem Cell 2009;5(6):659–65.
[161] Miharada K, Karlsson G, Rehn M, Rorby E, Siva K, Cammenga J, et al. Cripto regulates hematopoietic stem cells as a hypoxic-niche-related factor through cell surface receptor GRP78. Cell Stem Cell 2011;9(4):330–44.
[162] Miharada K, Karlsson G, Rehn M, Rorby E, Siva K, Cammenga J, et al. Hematopoietic stem cells are regulated by Cripto, as an intermediary of HIF-1alpha in the hypoxic bone marrow niche. Ann N Y Acad Sci 2012;1266:55–62.
[163] Dioum EM, Chen R, Alexander MS, Zhang Q, Hogg RT, Gerard RD, et al. Regulation of hypoxia-inducible factor 2alpha signaling by the stress-responsive deacetylase sirtuin 1. Science 2009;324(5932):1289–93.
[164] Zhong L, D'Urso A, Toiber D, Sebastian C, Henry RE, Vadysirisack DD, et al. The histone deacetylase Sirt6 regulates glucose homeostasis via Hif1alpha. Cell 2010;140(2):280–93.
[165] Lim JH, Lee YM, Chun YS, Chen J, Kim JE, Park JW. Sirtuin 1 modulates cellular responses to hypoxia by deacetylating hypoxia-inducible factor 1alpha. Mol Cell 2010;38(6):864–78.
[166] Taguchi A, Yanagisawa K, Tanaka M, Cao K, Matsuyama Y, Goto H, et al. Identification of hypoxia-inducible factor-1 alpha as a novel target for miR-17-92 microRNA cluster. Cancer Res 2008;68(14):5540–5.
[167] Brugarolas JB, Vazquez F, Reddy A, Sellers WR, Kaelin Jr WG. TSC2 regulates VEGF through mTOR-dependent and -independent pathways. Cancer Cell 2003;4(2):147–58.
[168] Mottet D, Dumont V, Deccache Y, Demazy C, Ninane N, Raes M, et al. Regulation of hypoxia-inducible factor-1alpha protein level during hypoxic conditions by the phosphatidylinositol 3-kinase/Akt/glycogen synthase kinase 3beta pathway in HepG2 cells. J Biol Chem 2003;278(33):31277–85.
[169] Brugarolas J, Kaelin Jr WG. Dysregulation of HIF and VEGF is a unifying feature of the familial hamartoma syndromes. Cancer Cell 2004;6(1):7–10.
[170] Bernardi R, Guernah I, Jin D, Grisendi S, Alimonti A, Teruya-Feldstein J, et al. PML inhibits HIF-1alpha translation and neoangiogenesis through repression of mTOR. Nature 2006;442(7104):779–85.
[171] Irani K, Xia Y, Zweier JL, Sollott SJ, Der CJ, Fearon ER, et al. Mitogenic signaling mediated by oxidants in Ras-transformed fibroblasts. Science 1997;275(5306):1649–52.
[172] Gerald D, Berra E, Frapart YM, Chan DA, Giaccia AJ, Mansuy D, et al. JunD reduces tumor angiogenesis by protecting cells from oxidative stress. Cell 2004;118(6):781–94.
[173] Bernstein BE, Meissner A, Lander ES. The mammalian epigenome. Cell 2007;128(4):669–81.
[174] Boyer LA, Mathur D, Jaenisch R. Molecular control of pluripotency. Curr Opin Genet Dev 2006;16(5):455–62.

[175] Meshorer E, Misteli T. Chromatin in pluripotent embryonic stem cells and differentiation. Nat Rev Mol Cell Biol 2006;7(7):540–6.
[176] Lee JH, Hart SR, Skalnik DG. Histone deacetylase activity is required for embryonic stem cell differentiation. Genesis 2004;38(1):32–8.
[177] Azuara V, Perry P, Sauer S, Spivakov M, Jorgensen HF, John RM, et al. Chromatin signatures of pluripotent cell lines. Nat Cell Biol 2006;8(5):532–8.
[178] Bernstein BE, Mikkelsen TS, Xie X, Kamal M, Huebert DJ, Cuff J, et al. A bivalent chromatin structure marks key developmental genes in embryonic stem cells. Cell 2006;125(2):315–26.
[179] Jackson M, Krassowska A, Gilbert N, Chevassut T, Forrester L, Ansell J, et al. Severe global DNA hypomethylation blocks differentiation and induces histone hyperacetylation in embryonic stem cells. Mol Cell Biol 2004;24(20):8862–71.
[180] Carlone DL, Lee JH, Young SR, Dobrota E, Butler JS, Ruiz J, et al. Reduced genomic cytosine methylation and defective cellular differentiation in embryonic stem cells lacking CpG binding protein. Mol Cell Biol 2005;25(12):4881–91.
[181] Sawarkar R, Paro R. Interpretation of developmental signaling at chromatin: the Polycomb perspective. Dev Cell 2010;19(5):651–61.
[182] Kolybaba A, Classen AK. Sensing cellular states–signaling to chromatin pathways targeting Polycomb and Trithorax group function. Cell Tissue Res 2014;356(3):477–93.
[183] Breiling A, Turner BM, Bianchi ME, Orlando V. General transcription factors bind promoters repressed by Polycomb group proteins. Nature 2001;412(6847):651–5.
[184] Lagarou A, Mohd-Sarip A, Moshkin YM, Chalkley GE, Bezstarosti K, Demmers JA, et al. dKDM2 couples histone H2A ubiquitylation to histone H3 demethylation during Polycomb group silencing. Genes Dev 2008;22(20):2799–810.
[185] Wang H, Wang L, Erdjument-Bromage H, Vidal M, Tempst P, Jones RS, et al. Role of histone H2A ubiquitination in Polycomb silencing. Nature 2004;431(7010):873–8.
[186] Francis NJ, Kingston RE, Woodcock CL. Chromatin compaction by a polycomb group protein complex. Science 2004;306(5701):1574–7.
[187] Dorighi KM, Tamkun JW. The trithorax group proteins Kismet and ASH1 promote H3K36 dimethylation to counteract Polycomb group repression in *Drosophila*. Development 2013;140(20):4182–92.
[188] Schuettengruber B, Martinez AM, Iovino N, Cavalli G. Trithorax group proteins: switching genes on and keeping them active. Nat Rev Mol Cell Biol 2011;12(12):799–814.
[189] Buszczak M, Spradling AC. Searching chromatin for stem cell identity. Cell 2006; 125(2):233–6.
[190] Herz HM, Mohan M, Garruss AS, Liang K, Takahashi YH, Mickey K, et al. Enhancer-associated H3K4 monomethylation by Trithorax-related, the *Drosophila* homolog of mammalian Mll3/Mll4. Genes Dev 2012;26(23):2604–20.
[191] Iovino N, Ciabrelli F, Cavalli G. PRC2 controls *Drosophila* oocyte cell fate by repressing cell cycle genes. Dev Cell 2013;26(4):431–9.
[192] Sen N, Satija YK, Das S. PGC-1alpha, a key modulator of p53, promotes cell survival upon metabolic stress. Mol Cell 2011;44(4):621–34.
[193] Cha TL, Zhou BP, Xia W, Wu Y, Yang CC, Chen CT, et al. Akt-mediated phosphorylation of EZH2 suppresses methylation of lysine 27 in histone H3. Science 2005;310(5746):306–10.
[194] Kaneko S, Li G, Son J, Xu CF, Margueron R, Neubert TA, et al. Phosphorylation of the PRC2 component Ezh2 is cell cycle-regulated and up-regulates its binding to ncRNA. Genes Dev 2010;24(23):2615–20.

[195] Wu SC, Zhang Y. Cyclin-dependent kinase 1 (CDK1)-mediated phosphorylation of enhancer of zeste 2 (Ezh2) regulates its stability. J Biol Chem 2011;286(32):28511–9.
[196] Nacerddine K, Beaudry JB, Ginjala V, Westerman B, Mattiroli F, Song JY, et al. Akt-mediated phosphorylation of Bmi1 modulates its oncogenic potential, E3 ligase activity, and DNA damage repair activity in mouse prostate cancer. J Clin Invest 2012;122(5):1920–32.
[197] Nakamura S, Oshima M, Yuan J, Saraya A, Miyagi S, Konuma T, et al. Bmi1 confers resistance to oxidative stress on hematopoietic stem cells. PloS One 2012;7(5):e36209.
[198] Prickaerts P, Niessen HE, Mouchel-Vielh E, Dahlmans VE, van den Akker GG, Geijselaers C, et al. MK3 controls Polycomb target gene expression via negative feedback on ERK. Epigenetics Chromatin 2012;5(1):12.
[199] Schwermann J, Rathinam C, Schubert M, Schumacher S, Noyan F, Koseki H, et al. MAPKAP kinase MK2 maintains self-renewal capacity of haematopoietic stem cells. EMBO J 2009;28(10):1392–406.
[200] Mukherji S, Ebert MS, Zheng GX, Tsang JS, Sharp PA, van Oudenaarden A. MicroRNAs can generate thresholds in target gene expression. Nat Genet 2011;43(9):854–9.
[201] Stark A, Brennecke J, Bushati N, Russell RBB, Cohen SM. Animal MicroRNAs confer robustness to gene expression and have a significant impact on 3'UTR evolution. Cell 2005;123(6):1133–46.
[202] Bartel DP. MicroRNAs: genomics, biogenesis, mechanism, and function. Cell 2004; 116(2):281–97.
[203] Cheung TH, Quach NL, Charville GW, Liu L, Park L, Edalati A, et al. Maintenance of muscle stem-cell quiescence by microRNA-489. Nature 2012;482(7386):524–8.
[204] Arnold CP, Tan R, Zhou B, Yue SB, Schaffert S, Biggs JR, et al. MicroRNA programs in normal and aberrant stem and progenitor cells. Genome Res 2011;21(5):798–810.
[205] Lechman ER, Gentner B, van Galen P, Giustacchini A, Saini M, Boccalatte FE, et al. Attenuation of miR-126 activity expands HSC in vivo without exhaustion. Cell Stem Cell 2012;11(6):799–811.
[206] Crist CG, Montarras D, Buckingham M. Muscle satellite cells are primed for myogenesis but maintain quiescence with sequestration of Myf5 mRNA targeted by microRNA-31 in mRNP granules. Cell Stem Cell 2012;11(1):118–26.
[207] Esau C, Kang X, Peralta E, Hanson E, Marcusson EG, Ravichandran LV, et al. MicroRNA-143 regulates adipocyte differentiation. J Biol Chem 2004;279(50):52361–5.
[208] Mohamed Ariff I, Mitra A, Basu A. Epigenetic regulation of self-renewal and fate determination in neural stem cells. J Neurosci Res 2012;90(3):529–39.
[209] Lin SL, Chang DC, Chang-Lin S, Lin CH, Wu DT, Chen DT, et al. Mir-302 reprograms human skin cancer cells into a pluripotent ES-cell-like state. RNA 2008;14(10):2115–24.
[210] Folmes CD, Nelson TJ, Martinez-Fernandez A, Arrell DK, Lindor JZ, Dzeja PP, et al. Somatic oxidative bioenergetics transitions into pluripotency-dependent glycolysis to facilitate nuclear reprogramming. Cell Metab 2011;14(2):264–71.
[211] Sadhu MJ, Guan Q, Li F, Sales-Lee J, Iavarone AT, Hammond MC, et al. Nutritional control of epigenetic processes in yeast and human cells. Genetics 2013;195(3):831–44.
[212] Lightfoot TJ, Barrett JH, Bishop T, Northwood EL, Smith G, Wilkie MJ, et al. Methylene tetrahydrofolate reductase genotype modifies the chemopreventive effect of folate in colorectal adenoma, but not colorectal cancer. Cancer Epidemiol Biomarkers Prev 2008; 17(9):2421–30.
[213] Yuan HX, Xiong Y, Guan KL. Nutrient sensing, metabolism, and cell growth control. Mol Cell 2013;49(3):379–87.
[214] Chervona Y, Costa M. The control of histone methylation and gene expression by oxidative stress, hypoxia, and metals. Free Radic Biol Med 2012;53(5):1041–7.

[215] Hino S, Sakamoto A, Nagaoka K, Anan K, Wang Y, Mimasu S, et al. FAD-dependent lysine-specific demethylase-1 regulates cellular energy expenditure. Nat Commun 2012;3:758.
[216] Hanover JA, Krause MW, Love DC. Bittersweet memories: linking metabolism to epigenetics through O-GlcNAcylation. Nat Rev Mol Cell Biol 2012;13(5):312–21.
[217] Jang H, Kim TW, Yoon S, Choi SY, Kang TW, Kim SY, et al. O-GlcNAc regulates pluripotency and reprogramming by directly acting on core components of the pluripotency network. Cell Stem Cell 2012;11(1):62–74.
[218] Choudhary C, Kumar C, Gnad F, Nielsen ML, Rehman M, Walther TC, et al. Lysine acetylation targets protein complexes and co-regulates major cellular functions. Science 2009;325(5942):834–40.
[219] Wellen KE, Hatzivassiliou G, Sachdeva UM, Bui TV, Cross JR, Thompson CB. ATP-citrate lyase links cellular metabolism to histone acetylation. Science 2009;324(5930):1076–80.
[220] Cedar H, Bergman Y. Linking DNA methylation and histone modification: patterns and paradigms. Nat Rev Genet 2009;10(5):295–304.

Part Two

Anaerobic-to-Aerobic Eukaryote Evolution: A Paradigm for Stem Cell Self-Renewal, Commitment and Differentiation?

Evolution of Eukaryotes with Respect to Atmosphere Oxygen Appearance and Rise: Anaerobiosis, Facultative Aerobiosis, and Aerobiosis[1]

8

Chapter Outline

8.1 From the First Prokaryotes to the Great Oxidation Event

According to the actual evidence-based knowledge, the first simple organisms (prokaryotes) appeared on Earth about 3.8 billion years ago and evolved in completely anoxic conditions. Initially they were chemoautotrophs, used CO_2 as a carbon source, and oxidized organic substrates to extract energy. In anaerobic conditions, prokaryotes also developed another way to generate energy: transformation of glucose to pyruvate (lactate) (i.e., anaerobic). In this process, adenosine triphosphate (ATP) appeared as a short-term energy currency, which has been in use by almost all organisms up to now. The geologic evidences revealed that the first oxygen appearance was not the result of a life process but rather of the water photolysis [1]. These were, however, extremely low oxygen amounts that were largely, if not completely, reduced by

[1] We adopted the habitual use of the terms "anaerobic" and "aerobic" to say "anoxic" and "oxic" in spite of the fact that these meanings do not match etymologically (aero = air and not oxygen); to characterize more precisely the ecological niches with respect to dissolved oxygen concentration, we use the terms "anoxic"—absence of oxygen; "nanooxic"—oxygen present in nanomolar concentrations; "microoxic"—oxygen present in concentrations of 1–10 μM; "oxic"—oxygen present in concentrations >10 μM.

Anaerobiosis and Stemness. http://dx.doi.org/10.1016/B978-0-12-800540-8.00008-9

inorganic molecules. However, for further discussion it is important to mention that a modeling study [2] pointed to the extremely low (approximately 10^{-8}–10^{-4} ppmv) atmospheric O_2 levels potentially generated by photochemical reactions prior to the advent oxygenic photosynthesis.

About 3 billion years ago, the first prokaryotic organisms appeared, able to perform the photosynthesis (cyanobacteria). This is related to development of protochlorophyll, which allowed utilization of sunlight energy for the reduction of CO_2 and its fixation in the form of carbohydrates [3]. This process results in production of oxygen during a reaction that allows the splitting of water into oxygen, protons, and electrons [4]. During one reaction cycle of oxygenic photosynthesis, approximately 0.1% of oxygen produced is released into the atmosphere (reviewed in Ref. [5]). Between appearance of the first organisms capable of photosynthesis and the rise of atmosphere oxygen concentration to approximately 0.21–2.1% (1–10% of present atmosphere level (PAL)) [6], a lag of several 100 million years was suggested (the duration of this period is still controversial; reviewed in Lyons et al. [7]). This lag was explained by the facts that the newly produced oxygen was consumed in various chemical reactions in the oceans (reaction with iron was probably predominant), similar to that produced by water photolysis. The other reason for slowing down the increase in atmosphere oxygen was related to the methane production by methanogenic and photosynthetic prokaryotes. This is due to the fact that in the presence of UV radiation (at that time, ozone atmosphere layer was not established), methane oxidizes to CO_2. The ozone layer, once the O_2 levels were sufficient to support its formation, seems to have been, in fact, a very important factor for the life evolution because the UV shielding by ozone decreased the rate of methane oxidation.

A hypothesis [6] proposes that the two steady states concerning the O_2 atmosphere concentration were established: the first one being 0.02% before, and the second one (21% or more) after ozone layer formation. It is believed that the rise in O_2 to 10% PAL (2% O_2) (this level is questioned today), which occurred between approximately 2.4 and 2.1 billion years ago, in the early Proterozoic period (Paleoproterozoic era of the Precambrian period), induced a phenomenon called "oxygen catastrophe," "oxygen crisis," "oxygen revolution," or the "great oxygenation event" ("GOE" in further text) [8–10] claim that the rise of atmospheric oxygen had occurred by 2.32 Gyr ago. It was probably preceded by a critical transition period, which took place between about 2.5 and 2.3 billion years ago (reviewed in Ref. [7]). It is believed that, due to this oxygen rise (O_2 acting as a poison), many species became extinct while the others possessing the mechanisms for detoxification of oxygen survived and evolved further. Therefore, at the beginning of oxygen rise, the organisms exhibiting some mechanisms of defense against the corrosive action of oxygen were involved in further evolution, multiplying these protective mechanisms including scavenging oxygen with heme-containing proteins and performing its metabolism by oxidases to form water and CO_2. The fact that these oxygen detoxifying reactions produce more ATP molecules than anaerobic glycolysis was very fortunate for further evolution, because it allowed a higher rate of energy production, dramatically enhancing the number of biochemical and molecular reactions that can take place in a cell.

8.2 Appearance of Eukaryotes, First Eukaryotic Common Ancestor, Last Eukaryotic Common Ancestor, and Diversification of Eukaryotes

It is estimated that the first eukaryote appeared about 2 billion years ago. Some organic biomarker data [11] suggested even earlier appearance both of eukaryotes and cyanobacteria, 2.7 billion years ago (integrity of biomarker data has been challenged in recent years (reviewed in Ref. [5])); the other, also controversial, arguments pointing to oxygenic synthesis even 3 billion years ago appeared recently [12]. The origin of eukaryotes was not clearly established up to now. While several hypotheses exist [13] (either a nucleus-bearing amitochondriate cell first followed by the acquisition of mitochondria in an eukaryotic host or acquisition of mitochondria in a prokaryotic host followed by acquisition of eukaryotic specific features), the integration of a symbiont, a hypothesis consolidated by Lynn Margulis almost five decades ago [14], is the one that is generally accepted [15]. The appearance of the first eukaryotic common ancestor (FECA) is probably related to the gene duplication, considered to be a crucial mechanism of evolutionary innovation [16]. A major increase in the level of gene paralogy as a hallmark of the early evolution of eukaryotes was pointed out [16], but it is possible that the early phase of eukaryotic evolution was particularly conductive to pseudoparalogy through gene transfer to the nuclear genome from the proto-mitochondrial endosymbiont, to the other, transient endosymbionts and, possibly, via other routes as well [17–19].

Since Mereschowsky proposed in 1910 that the nucleus was formed from bacteria that had found a home in an entity that was composed of "amoebaplasm" and was not a bacterium [20], the numerous theories tried to explain the origin of eukaryotes (reviewed in Ref. [21]). It is believed that the origin of the nucleus (a place that enables stabilization of the conditions for gene preservation) basically marked the beginning of the eukaryotic cell (for review of eukaryotic evolution, [21]). The most known hypothesis proposes that the nucleus was derived in the symbiosis of archaea and bacteria [22]. This hypothesis is upgraded by a theory proposing that the structure formed in such way became the nucleus after being engulfed by a hypothetic cell that the authors call "chronocyte" (this is the so-called "ABC hypothesis"). This formation of the nucleus would restore the three cellular domains as the chronocyte was not a cell that belonged to the archaea or to the bacteria [23]. This host cell ("chronocyte") was not a prokaryotic cell but one that had a cytoskeleton composed of actin and tubulin and an extensive membrane system. The chronocyte donated to the resulting eukaryotic cell, its cytoskeleton, ER, Golgi apparatus, and major intracellular control systems, such as calmodulin, ubiquitin, inositol phosphates, cyclin, and the GTP-binding proteins. Some authors [23,24] propose that the nucleus originated by recombination of eu- and archaebacterial DNA that remained attached to eubacterial motility structures and became the microtubular cytoskeleton, including the mitotic apparatus. In fact, the earliest symbiogenetic fusion that integrated thermoacidophilic archaebacterial thermoplasmas with aerotolerant spirochetes produced the first protist, a swimming chimera that evolved into a stable nucleated protist cell: the last eukaryotic common ancestor, LECA (or that would be, in fact, FECA?—*author remark*). These authors propose that the integration of mitochondrial symbiont came later. Indeed, the lateral gene transfer phenomenon [19] or at least a major part of genes (nucleus) considered through the prism of spirochete-archeobacterial thermoplasma symbiogenetic fusion seems to be similar to the image offered by the ovule fecundation; that is, fusion of spermatozoid and ovule (duplication of genes by fusion of two haploid cells). However, this supposed "model" would suggest that one of two prokaryotes already integrated the mitochondrial endosymbiont (as only the ovule has the mitochondria).

Another hypothesis seems to be compatible with the general frame of the fusion scenario: "the libertine bubble theory" [25]. This theory proposes that the proto-eukaryotes developed the "relations" of DNA exchange since the membrane of the first "bubbles" derived from bacterias was essentially permeable. This membrane permeability specialized the "bubbles" through the evolution to favor the exchanges of DNA (instead to "eat" each other), which will give the basis for sexual reproduction (in its most developed form the molecular recognition between the ovule and spermatozoide). Therefore, the DNA exchange process enabled the renewal of the metabolic capacities (enzymes are coded by DNA) and evolutionary advantage for the most successful and most stable products of DNA exchanges. If we continue this reasoning, then the meiosis would be preparation, a step back to roughly repeat (in form of the fusion of spermatozoid and ovule) a basic ancestral event that resulted in the first eukaryote appearance.

However, the most recent discovery of *Lokiarcheota* [76], a monophyletic complex archaea bridge the gap between prokaryotes and eukaryotes: its genome encode an expanded repertoire of eukaryotic signature proteins that are suggestive of sophisticated membrane remodeling capabilities—a dynamic actin cytoskeleton and potentially endo- and/or phagocytic capabilities, which would have facilitated the invagination of the mitochondrial ancestor [76]. Thus, the abovementioned "chronocite" [23] probably was an archaea.

Whatever the case, the increase in ploidy/genome size resulted in gene redundancy that allowed novel sequences/functions to arise in duplicate genes without compromising existing pathways. Furthermore, the recessive nonadaptive alleles, masked but carried through time, may prove to be adaptive in future contexts. The most recent considerations agree that the cell fusion may be a plausible explanation as a step in the appearance of the FECA from a variety of proto-eukaryotes since their fusions would be expected to have yielded more gene-rich and more successful lineages [26]. At the same time, the successful mitosis events were positively selected, leading to the stabilization of the genome balance with respect to the copy number and capacity of repair and expression regulation. Indeed, it has been shown with the modern eukaryotes (yeasts) that a cross-species cell–cell fusion may give rise to new species [27,28]. However, the fusion of two proto-eukaryotes (each one presenting a paradigm of a future haploid entity) can transitorily result in a diploid cell entity to finally come back to "haploid state" by a loss of chromosomes. This ancestral event, conserved in some modern mono-cellular eukaryotes [29], is known as the "parasexual cycle." The transition haploid-diploid-haploid state could have evolved to meiosis.

The meiotic divisions were probably messy and inaccurate at the beginning to finally become an efficient mechanism to purge the deleterious mutations from the genome and reveal (unmask) the advantageous recessive alleles in haploid progeny of diploid heterozygots.

By mutual adaptation and compatibilization of the genetic material, evolution of meiosis resulted in a homolog chromosome pairing. Early steps in meiosis evolution existed in proto-eukaryotes and the transition from parasexual to sexual experimentation certainly antedates LECA, which had the capacity of a well done meiosis (reviewed in Ref. [26]). While the transition of haploid versus diploid entities (or vice versa) could provide the advantages with respect to the ecological niche environment [30], the evolutionary usefulness definitively confirmed the haploid-diploid-haploid transition, which finally becomes what we know as the sexual reproduction, perceived as diploid-haploid-diploid transition. At the level of single-celled eukaryotes it can still operate or not in response to the conditions of ecological niche leading to the coexistence of both mitotic and meiotic divisions at the population level. With the appearance of first complex colonies and first metazoan, some colony members started to ensure primarily meiotic divisions to evolve finally in specialized germ-line cells in animals. However, most metazoan entities conserved the possibility of asexual reproduction of the whole organism or of a "partial reproduction"; that is, organ and member regeneration (reviewed in Ref. [31]). This capacity is ensured by the primitive cells (stem cells), which, by the mitotic divisions combined or not by commitment and differentiation, allow reconstructing of either organisms or the complex biological structures in parallel with self-renewing. This point will be discussed in detail in Chapter 9.

The FECA was preceded by archaeoprotist motile nucleated cells with Embden-Meyerhof glycolysis and substrate-level phosphorylation that lack the symbiont that became the mitochondrion [24]. Until recently, it was widely believed that cyanobacteria exhibiting aerobic properties took up residence inside more complex single-cell organisms, which led to establishment of the symbiotic state. The recent analyses suggest, however, that the first mitochondrial endosymbiont was a facultative anaerobic alpha proteobacteria that could generate ATP in aerobic environments (using PHD, the citric acid cycle, and the respiratory chain) as well as under anoxic conditions (using anaerobic respiration) [32]. This scenario is not incompatible with the "Canfield concept" [33] considering that the O_2 concentration rise did not follow the same kinetics in the ocean as in the atmosphere—the ocean remained "euxinic" (waters free of oxygen and rich in hydrogen-sulphide) for a billion years of mid-Proterozoic (from 1.8 to 0.8 billion years ago), highlighting the essential lag between atmospheric (GOE) and oceanic oxygenation. More recent data, proposing more nuanced chemistry models, agree with the Canfield concept with respect to oxygenation of the ocean, which remained dominantly anoxic (and not euxinic everywhere) [34] (reviewed in Ref. [7]). This is even more probable in light of the most recent data based on chromium isotope, implying that the atmosphere O_2 concentrations in the period from 1.8 to 0.8 billion years ago were at most 0.21% (0.1% of the present atmospheric levels; PAL) [35], which is much lower compared to the earlier estimations (see Ref. [9]). These data strongly support the idea that the oceans were anoxic or nearly anoxic at least until 0.8 billion years ago. In other words, the first eukaryotes had to be "equipped" with anaerobic mitochondria long before entering the aerobic evolution.

The argument in favor of the anaerobic nature of the first eukaryotes is the discovery of mitosomes and hydrogenosomes representing the anaerobic mitochondria operating by basic mechanisms; these structures are still present in contemporary mitochondria. However, the most important argument in favor of anaerobic nature of the first eukaryote respiration is the convincing data that the mentioned anoxic/extremely low oxygenation state persisted in the oceans until about 600 million years ago, providing several ecologic niches where eukaryotes persisted on the margins of the aerobic (very low) atmospheric O_2 levels with respect to the PAL world.

In the past years, a lot of effort was invested to date the last eukaryotic common ancestor (LECA) combining the different approaches, including the so-called molecular-clock model. Depending on the model and the methods, these studies provided two estimations of LECA age 1.007 (0.943–1.102) and 1.898 (1.655–2.094) billion years. All models, however, agree that the eukaryotic supergroups diverged rapidly within 300 million years after appearance of LECA [36]. Of note is that the most ancient nondisputable primitive eukaryote fossils were found so far in the geologic strata dating 1.45 billion years, which provides a minimum age for the group [37,38]. In view of the arguments implying that the major animal clades diverged tens of millions of years before their first appearance in the fossil records [39] (extrapolating this phenomenon to early eukaryotes), these findings are in favor of the second molecular-clock estimation proposing the mean age for LECA as 1.9 billion years (1.6–2.1) [36]. These analyses are corroborated by the data of Parfrey et al. [40] suggesting that the last common eukaryotic ancestor lived between 1.866 and 1.679 billion years ago. Furthermore they estimated that most or all major clades (six major lineages recognized today) of eukaryotes diverged before 1.2 billion years ago. These considerations are, of course, troubled by the controversy related to the complex eukaryote fossils dated 2.1 billion years ago [41,42].

Thus, in view of geological data concerning the Proterozoic atmosphere and ocean chemistry exposed above, it is clear that not only LECA inhabited the anoxic or nearly anoxic conditions, but also eukaryotes diverged in the same conditions. Here it should be stressed that the dissolved O_2 concentration at which facultative anaerobic eukaryotes start to respire is about 2 μM ("Pasteur point") [43]. Pasteur point value varies from organism to organism, but a value of approximately 0.01 (1%) of the present atmospheric oxygen level (PAL) (i.e., 0.21% atmospheric O_2 concentration) is typical for modern prokaryotes and single-celled eukaryotes [44]. It is interesting to mention that the Pasteur point is also used to mark the level of oxygen in the early atmosphere of the Earth that is believed to have led to major evolutionary changes, and that this value for the modern eukaryotes fits precisely with the recently estimated O_2 concentration in Proterozoic atmosphere [35]. The current atmosphere O_2 concentration (21% O_2) in the same conditions results in 250 μM dissolved O_2 concentration, which is more than 100-fold above the Pasteur point. At 0.21% O_2, concentration in gas phase at 25 °C has a dissolved oxygen concentration in seawater at 2.2 μM, which is comparable to the Km values (Michaelis–Menten constants) for oxygen utilization for many aerobic bacteria isolated from marine environments [45]. The microorganisms, capable of performing aerobic metabolism in micromolar-dissolved O_2 concentrations (i.e., atmosphere O_2 concentrations varying from one to several percent; hence "microaerofiles") are not, however, the "champions" in O_2 utilization. A number of facultative aerobic microorganisms may subsist in nanomolar-dissolved O_2 concentrations, as low as 3–8 nM [46–48] (in the further text we will use the term "nanaoaerophiles" for organisms capable of metabolizing the nanomolar O_2 concentrations). Furthermore, the *Escherichia coli* K-12 (exhibiting a low-oxygen-affinity terminal oxidase) grows at dissolved oxygen levels of ≥3 nM [49]. Of note this value is three orders of magnitude lower than the abovementioned values for microaerophylic/aerobic microorganisms including the eukaryotes.

In view of these data, it is interesting to come back to the abovementioned problematic dating with sterol biomarkers [11]. The critical evaluation of the sterol biomarkers-based dating showed its irrelevance for the dating of the eukaryotes not only due to the methodological flaws (obtained results are inconsistent with an indigenous origin for the biomarkers, which are, in fact, much younger than the rocks studied [50,51]), but also to the fact that some prokaryotes, including proteobacteria and synthesize sterols [52]. Now comes the interesting point: it is established that only extremely low, trace amounts of oxygen are required for sterol synthesis [53]. Namely, the unimpaired sterol synthesis occurs at the lowest dissolved O_2 concentration tested (7 nM). Since the concentrations lower than 7 nM were not tested, the real minimal O_2 threshold might be even lower, although it is uncertain at which extent. These data, in fact, are strongly in favor of the hypothesis that the nanoaerobic (the authors called it "microaerobic") marine environments (where steroid biosynthesis was possible) could have existed for long periods of time prior to the earliest geologic and isotopic evidence for atmospheric O_2 [53]. The authors propose that "In the late Archean, molecular oxygen likely cycled as a biogenic trace gas, much as compounds such as dimethyl-sulfide do today" [53].

In view of all these considerations it can be concluded that the oxygenation of Proterozoic ocean after appearance in diverging of eukaryote supergroups was not homogenous and the different ecologic niches existed: euxinic, anoxic, nanoaerobic, and, possibly, microaerobic. If so, for the life at the "aerobic margin" a capacity of both aerobic and anaerobic respiration would be an advantage, which is fitting well with facultative anaerobic mitochondria origin. It should be mentioned, however, that some authors propose that the earliest eukaryotes were strict aerobes that adapted to microaerobic and anoxic conditions [54,55]. This adaptation is proposed to have occurred by lateral gene transfer. As already said, the other, more accepted point of view is that the ancestors of mitochondria were themselves facultative anaerobes with a diversity of bioenergetics enzymes [56] and that the early eukaryotes were equipped with facultative anaerobic mitochondria [57]. Furthermore, as Rutten remarked 45 years ago, "we had better speak of 'facultative respirators' than of facultative aerobes" [43].

Indeed, if today the highly "hypoxic" sulfidic waters of hypersaline Mediterranean basins and the Black Sea are abound with eukaryotic microorganism from the six supergroups [58,59], why would the ancestral eukaryotes not be able to appear and diverge in the similar conditions?

8.3 Neoproterozoic Oxygenation Event and Metazoan Controversy

Between 0.8 and 0.55 billion years ago, the O_2 concentration in atmosphere raised again. This phenomenon is known as Neoproterozoic Oxygenation Event (NOE) [60]. This increase in O_2 concentration (at the time dated 0.6 billion years ago) was considered by Nursall [3] "as a prerequisite to the origin of metazoa" and this idea was generally accepted. However, Canfield et al. [61] claim that the deep ocean was anoxic and ferruginous before and during the Gaskiers glaciation (580 million years ago), and that it became "oxic" afterward (see previous discussion concerning recent modifications of this concept). Since the first known members of the Ediacara biota arose shortly after the Gaskiers glaciations, a causal link between their evolution and this oxygenation event was suggested. A prolonged stable oxic environment may have permitted the emergence of bilateral motile animals some 25 million years later. Recent serious arguments appeared in favor of another viewpoint. Raising the arguments that the long delay in animal emergence is rather related to the intrinsic kinetics in evolution of gene expression and cell signaling in animals, Butterfield [62] suggests that the simultaneous emergence of animals and NOE might be either a coincidence or even an inverse feedback: the animal evolution itself triggered NOE.

It is interesting to mention that fossil records of at least basal animals (sponges and perhaps cnidarians) point to a period of time significantly before the beginning of the Cambrian, while the "Cambrian explosion" itself still seems to represent the arrival of the bilaterians [63]. It seems that the evolution of the earliest animals (sponges and eumetazoans) was probably not limited by the low absolute oxygen levels that may have characterized Neoproterozoic oceans, although these inferred levels would constrain animals to very small sizes and low metabolic rates [64]. Indeed, even today's successors of these Neoproterozoic simple animals do not need a lot of oxygen and can survive and proliferate in microoxyc conditions. Some sponges, for example, require roughly 0.1 ml/l (4.46 μM) of dissolved O_2; a recent study [65] shows that *Halichondria panicea*, a modern marine demosponge inhabiting well-oxygenated waters, can survive at least 24 days in the water equilibrated with only 3–4% PAL oxygen concentrations (0.63–0.84% O_2). Also, some bilaterian worms need only around 0.02 ml/l (8.9 μM) O_2 (cited in Ref. [64]) (see information concerning the Pasteur point, above). So, as can be concluded from the previous discussion, there has been sufficient oxygen in the atmosphere to ventilate the ocean surface above these O_2 concentrations since the GOE (~2.4 or at least 2.1 billion years ago). Even before GEO, some localized oxygenated niches probably existed since, as previously mentioned, oxygenic photosynthesis largely predated GOE. Furthermore, a possibility of a much larger increase in atmospheric O_2 concentration early on and then a deep plunge to the levels maintained during the GOE period has been seriously considered recently [7] (Figure 8.1). Perhaps the discoveries of a nearly 2-Gyr–old coil-shaped fossil *Grypania spiralis*, which may have been eukaryotic [41], as well as of centimeter-sized structures from the 2.1-Gyr interpreted as highly organized and spatially discrete populations of colonial organisms [42], should be considered in view of these recent ideas. Indeed, it is possible that these complex eukaryotic organisms appeared when favorable ecologic

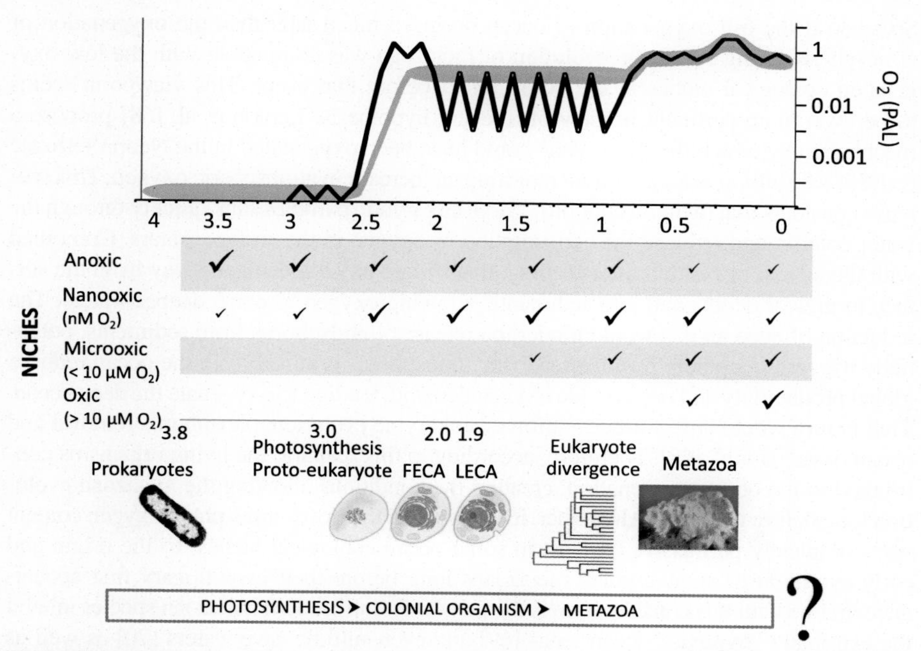

Figure 8.1 Oxygenation of atmosphere, the supposed ocean ecological niches with respect to dissolved O_2 and apparition end evolution of eukaryotes. Two curbs are presented: the "traditional one" [9] (blue) and one based on recent data [7] (black). In view of heterogeneity of ecological niches, the probable existence of "anoxic," "nanooxic" (nanomolar O_2 concentrations), "microoxic" (O_2 concentrations 1–10 μM), and oxic (O_2 concentrations >10 μM) were presented on the basis of synthesis of literature data for each geological period in parallel with the crucial evolution events (illustration). Note that, following this model, the anoxic niches coexisted with nanooxic through the whole process of evolution until today. The bottom frame (question mark) represents another point of view proposing the much earlier appearance of photosynthesis and other consequent forms of life [40,41] with respect to the "mainstream" model.

niches became disposable and then disappeared (or were extremely rarified) when these niches disappeared (or were rarified).

It seems that in the early Ediacaran period the pulsed oxidation events provoked a fluctuation of redox conditions and heterogeneity of eukaryote distribution in ocean [66]. These authors hypothesize that the distribution of early Ediacaran eukaryotes likely tracked redox conditions and that only after approximately 551 Ma (when Ediacaran oceans were pervasively oxidized) did evolution of oxygen-requiring taxa reach global distribution [66].

The convincing evidences were provided [67] that throughout the Ediacaran period (635–542 million years ago) several different ecologic niches existed in oceans; that is, "a stratified ocean with coeval oxic, sulfidic, and ferruginous zones, favored by overall low oceanic sulfate concentrations, was maintained dynamically throughout the Ediacaran Period." So the fluctuating oceanic redox conditions could explain the generally patchy record of metazoans observed through the Ediacaran might be explained by Li et al. [67].

So, even if the full oxygenation of ocean occurred much later than the oxygenation of atmosphere during NOE, the evolution of metazoan was compatible with the low oxygenated ecological niches in the ocean much before that event. This viewpoint seems to be even more pertinent in view of a recent hypothesis: Lenton et al. [68] propose a mechanism by which the deep ocean could have been oxygenated in the Neoproterozoic (c.1.85–0.85 billion years), without requiring an increase in atmospheric oxygen. This scenario considers that (a literal citation): "large eukaryotic particles sank quickly through the water column and reduced the consumption of oxygen in the surface waters. Combined with the advent of benthic filter-feeding, this shifted oxygen demand away from the surface to greater depths and into sediments, allowing oxygen to reach deeper waters. The reduction of deep anoxia would hinder the release of phosphorus from sediments, potentially triggering a potent positive feedback: phosphorus removal from the ocean reduces global productivity and ocean-wide oxygen demand, tending to oxygenate the deep ocean. That in turn would have further reinforced eukaryote evolution, phosphorus removal and ocean oxygenation" [68]. Therefore, according to this scenario the living organisms contributed to the ocean oxygenation, creating the conditions allowing the metazoan evolution long before the NOE. Altogether, it appears that the (low) atmosphere oxygen content was completely permissive (at least in some ocean ecological niches) to the origin and early evolution of sponge-grade metazoans long before their evolutionary first appearance. In line with this consideration is the fact that some modern metazoan species inhabit the extremely "hypoxic" ocean areas [69] and even sulfidic cave waters [70] as well as environments 1–4 km deep within the terrestrial crust (some nematodes) [71].

It is believed that the mitochondrial sulphide-oxidizing metabolism of some modern animals (for instance *Arenicola* or *Urechis*) might be a relic from early metazoan evolution [72,73]. It was proposed [65] that the accumulation of requirements to build a multicellular organism (i.e., adhesion molecules, signaling proteins for cell–cell communication, transcription factors essential for coordinated development) conducted to the onset of animal evolution and not an enhanced availability of oxygen beyond the one already existing during GOE. However, the increased availability of oxygen that allowed the O_2-dependent protein and other complex molecules synthesis (as that of collagen [74]) certainly led to more rigid bodies, increase in size [75], and further diversification. Using modern oxygen minimum zones as an analog for Proterozoic oceans, Sperling et al. [64] clearly demonstrated the effect of low oxygen levels on the feeding ecology of polychetes, the dominant macrofaunal animals in deep-sea sediments. The low oxygen is clearly linked to low proportions of carnivores in a community and low diversity of carnivorous taxa, whereas higher oxygen levels support more complex food webs.

8.4 Cambrian Period, Further Increase in Atmospheric O_2 Concentration, Paleozoic Era, and Definitive Stabilization 650 Million Years Ago

During the Cambrian period of the Paleozoic era (542–488 million years ago) characterized by the presence of plants (performing intensive photosynthesis), atmosphere oxygen increased to 10–20% (0.5–1 PAL). The rise in O_2 concentration in this period was clearly related to the Cambrian animal radiation [64]. In the further period of the Paleozoic era (until 251 million years), oxygen reached the physical maximum of 35% (1.6 PAL). It is believed that this is the highest value possible, since beyond such oxygen and nitrogen levels (nitrogen remains constant), combustion of biosphere would occur. In the next geological period, the oxygen level fluctuated from 15% to 27% to reach a steady state of 21% 350 million years ago, which still persists (reviewed in Ref. [7]).

8.5 Conclusions

On the basis of available data, it turns out that the early eukaryotic evolution occurred either in completely anaerobic conditions, or in conditions of extremely low oxygenation during a period that lasted roughly 1.4 billion years. The important point to stress is the heterogeneity of ecological niches in the Proteozoric Ocean with respect to O_2 concentration (Figure 8.1), providing the nanooxic and microoxic conditions in some of them during this period. It is interesting that the anoxic, nanooxic, and microoxic conditions exist even today in some ocean areas, as well as that the modern protists and some metazoan conserved the capacity to survive and even to proliferate in anoxic or nanooxic conditions. In the period before, during, and after GEO conditions of pulsation between anoxic, nanooxic, and microoxic conditions, were permissive for appearance and evolution of not only the first eukaryotes but also of the first metazoan. These organisms were facultative respirators and in addition to the cytoplasmic ways of energy production inherited from prokaryotes (fermentation, glycolisis), they were able to perform an anaerobic mitochondrial respiration as well as aerobic mitochondrial respiration if enough oxygen was available. The mitosis (asexual reproduction) and meiosis (leading to the sexual reproduction) as well as the further specialization into somatic (mitotic) and germ-line (meiotic) cells in the context of a multicellular organism, were established in these anaerobic/nanoaerophilic (and possibly microaerophilic in some ecological niches) conditions. These basic, primitive oxygenation conditions during the period somewhere between FECA and LECA and until divergence from LECA, in which the mitosis was stabilized (representing the simple asexual reproduction of the first eukaryotes), seem to be reproduced by the stem cell niches as well as being the most simple mitotic cell division *à l'identique*—self-renewal, which will be discussed in the further chapters.

References

[1] Kasting JF, Pollack JB, Crisp D. Effects of high CO_2 levels on surface temperature and atmospheric oxidation state of the early Earth. Effects of high CO_2 levels on surface temperature and atmospheric oxidation state of early Earth. J Atmos Chem 1984;1:403–28.

[2] Haqq-Misra J, Kasting J, Lee S. Availability of O_2 and H_2O_2 on pre-photosynthetic earth. Astrobiology 2011;11:293–302.

[3] Nursall JR. Oxygen as a prerequisite to the origin of the metazoa. Nature 1959;183:1170–2.
[4] Dismukes GC, Klimov VV, Baranov SV, Kozlov YN, DasGupta J, Tyryshkin A. The origin of atmospheric oxygen on Earth: the innovation of oxygenic photosynthesis. Proc Natl Acad Sci USA 2001;98:2170–5.
[5] Stamati K, Mudera V, Cheema U. Evolution of oxygen utilization in multicellular organisms and implications for cell signalling in tissue engineering. J Tissue Eng 2011;2: 2041731411432365.
[6] Goldblatt C, Lenton TM, Watson AJ. Bistability of atmospheric oxygen and the great oxidation. Nature 2006;443:683–6.
[7] Lyons TW, Reinhard CT, Planavsky NJ. The rise of oxygen in Earth's early ocean and atmosphere. Nature 2014;506:307–15.
[8] Holland HD. The oxygenation of the atmosphere and oceans. Philos Trans R Soc Lond B Biol Sci 2006;361:903–15.
[9] Kump LR. The rise of atmospheric oxygen. Nature 2008;451:277–8.
[10] Bekker A, Holland HD, Wang PL, Rumble 3rd D, Stein HJ, Hannah JL, et al. Dating the rise of atmospheric oxygen. Nature 2004;427:117–20.
[11] Brocks JJ, Logan GA, Buick R, Summons RE. Archean molecular fossils and the early rise of eukaryotes. Science 1999;285:1033–6.
[12] Crowe SA, Døssing LN, Beukes NJ, Bau M, Kruger SJ, Frei R, et al. Atmospheric oxygenation three billion years ago. Nature 2013:501535–8.
[13] Embley TM, Martin W. Eukaryotic evolution, changes and challenges. Nature 2006;440: 623–30.
[14] Margulis L. On the origin of mitosing cells. J Theor Biol 1967;14(3):IN1–6.
[15] Schaechter M. Retrospective. Lynn margulis (1938–2011). Science 2012;335:302.
[16] Makarova KS, Wolf YI, Mekhedov SL, Mirkin BG, Koonin EV. Ancestral paralogs and pseudoparalogs and their role in the emergence of the eukaryotic cell. Nucleic Acids Res 2005;33:4626–38.
[17] Doolittle WF. You are what you eat: a gene transfer ratchet could account for bacterial genes in eukaryotic nuclear genomes. Trends Genet 1998;14:307–11.
[18] Doolittle WF, Boucher Y, Nesbø CL, Douady CJ, Andersson JO, Roger AJ. How big is the iceberg of which organellar genes in nuclear genomes are but the tip? Philos Trans R Soc Lond B Biol Sci 2003;358:39–57.
[19] Andersson JO, Sjögren AM, Davis LA, Embley TM, Roger AJ. Phylogenetic analyses of diplomonad genes reveal frequent lateral gene transfers affecting eukaryotes. Curr Biol 2003;13:94–104.
[20] Mereschowsky C. Theorie der zwei Plasmaarten als Grundlage der Symbiogenesis, eine neue Lehre von der Entstehung der Organismen. Biol Zentralbl 1910;30:278–367.
[21] Lodé T. For quite a few chromosomes more: the origin of eukaryotes. J Mol Biol 2012;423:135–42.
[22] Horiike T, Hamada K, Kanaya S, Shinozawa T. Origin of eukaryotic cell nuclei by symbiosis of Archaea in Bacteria is revealed by homology-hit analysis. Nat Cell Biol 2001;3:210–4.
[23] Hartman H, Fedorov A. The origin of the eukaryotic cell: a genomic investigation. Proc Natl Acad Sci USA 2002;99:1420–5.
[24] Margulis L, Chapman M, Guerrero R, Hall J. The last eukaryotic common ancestor (LECA): acquisition of cytoskeletal motility from aerotolerant spirochetes in the Proterozoic Eon. Proc Natl Acad Sci USA 2006;103:13080–5.
[25] Lodét T. Sex is not a solution for reproduction: the libertine bubble theory. BioEssays 2011;33:419–22.

[26] Goodenough U, Heitman J. Origins of eukaryotic sexual reproduction. Cold Spring Harb Perspect Biol 2014 March 1;6(3). http://dx.doi.org/10.1101/cshperspect.a016154.
[27] Wolfe KH, Shields DC. Molecular evidence for an ancient duplication of the entire yeast genome. Nature 1997;387:708–13.
[28] Scannell DR, Byrne KP, Gordon JL, Wong S, Wolfe KH. Multiple rounds of speciation associated with reciprocal gene loss in polyploid yeasts. Nature 2006;440:341–5.
[29] Schoustra SE, Debets AJ, Slakhorst M, Hoekstra RF. Mitotic recombination accelerates adaptation in the fungus *Aspergillus nidulans*. PLoS Genet 2007;3:e68.
[30] Zörgö E, Chwialkowska K, Gjuvsland AB, Garré E, Sunnerhagen P, Liti G, et al. Ancient evolutionary trade-offs between yeast ploidy states. PLoS Genet 2013;9:e1003388.
[31] Li Q, Yang H, Zhong TP. Regeneration across metazoan phylogeny: lessons from model organisms. J Genet Genomics 2015;20(42):57–70.
[32] Hampl V, Stairs CW, Roger AJ. The tangled past of eukaryotic enzymes involved in anaerobic metabolism. Mob Genet Elements 2011;1:71–4.
[33] Canfield DE. A new model for Proterozoic ocean chemistry. Nature 1998;396:450–3.
[34] Shen Y, Canfield DE, Knoll AH. Middle Proterozoic ocean chemistry: evidence from the McArthur Basin, northern Australia. Am J Sci 2002;302:81–109.
[35] Planavsky NJ, Reinhard CT, Wang X, Thomson D, McGoldrick P, Rainbird RH, et al. Low mid-Proterozoic atmospheric oxygen levels and the delayed rise of animals. Science 2014;346:635–8.
[36] Eme L, Sharpe SC, Brown MW, Roger AJ. On the age of eukaryotes: evaluating evidence from fossils and molecular clocks. Cold Spring Harb Perspect Biol 2014;1:6(8).
[37] Javaux EJ, Knoll AH, Walter MR. Morphological and ecological complexity in early eukaryotic ecosystems. Nature 2001;412:66–9.
[38] Knoll AH, Javaux EJ, Hewitt D, Cohen P. Eukaryotic organisms in Proterozoic oceans. Philos Trans R Soc Lond B Biol Sci June 29, 2006;361(1470):1023–38.
[39] Erwin DH, Laflamme M, Tweedt SM, Sperling EA, Pisani D, Peterson KJ. The Cambrian conundrum: early divergence and later ecological success in the early history of animals. Science 2011;334:1091–7.
[40] Parfrey LW, Lahr DJ, Knoll AH, Katz LA. Estimating the timing of early eukaryotic diversification with multigene molecular clocks. Proc Natl Acad Sci USA 2011;108:13624–9.
[41] Han TM, Runnegar B. Megascopic eukaryotic algae from the 2.1-billion-year-old Negaunee Iron-Formation, Michigan. Science 1992;257:232–5.
[42] El Albani A, Bengtson S, Canfield DE, Bekker A, Macchiarelli R, Mazurier A, et al. Large colonial organisms with coordinated growth in oxygenated environments 2.1 Gyr ago. Nature 2010;466:100–4.
[43] Rutten MG. The history of atmospheric oxygen. Space Life Sci 1970;2:5–17.
[44] Fenchel T, Finlay BF. Ecology and evolution in anoxic worlds. Oxford: Oxford University Press; 1995.
[45] Devol AH. Bacterial oxygen uptake kinetics as related to biological processes in oxygen deficient zones of the oceans. Deep Sea Res 1978;5:137–46.
[46] Rice CW, Hempfling WP. Oxygen-limited continuous culture and respiratory energy-conservation in *Escherichia coli*. J Bacteriol 1978;134:115–24.
[47] Kuzma MM, Hunt S, Layzell DB. Role of oxygen in the limitation and inhibition of nitrogenous activity and respiration rate in individual soybean nodules. Plant Physiol 1993;101:161–9.
[48] Bergersen FJ, Turner GL. Bacteroids from soybean root nodules; respiration and N_2 fixation inflow-chamber reactions with oxyleghemoglobin. Proc R Soc Lond Ser B 1990;238:295–320.

[49] Stolper DA, Revsbech NP, Canfield DE. Aerobic growth at nanomolar oxygen concentrations. Proc Natl Acad Sci USA 2010;107:18755–60.
[50] Fischer WW. Biogeochemistry: life before the rise of oxygen. Nature 2008;455:1051–2.
[51] Rasmussen B, Fletcher I, Brocks JJ, Kilburn MR. Reassessing the first appearance of eukaryotes and cyanobacteria. Nature 2008;455:1101–4.
[52] Pearson A, Budin M, Brocks JJ. Phylogenetic and biochemical evidence for sterol synthesis in the bacterium *Gemmata obscuriglobus*. Proc Natl Acad Sci USA 2003;100: 15352–7.
[53] Waldbauer JR, Newman DK, Summons RE. Microaerobic steroid biosynthesis and the molecular fossil record of Archean life. Proc Natl Acad Sci USA 2011;108:13409–14.
[54] Andersson SG, Kurland CG. Origins of mitochondria and hydrogenosomes. Curr Opin Microbiol 1999;2:535–41.
[55] Hug LA, Stechmann A, Roger AJ. Phylogenetic distributions and histories of proteins involved in anaerobic pyruvate metabolism in eukaryotes. Mol Biol Evol 2010;27: 311–24.
[56] Degli Esposti M, Chouaia B, Comandatore F, Crotti E, Sassera D, Lievens PM, et al. Evolution of mitochondria reconstructed from the energy metabolism of living bacteria. PLoS One 2014;9(5):e96566.
[57] Müller M, Mentel M, van Hellemond JJ, Henze K, Woehle C, Gould SB, et al. Biochemistry and evolution of anaerobic energy metabolism in eukaryotes. Microbiol Mol Biol Rev 2012;76:444–95.
[58] Edgcomb V, Orsi W, Leslin C, Epstein SS, Bunge J, Jeon S, et al. Protistan community patterns within the brine and halocline of deep hypersaline anoxic basins (DHABs) in the eastern Mediterranean Sea. Extremophiles 2009;13:151–67.
[59] Wylezich C, Jürgens K. Protist diversity in suboxic and sulfidic waters of the Black Sea. Environ Microbiol 2011;13:2939–56.
[60] Och LM, Shields-Zhou GA. The Neoproterozoic oxygenation event: environmental perturbations and biogeochemical cycling. Earth Sci Rev 2012;110(1):26–57.
[61] Canfield DE, Poulton SW, Narbonne GM. Late-Neoproterozoic deep-ocean oxygenation and the rise of animal life. Science 2007;315:92–5.
[62] Butterfield NJ. Oxygen, animals and oceanic ventilation: an alternative view. Geobiology 2009;7:1–7.
[63] Budd GE. The earliest fossil record of the animals and its significance. Philos Trans R Soc Lond B Biol Sci 2008;363:1425–34.
[64] Sperling EA, Frieder CA, Raman AV, Girguis PR, Levin LA, Knoll AH. Oxygen, ecology, and the Cambrian radiation of animals. Proc Natl Acad Sci USA 2013;110:13446–51.
[65] Mills DB, Ward LM, Jones C, Sweeten B, Forth M, Treusch AH, et al. Oxygen requirements of the earliest animals. Proc Natl Acad Sci USA 2014;111:4168–72.
[66] McFadden KA, Huang J, Chu X, Jiang G, Kaufman AJ, Zhou C, et al. Pulsed oxidation and biological evolution in the Ediacaran Doushantuo Formation. Proc Natl Acad Sci USA 2008;105:3197–202.
[67] Li C, Love GD, Lyons TW, Fike DA, Sessions AL, Chu X. A stratified redox model for the Ediacaran Ocean. Science 2010;328:80–3.
[68] Lenton TM, Boyle RA, Poulton SM, Shields-Zhou GA, Butterfield NJ. Co-evolution of eukaryotes and ocean oxygenation in the Neoproteozoric era. Nat Geosci 2014;7:257–65.
[69] Danovaro R, Dell'Anno A, Pusceddu A, Gambi C, Heiner I, Kristensen RM. The first metazoa living in permanently anoxic conditions. BMC Biol 2010;8:30.
[70] Engel AS. Observation on the biodiversity of sulfidic karst habitats. J Cave Karst Stud 2007;69:187–206.

[71] Borgonie G, García-Moyano A, Litthauer D, Bert W, Bester A, van Heerden E, et al. Nematoda from the terrestrial deep subsurface of South Africa. Nature 2011;474:79–82.
[72] Theissen U, Hoffmeister M, Grieshaber M, Martin W. Single eubacterial origin of eukaryotic sulfide:quinone oxidoreductase, a mitochondrial enzyme conserved from the early evolution of eukaryotes during anoxic and sulfidic times. Mol Biol Evol 2003;20:1564–74.
[73] Ma ZJ, Bao ZM, Wang SF, Zhang ZF. Sulfide-based ATP production in Urechis unicintus. Chin J Oceanol Limnol 2010:521–6.
[74] Towe KM. Oxygen-collagen priority and the early metazoan fossil record. Proc Natl Acad Sci USA 1970;65:781–8.
[75] Knoll AH, Carroll SB. Early animal evolution: emerging views from comparative biology and geology. Science 1999;284:2129–37.
[76] Spang A, Saw JH, Jørgensen SL, Zaremba-Niedzwiedzka K, Martijn J, Lind AE, et al. Complex archaea that bridge the gap between prokaryotes and eukaryotes. Nature 2015;521:173–9. http://dx.doi.org/10.1038/nature14447. [Epub ahead of print].

Evolution of Mitochondria in Eukaryotes versus Mitochondria "Maturing" from the Stage of Stem Cells to Committed Progenitors and Mature Cells[1]

Chapter Outline

9.1 Integration of Bacterial Endosymbiont and the Acquisition of Aerobic Respiration: A Simultaneous or Two-Step Process?

As mentioned in Chapter 8, the classical Margulis hypothesis on the origin of mitochondria proposes that an anaerobic organism engulfed a smaller aerobic organism (ancestor of mitochondrion) [1,2], which was a mutually beneficial arrangement since the small anaerobe reduced oxygen-generating ATP. According to this hypothesis, the ancestor of "nucleocytoplasm" was an O_2-intolerant fermentative cell, which became associated with aerobic bacteria able to consume any O_2 present. In turn, the host provided the nutrients (i.e. pyruvate) to bacteria ("OxTox model" [3]). The "endosymbiont theory" is now generally accepted, but the viewpoint that the first eukaryotes were aerobes is seriously questioned [4]. While nobody contests either

[1] We adopted the habitual use of the terms "anaerobic" and "aerobic" to say "anoxic" and "oxic" in spite of the fact that these meanings do not match etymologically (aero = air and not oxygen). In order to characterize more precisely the ecological niches in respect to dissolved oxygen concentration, we use the terms "anoxic"—absence of oxygen; "nanooxic"—oxygen present in nanomolar concentrations; "microoxic"—oxygen present in concentrations of 1–10 μM; oxic—oxygen present in concentrations >10 μM.

Anaerobiosis and Stemness. http://dx.doi.org/10.1016/B978-0-12-800540-8.00009-0

the important, if not decisive role of oxygen in evolution of multicellular complex life or the energetic advantage when O_2 appears as the terminal electron acceptor, the idea that oxygen was the trigger for the integration of mitochondrial symbiont is currently being challenged. First, this proposal does not hold in light of recent geological arguments, concerning the oxygen presence in the ocean environment (so-called "Canfield ocean") at the time when this event happened. The newer view of "Proterozoic ocean chemistry" that emerged in the mid-1990s provides the data that show ecological niches during a billion-year-long period (from 2.5 to 1.5 billion years ago) to have been mainly anoxic or nanooxic at best (see Chapter 8 and Figure 8.1 for further details). In fact, there is a significant lag over a billion years long between the time when O_2 started accumulating in the atmosphere and the time when the oceans became fully oxidized [5]. Thus, these environmental imperatives suggest that the first eukaryotes had to be "equipped" with mitochondria capable of anaerobic respiration long before entering the aerobic evolution. In fact, it is generally accepted now that the first mitochondrial endosymbiont was an α-proteobacterium able to live with or without the oxygen [6] (reviewed in Refs. [7,8]). Second, the data suggesting that the so-called "mitochondrion-related organelles" (currently termed "organelles of mitochondrial origin" [8]), which are vertical descendants from the original mitochondrial symbiont, seem to prevail (reviewed in Ref. [9]) over hypotheses proposing two serial endosymbiotic events [10]. Third, the thesis that the role of mitochondria is to protect the cell (host) from the dangers of molecular oxygen [3] is very problematic: the reactive oxygen species (ROS) generated in the course of oxidative phosphorylation, exhibiting a high-energy nature and reactivity, result in degradation, inactivation, and destruction of enzymes and nucleic acids. These effects are considered to be much more pronounced than the effects of molecular O_2 (note that the lowest cellular level of molecular O_2 is in the mitochondrial matrix). In that respect, Lane [11] compared the proposed strategy of mitochondrial aerobic respiration aimed to save the cell as "to try to save oneself from drowning by drinking the ocean" and Van der Giezen and Lenton [12] said figuratively, "No one ingests an umbrella to protect themselves from the rain."

9.2 Organelle of Mitochondrial Origin

It seems now that the term "amitochondriate" (initially describing the eukaryotes lacking typical mitochondria) [13,14] is obsolete since none of the extant protist is amitochondriate in the sense of ancestrally lacking mitochondria [15]. In fact, although many lack cytochromes, the "amitochondriates" contain organelles of mitochondrial origin differing in their functional capacities and structure. In general, all organelles of mitochondrial origin can be, according to Muller et al. [8], classified into five types:

1. *The "canonical"-type mitochondrion*: uses oxygen as the terminal receptor (produces ATP by oxidative phosphorylation; i.e., possesses a proton-pumping electron transport chain composed of subunits).
2. *Anaerobic mitochondrion*: does not use oxygen as the final electron acceptor; instead, it uses either, an endogenously produced molecule (mostly fumarate generating succinate as excreted end product), or environmental acceptors (i.e. nitrate) [16].

3. *Hydrogen-producing mitochondria*: use protons as terminal electron acceptors, producing hydrogen. Their membrane-associated proton-pumping electron transport chains do not transfer electrons to oxygen.
4. *Hydrogenosomes*: lack a membrane-associated electron transport chain and a genome. They generate ATP via substrate-level phosphorylation (hydrogen-producing fermentations) [17].
5. *Mitosomes*: do not produce ATP but have retained either components of FeS cluster assembly or of sulfate activation [18–20].

9.3 Anaerobic Respiration Is More Primitive Than Aerobic

As already suggested in previous chapters, the actual hypothesis concerning the origin of eukaryotes and the mitochondria is represented by the "symbiogenesis scenario": the host for mitochondrial synthesis was not a eukaryote but a prokaryote, an archeon [21]. This hypothesis fits very well with the scenario on metabolic integration of mitochondrial symbiont known as "hydrogen hypothesis" [22]. According to this hypothesis the host was a strictly anaerobic hydrogen-dependent autotrophic archaebacterium and the symbiont was a eubacterium that was able to respire, but generated molecular hydrogen as a waste product in anaerobic conditions [22]. In this way, both the first complete eukaryote and the ancestral mitochondrion appeared in parallel. The authors consider that the three forms of energy metabolism found in eukaryotes today were inherited from the common ancestor of hydrogenosomes and mitochondria. The opposed view is that the earliest eukaryotes were strict aerobes [23–27].

Endocytosis has long been considered as an essential function for incorporating a bacterial symbiont as well as a eukaryotic hallmark. So, if the host was not a eukaryote, how did it engulf the eubacterium? These arguments can be declined on the basis of two facts: (1) bacterial endosymbiosis (e.g., g-Proteobacteria inside β-Proteobacteria) have been documented [28,29] and (2) newly discovered *Lokiarchaeota* (i.e., complex archaea, the best prokaryote candidate for eukaryote precursor, or mitochondrial endosymbiont host) seems to have a dynamic actin cytoskeleton [30], hence, a priori, the capacity for phagocytosis. For our discussion here, it is important to stress that the "hydrogen hypothesis" implies that both anaerobic and aerobic metabolic pathways were contained in the first bacterial endosymbionts and have been alternatively expressed depending on the ecological niche (O_2 availability). In eukaryote lines that were permanently exposed to anaerobic conditions, the genes for anaerobic respiration were maintained while those for aerobic respiration were either lost or substantially reduced (explaining the existence of anaerobic mitochondria, hydrogen-producing mitochondria, hydrogenosomes, and mitosomes—see above). In contrast, the eukaryote lines that were frequently or permanently exposed to O_2 maintained or even developed [31] the aerobic metabolic pathways that became more complex. This adaptation first resulted in the complex life cycle of protists (to protect the reproduction of primitive lineage), as an adaptation to the increase in O_2 concentration (see Chapter 10); that is, acquisition of new functional extensions. In general, the appearance of new molecular, functional, and morphologic extensions is related to the availability of energy; that is, to the increase in oxidative phosphorylation part in general metabolic rate. We believe that this was the main factor enabling the divergence of eukaryotes and diversification of metazoans.

The differentiation of certain cell lineages in complex organisms is reminiscent of this evolutionary process. The eukaryotes living on the edge of aerobic and anaerobic environments remained the facultative respirators (nanoaerophiles and microaerophiles) with mitochondria arrangements allowing both anaerobic and aerobic mitochondrial respiration. The same options are present for multicellular organisms, but some of them are strictly anaerobes possessing the anaerobic mitochondria [8]. It is necessary to underline that the eukaryote anaerobes occur through the full spectrum of eukaryote lineages, including the animals [5,8]. Indeed, the various forms of worms, mussels, and crustaceans reside in anaerobic conditions most of the time. Their metabolism relies on the *anaerobic mitochondrial respiration* using, as the final electron receptor, fumarate instead of oxygen. Fumarate-reductase donates electrons from glucose oxidation to fumarate, yielding succinate. Succinate is either excreted as the end product of this reaction or can participate in further reactions resulting in ATP synthesis through substrate-level phosphorylation yielding two excreted products: acetate and propionate. The fumarate itself is a result of conversion from malate, which enters mitochondrion from cytoplasm [5,8].

All these arguments are in favor of the concept that the functional anaerobic respiration preceded the aerobic one in mitochondria of the first eukaryotes [32]. In that respect, Mentel and Martin consider that "The broad phylogenic distribution of a uniform and recurrently conserved energy metabolic repertoire among anaerobic eukaryotes no longer has to be explained away as some adaptation to anaerobic niches from an assumed aerobic-specialized ancestral state" and concluded "…the anaerobic lifestyle in eukaryotes can be readily understood as a direct holdover among diversified descendants of a kingdom's anaerobic youth" [5].

Even in strictly aerobic animals, the multicellular tridimensional architecture results in the forming of the microenvironmental niches reminiscent of the ecological niches during the ancestral eukaryotic evolution. In these organisms, some cells may be sequestrated in anoxic, nanooxic, or microoxic niche and exhibit anaerobic, nanoaerophylic, and microaerophylic metabolic character. These cells, exhibiting a primitive character (nondifferentiated, self-renewing capacity, high differentiation capacity, high proliferative potential) are the stem cells. Based on the above, the logical question is whether we can establish any analogies between the evolution of first eukaryotes' mitochondria and the mitochondria in stem cells in their primitive state and during their commitment to the specific lineages.

9.4 Mitochondria Issue and Stem Cells

Embryonic stem cells (ESCs), derived from the inner cell mass of human and mouse preimplantation stage of blastocysts have few mitochondria with poorly developed cristae [33–37]. Few mitochondria in human ESCs are organized in small perinuclear groups and exhibit low mitochondrial DNA (mtDNA) content [36–38]. Furthermore, the clustering of mitochondria around pronuclei as well as the mitochondrial characteristics of fertilized oocytes and embryos in the cleavage stage in several species [39–41], including human [42], seem to be similar to those of ESCs. Therefore, it was suggested that a perinuclear arrangement of mitochondria might be a cellular marker for stemness [43]. In parallel with a commitment to the specific cell fate, the number of mitochondria increases, they are larger, they exhibit the distinct cristae, and the number of mtDNA copies increases as well. Upon cellular differentiation, mitochondria acquire an elongated morphology with swollen cristae and dense matrices [36–38]. The low activity of mitochondria is in direct relation with the maintenance of stemness as demonstrated by a functional approach: the inhibition of mitochondrial respiratory chain (aerobic respiration) enhances human ESC pluripotency [44].

Concerning induced pluripotent stem cells (iPSCs) quite a similar situation was found: the somatic mitochondria iPSCs revert to an immature ESC-like state with respect to organelle morphology and distribution, expression of nuclear factors involved in mitochondrial biogenesis, and content of mtDNA [45,46]. Furthermore, bioenergetic transition from an oxidative to an anaerobic glycolytic state precedes inducing of pluripotency [47,48]. Conversely, upon differentiation, mitochondria within iPSCs and ESCs exhibited analogous maturation and anaerobic to aerobic metabolic modifications [37,45,49,50].

9.4.1 Mitochondria and Adult Stem Cell Systems

In mammals, the number, morphology, functional activity, and position of mitochondria in the cell vary not only from one tissue to another, but also between the cells of the same tissue [51]. Although in terms of its functional definition, a direct approach to "real" stem cells in adult tissues is very difficult and always related to a risk of the misinterpretations due to "elusive" nature of cell target cell (see Chapter 1), the low mitochondrial content and their decreased function in stem cells was pointed out [52,53] and further confirmed by other approaches. Also, the perinuclear arrangement (proposed as a hallmark of stemness) was described in some adult human $CD34^{+}$ cells [54], and in the rhesus monkey adult mesenchymal cell line (ATSC line) [43].

Long ago, it was shown that murine and human bone marrow morphologically recognizable hematopoietic precursors completely rely on oxidative phosphorylation—mitochondrial aerobic respiration—since they are, unlike some undifferentiated progenitors, highly vulnerable to mitochondrial toxic treatment [55]. The classic experimental hematology studies revealed that the more primitive hematopoietic stem cells (HSCs) were enriched in Rhodamine-$123^{low(dull)}$ cell fraction, and vice versa, less primitive HSCs and committed progenitors in Rhodamine-123^{high} fraction [56–59]. However, the Rhodamine-123 (Rho-123) is also substrate for ABC transporters (P-glycoprotein and ABCG2), responsible for the so-called "multidrug resistance" (MDR) phenotype (see Chapter 12) conferring to the cell the capacity to reject certain molecules (including Rho-123). So, even if Rho-123 did label active mitochondria, these data were not specific enough to derive a conclusion concerning the low mitochondrial mass and activity in primitive HSC. By using a complex approach with Thy-1.1^{low} Sca-1^{+}, Lin^{-} cells, Kim et al. [60] demonstrated that, at least in young mice, Rho-123 labeling is much more specific for mitochondrial activation than either to mitochondrial mass or MDR function. Thus, these results strongly suggested that in primitive HSC (Rho-123^{low}), the mitochondria are at the low activation level. If the same principle is valid for neonatal (cord blood) human cells, then the results of McKenzie et al. [61] would also act in favor of the low mitochondria activation state in the primitive HSC: purification based on Rho-123 uptake led to a fourfold enrichment of in vivo repopulating HSC (SCID-repopulating cells—SRCs) in the $Lin^-CD34^+CD38^-$ fraction, with a frequency of 1 SRC in 30 $Lin^-CD34^+CD38^-Rho^{low}$ cells. In $CD34^+CD38^-Rho^{low}$ fraction were present the HSC capable of long-term engraftment as measured by serial transplantation [61].

When MitoTracker green, a staining agent allowing to deduce the mitochondrial mass, was used to analyze the overall $CD34^+$ cell population, the mitochondrial mass was reported to be lower than in other cell types [54]. The important finding of this study was that the $CD34^+$ cells were heterogeneous with respect to the mitochondrial mass per cell. However, the authors claim that there is a negative correlation between the cellular mitochondrial content and $CD34^+$ expression attributed to the more primitive character of highly positive $CD34^+$ cells. This point generated confusion. It should be noted that the authors use the term "HSC" as a synonym for "$CD34^+$ cell," which does not stand: an extremely low fraction of $CD34^+$ cells mobilized in peripheral blood (used in Piccoli et al. [54] study ($\ll$1%)) may be considered as HSC; roughly 20–50% of these $CD34^+$ cells represent the committed progenitors, while the other $CD34^+$ cells exhibit neither HSC nor hematopoietic progenitor cell activity. In addition, the primitive HSC are either CD34 negative or exhibit a low CD34 molecule expression, and the highly positive $CD34^+$ cells are composed mainly of committed progenitors (see Chapter 1). Therefore, the results obtained on total $CD34^+$ population cannot be ascribed to HSC, but rather to committed progenitors. Thus, on the basis of these results, if correctly interpreted, the finding of a lower mitochondrial content should be related to committed progenitors, which is not in line with most other studies providing the arguments for a totally opposite conclusion: when the same authors, in addition to CD34, explored the CD133 marker, they found an inverse correlation between the mitochondrial content (rather mitochondrial mass) and CD133 expression on $CD34^+$ cells mobilized in peripheral blood. While this finding can be a significant cue toward the suggestion that more primitive HSC contain less mitochondria, the claim that CD133 is "diagnostic for more primitive stem cells" (reviewed in Ref. [62]) is highly exaggerated. A $CD34^-$ minority of HSC express CD133, and hence, CD133 may help to detect this HSC subpopulation; also, a part of $CD34^+$ committed progenitors does not express CD133, implying that in the $CD133^+$ cell population the HSC are a little bit more enriched. However, the HSC in the $CD133^+$ cell population (especially $CD34^+$ $CD133^+$ population) still represent an absolute minority (for mobilized PB cells <1%); furthermore, the relative majority represents, again, the committed progenitors. For the same reason the number of mtDNA (565 ± 100), average number of mitochondria (50 ± 20), and the number of mtDNA molecules per mitochondria (13 ± 5) judged as "high" in comparison to some mature cell populations [62] cannot be related to HSC, which represent an infinitesimal minority in $CD34^+$ cell population but rather to committed progenitors known to rely on the oxidative phosphorylation (see Chapter 6).

This viewpoint is confirmed by the results of Romero-Moya et al. [63], who demonstrated that the low-mitochondria content human cord blood $CD34^+$ fraction is enriched in $CD34^+CD38^-$ fraction and, more importantly, in vivo repopulating HSC, while containing a relatively lower proportion of committed progenitors (colony-forming cells) compared to corresponding high-mitochondria content fraction. Of note, this "advantage" of low-mitochondria $CD34^+$ cells was lost upon their culture at atmospheric oxygen concentration (21% O_2) [63]. Using the Lin^- $Sca\text{-}1^+$ population from murine bone marrow Piro et al. [64] showed that the smaller the cell, the higher the Sca-1 expression and the lower the mitochondrial mass (as determined by MitoTracker fluorescence intensity). These results might be more specific to the HSC since they demonstrate that with the enrichment of HSC in a functionally heterogeneous population, the mitochondrial mass per cell decreases, indicating that the real HSC exhibit a low mitochondrial mass [64]. Of course, the much more specific argument for this claim would be analysis of the mitochondrial mass on side population cells in Lin-negative Sca-1 positive fraction that the authors report to represent ~10% of these fractions [64] (see Chapter 1). However, murine LSK population ($Lin^-Sca\text{-}1^+c\text{-}Kit^+$; Chapter 1), which is heterogeneous and relatively enriched in both long-term (LT-HSC) and short-term (ST-HSC) reconstituting cells, offers the possibility to additionally concentrate LTC-HSC on the basis of CD34 expression: appearance and expression of CD34 on murine bone marrow LSK cells is a sign of a loss in repopulating capacity [65]. While LSK $CD34^{-/low}$ cells exhibited much lower mitochondrial mass compared to LSK $CD34^{hi}$ cells [66], these authors did not find a relationship between the Sca-1 expression density and mitochondrial mass but did between mitochondrial mass and cell size—the smallest cells had lower mitochondrial mass, but also low mitochondrial membrane potential.

In parallel with the loss of long-term repopulating capacity and increase in $CD34^+$ expression, increase the size, mitochondrial mass, and mitochondrial membrane potential [66]. This feature was further supported by Mortensen et al. [67]. These authors, after provoking conditional deletion of the essential autophagy gene Atg7 (thus preventing the autophagy), observed an accumulation of mitochondria and ROS in the LSK cell compartment, as well as increased proliferation and DNA damage of these cells. In spite of an amplification of overall LSK compartment, Atg7-deficient LSK cells lost in vivo repopulating capacity (i.e., the functional HSC were lost) [67]. These data confirmed the association between the increase in mitochondrial mass with the loss of stemness. Furthermore, the correct interpretation of the HSC entity (functional one) inside heterogeneous phenotypic (LSK) cell population in this study, clearly points out the relative value of HSC "definition" based on membrane markers expression for an appropriate reasoning and attribution of critical results to the real HSC entity.

In line with these data are the results of Chen et al. [68], obtained on Flt2 Lin $Sca\text{-}1^+$ $c\text{-}Kit^+$ (FLSK) $CD48^-$ cells, highly concentrated in LT-HSC. In *Drosophila* and mammalian cells, in the culture exposed to a low O_2 concentration, an activation of tuberous sclerosis complex (TSC) occurs. This can inhibit the target of rapamycin (TOR), through REDD1 and AMP-activated protein kinase. Thus, the authors investigated the effect of targeted mutation of TSC1 on proliferation and functional properties of murine bone marrow $FLSKCD48^-$ cells. This mutation led to mTOR activation, which, in turn, resulted in increased mitochondria biogenesis and accumulation of ROS followed by a diminution in quiescent fraction and increased proliferation and $FLSKCD48^-$ cells, which were associated with the loss of LT-HSC activity, evidenced on the basis of the best approach available—a serial transplantation model [68]. Again, the commitment and differentiation events are associated with mitochondria proliferation (mitochondrial biogenesis), which is paralleled by increase of aerobic metabolic properties of cells and vice versa, repressing mitochondrial biogenesis and ROS by TSC-M tor maintains the HSC stemness [68]. Norddahl et al. [69] revealed, by MitoTracker green staining (in the presence of Verapamil to block its efflux), that the mitochondrial content was reduced at the stage of the transition of HSCs to bipotent megakaryocytic/erythroid common progenitors but gradually increased upon differentiation through the erythroid committed progenitors [69]. Their study proves that the intact mitochondrial function is required for appropriate multilineage differentiation of stem cell descendants.

In the steady state, the mitochondrial aerobic respiration (oxidative phosphorylation) does not play an important role in HSC, as shown by the murine bone marrow LSK $CD34^-$ and $Flk2^-$ population, highly enriched in HSC (LT-HSCs), that is characterized by low mitochondrial potential and low NADH fluorescence (which is an index of overall mitochondrial respiration) [70]. Furthermore, when these "low metabolic profile" cells were sorted and evaluated by functional assays, the in vivo hematopoietic reconstituting activity was highly enriched in this population including the long-term one, relaying again the stemness with the low mitochondrial aerobic activity. This metabolic phenotype may exhibit a protective character if the HSC are in microaerophilic and anoxic conditions, which is incompatible with the survival of more committed or differentiated cells that rely on oxidative phosphorylation [70]. The same conclusion that a stem cell entity relies on anaerobic metabolism can be derived from the experiments with the mesenchymal stromal cells (MStroCs), a population containing the mesenchymal stem cells (MSCs) (see Chapters 1 and 4). These cells are resistant to exposure to 0.5% O_2 atmosphere as well as to inhibition of mitochondrial respiration with 2,4-dinitrophenol for 72h, indicating that in the absence of oxygen, MSCs can survive using anaerobic ATP production [71]. In another experimental design, these cells were maintained for 12days in the cultures exposed to <0.2% O_2 (<1.5mmHg).

As an interesting point, we mention the recent results with a mammary stem–cell-like model cell line [72], showing that apportion older mitochondria to postdivision daughter cells is unequal and that the daughter containing younger mitochondria maintains the stem-like cell pool, evidenced by a functional assay (mammosphere-forming capacity). Indeed, asymmetric segregation of mitochondria during mitosis, observed in some species, was suggested to play a role in determination self-renewal versus differentiation [37,73,74].

It is interesting that in *Planaria*, a model animal extensively used for the regeneration studies the neoblasts, a nondifferentiated cell population containing several functional types of stem/progenitor cells, including the totipotent stem cells, the low number of mitochondria ("few mitochondria") was repeatedly observed since 1969 [75–77]. The complexity of the planarian neoblast cells system is reminiscent of the one of $CD34^+$ cells in rodents and primates with respect to the number of the cell types that the most primitive neoblast and $CD34^+$ cells can generate; of course, the most primitive neoblasts are totipotent (small primitive animal) and the most primitive $CD34^+$ cells only multipotent (hematopoietic tissue is only one morphophysiological unit of a complex animal system; for further details see Chapter 10). At the moment, the data concerning the mitochondrial mass and activity in different functional and phenotypic planarian neoblast subpopulations are not available.

We believe that the exploration of this model will provide further interesting data relaying the amount of mitochondria, their position, metabolic type and intensity to the level of primitiveness, self-renewal, and differentiation potentials.

9.4.2 Might Adult Stem Cells Mitochondria Revert to Anaerobic Respiration Mode?

As explained above, the anaerobic mitochondria do not use O_2 as electron acceptor but other substrates, mainly fumarate, producing succinate, and the hydrogen-producing mitochondria use protons as terminal electron acceptors, producing hydrogen. We believe that, in line with "hydrogen hypothesis" (see above), these metabolic features represent the ancestral mitochondrial mechanisms that operated before the eukaryotes were exposed to a microoxic and oxic environment. The discovery of the anaerobic mitochondria that produce hydrogen in *Nyctotherus ovalis* reveals that this microorganism possesses the hydrogenosomes with a rudimentary genome encoding components of the mitochondrial electron transport chain (only complex I and II and not complex III and IV) [78]. The authors stated that "The presence of respiratory chain activity mitochondrial complex I and II, in combination with hydrogen formation, characterizes the *N. ovalis* hydrogenosome as a true missing link in the evolution of mitochondria and hydrogenosomes" [78]. As already mentioned, a number of eukaryotes possess the anaerobic mitochondria (*Ascaris lumbricoides*, *Mytilus edulis*, *Arenicola marina*, *Sipunculus nudus*, etc.) with several options to avoid the O_2 as electron receptor [8]. This anaerobic mitochondrial respiration is known as the "fumarate respiration" takes on in some protists, but also in anaerobic bacteria [79], which support hypothesis of its ancestral evolutionary origin.

Concerning the mammalian cells, the anaerobic mitochondrial respiration was usually considered as specificity of cancer cells (see Chapter 14, entirely dedicated to this issue; reviewed in Ref. [79]). However, the generation of ATP by intramitochondrial substrate-level phosphorylation (citric acid metabolites) and maintenance of mitochondrial membrane potential via electron transport in complex I is also documented in normal cells, as, for example, in cells in freshly isolated kidney proximal tubules [80]. Malate, fumarate, and α-ketoglutarate can promote this anaerobic mitochondrial metabolism and its end product is succinate [81]. Accumulation of succinate is a well-known ischemic feature in heart tissue resulting from reversal of succinate dehydrogenase, which, in turn, is driven by fumarate overflow from purine nucleotide breakdown and partial reversal of the malate/aspartate shuttle [82].

To test the hypothesis that, in conditions favoring stemness, the mammalian stem cells may regress to functioning mode supposed to characterize the first eukaryotes, we cultured the human bone marrow MStroCs in anoxia (0% O_2 atmosphere) for 10 days. It turned out that a subpopulation of these cells (MSCs) continued not only to proliferate, but even amplified in this condition. Of course, in this condition the MSCs exhibit a pronounced glycolytic profile, but still an important proportion of ATP production belongs to the mitochondrial one. When the transformation fumarate-succinate was blocked by the inhibition of NADH-fumarate reductase, the proliferation of MSC was inhibited and their numerical amplification failed (our unpublished data). This shows that the human normal-tissue (noncancer) adult stem cells mitochondria may assume the functional feature of anaerobic mitochondria, which implies the ancestral functional set-up of stemness. We are studying now in detail the energetic aspects of this anaerobic MSC proliferation and anaerobic mitochondria respiration.

9.5 Conclusions

On the basis of presented literature data, although not yet complete, we believe that a parallel can be drawn between the morphofunctional evolution of mitochondria from the first eukaryotic common ancestor (FECA) through the eukaryote diversification and their further evolution in metazoans on one side, and morphofunctional development of mitochondria from the primitive stem cells through the committed progenitors and their differentiation on the other side. This parallel concerns energetic participation of mitochondria in stem cells (their low number/mass, relatively lower energetic participation, simplified morphology, and probably, reversal to anaerobic mitochondrial function; i.e., "fumarate" respiration), which increases with commitment and differentiation. These features are reminiscent of the different evolution stages starting from the FECA to the complex eukaryote organisms. The anoxic/nanooxic/microaerophilic stem and progenitor cell niches are reminiscent of ecological niches of different stages in eukaryote evolution. Figure 9.1 represents a simplified model of this parallel; the possibility that mitochondria in mammalian stem cells may assume the function of hydrogen-producing mitochondria remains to be explored.

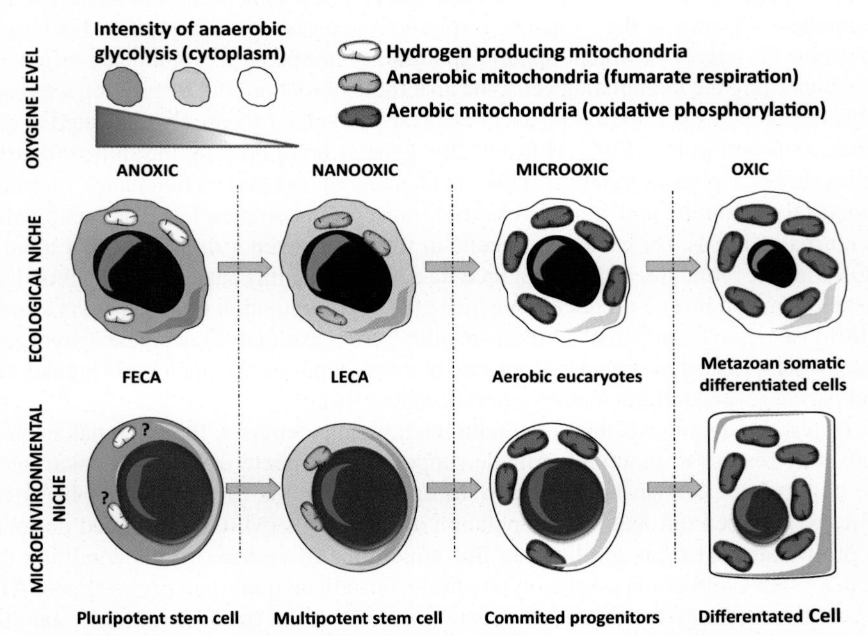

Figure 9.1 The evolution of energetic metabolism in eukaryotes versus changes in energetic metabolism during commitment and differentiation of stem and progenitor cells. Relation with the ecological niche (microenvironmental niche) is presented.

References

[1] Margulis L. Symbiosis in cell evolution. Life and its environment on the early earth. New York: WH Freeman and Co.; 1981.
[2] Gray MW. The evolutionary origins of organelles. Trends Genet 1989;5:294–9.
[3] Kurland CG, Andersson SG. Origin and evolution of the mitochondrial proteome. Microbiol Mol Biol Rev 2000;64:786–820.
[4] Searcy DG. Metabolic integration during the evolutionary origin of mitochondria. Cell Res 2003;13:229–38.
[5] Mentel M, Martin W. Energy metabolism among eukaryotic anaerobes in light of Proterozoic ocean chemistry. Philos Trans R Soc Lond B Biol Sci 2008;27(363):2717–29.
[6] Gray MW, Burger G, Lang BF. Mitochondrial evolution. Science 1999;283:1476–81.
[7] Gray MW, Burger G, Lang BF. The origin and early evolution of mitochondria. Genome Biol 2001;2. Reviews 1018.
[8] Müller M, Mentel M, van Hellemond JJ, Henze K, Woehle C, Gould SB, et al. Biochemistry and evolution of anaerobic energy metabolism in eukaryotes. Microbiol Mol Biol Rev 2012;76:444–95.
[9] Van der Giezen M, Tovar J, Clark CG. Mitochondrion-derived organelles in protists and fungi. Int Rev Cytol 2005;244:175–225.
[10] Dyall SD, Brown MT, Johnson PJ. Ancient invasions: from endosymbionts to organelles. Science 2004;304:253–7.
[11] Lane N. Hydrogen bombshell: rewriting life's history. New Sci August 4, 2010:36–9.
[12] Van der Giezen M, Lenton TM. The rise of oxygen and complex life. J Eukaryot Microbiol 2012;59:111–3.
[13] Müller M. Energy metabolism of protozoa without mitochondria. Annu Rev Microbiol 1988;42:465–88.
[14] Müller M. Energy-metabolism of ancestral eukaryotes—a hypothesis based on the biochemistry of amitochondriate parasitic protists. Biosystems 1992;28:33–40.
[15] Embley TM, Hirt RP. Early branching eukaryotes? Curr Opin Genet Dev 1998;8:624–9.
[16] Tielens AGM, Rotte C, van Hellemond JJ, Martin W. Mitochondria as we don't know them. Trends Biochem Sci 2002;27:564–72.
[17] Müller M. The hydrogenosome. J Gen Microbiol 1993;139:2879–89.
[18] Goldberg AV, Molik S, Tsaousis AD, Neumann K, Kuhnke G, Delbac F, et al. Localization and functionality of microsporidian iron-sulphur cluster assembly proteins. Nature 2008;452:624–8.
[19] Tovar J, León-Avila G, Sánchez LB, Sutak R, Tachezy J, van der Giezen M, et al. Mitochondrial remnant organelles of Giardia function in iron-sulphur protein maturation. Nature 2003;426:172–6.
[20] Mi-ichi F, Yousuf MA, Nakada-Tsukui K, Nozaki T. Mitosomes in *Entamoeba histolytica* contain a sulfate activation pathway. Proc Natl Acad Sci USA 2009;106:21731–6.
[21] Koonin EV. The origin and early evolution of eukaryotes in the light of phylogenomics. Genome Biol 2010;11:209.
[22] Martin W, Müller M. The hydrogen hypothesis for the first eukaryote. Nature 1998; 392:37–41.
[23] Anderson SG, Kurland CG. Origins of mitochondria and hydrogenosomes. Curr Opin Microbiol 1999;2:535–41.
[24] Hug LA, Stechmann A, Roger AJ. Phylogenetic distributions and histories of proteins involved in anaerobic pyruvate metabolism in eukaryotes. Mol Biol Evol 2010;27:311–24.

[25] Hampl V, Stairs CW, Roger AJ. The tangled past of eukaryotic enzymes involved in anaerobic metabolism. Mob Genet Elements 2011;1:71–4.
[26] Leger MM, Gawryluk RM, Gray MW, Roger AJ. Evidence for a hydrogenosomal-type anaerobic ATP generation pathway in *Acanthamoeba castellanii*. PLoS One 2013;8:e69532.
[27] Stairs CW, Roger AJ, Hampl V. Eukaryotic pyruvate formate lyase and its activating enzyme were acquired laterally from a Firmicute. Mol Biol Evol 2011;28:2087–99.
[28] Von Dohlen CD, Kohler S, Alsop ST, McManus WR. Mealybug β-proteobacterial endosymbionts contain γ-proteobacterial symbionts. Nature 2001;412:433–6.
[29] Thao ML, Gullan PJ, Baumann P. Secondary (γ-proteobacteria) endosymbionts infect the primary (β-proteobacteria) endosymbionts of mealybugs multiple times and coevolve with their hosts. Appl Environ Microbiol 2002;68:3190–7.
[30] Spang A, Saw JH, Jørgensen SL, Zaremba-Niedzwiedzka K, Martijn J, Lind AE, et al. Complex archaea that bridge the gap between prokaryotes and eukaryotes. Nature 2015;14(521):173–9.
[31] Kitazoe Y, Tanaka M. Evolution of mitochondrial power in vertebrate metazoans. PLoS One 2014;9(6):e98188.
[32] Mentel M, Röttger M, Leys S, Tielens AG, Martin WF. Of early animals, anaerobic mitochondria, and a modern sponge. Bioessays 2014;36:924–32.
[33] Sathananthan H, Pera M, Trounson A. The fine structure of human embryonic stem cells. Reprod Biomed 2001;4:56–61.
[34] Baharvand H, Matthaei KI. The ultrastructure of mouse embryonic stem cells. Reprod Biomed Online 2003;7:330–5.
[35] Oh SK, Kim HS, Ahn HJ, Soel HW, Kim YY, Park YB, et al. Derivation and characterization of new human embryonic stem cell lines: SNUhES1, SNUhES2, and SNUhES3. Stem Cells 2005;23:211–9.
[36] Cho YM, Kwon S, Pak YK, Seol HW, Choi YM, Park DJ, et al. Dynamic changes in mitochondrial biogenesis and antioxidant enzymes during the spontaneous differentiation of human embryonic stem cells. Biochem Biophys Res Commun 2006;348:1472–8.
[37] Facucho-Oliveira JM, St John JC. The relationship between pluripotency and mitochondrial DNA proliferation during early embryo development and embryonic stem cell differentiation. Stem Cell Rev 2009;5:140–58.
[38] St John J, Ramalho-Santos J, Gray H, Petrosko P, Rawe VY, Navara CS, et al. The expression of mitochondrial DNA transcription factors during early cardiomyocyte in vitro differentiation from human embryonic stem cells. Cloning Stem Cells 2005;7:141–53.
[39] Barnett DK, Kimura J, Bavister BD. Translocation of active mitochondria during hamster preimplantation embryo development studies by confocal laser scanning microscopy. Dev Dyn 1996;205:64–72.
[40] Batten BE, Albertini DF, Ducibella T. Patterns of organelle distribution in mouse embryos during preimplantation development. Am J Anat 1987;178:204–13.
[41] Squirrell JM, Schramm RD, Paprocki AM, Wokosin DL, Bavister BD. Imaging mitochondrial organization in living primate oocytes and embryos using multiphoton microscopy. Microsc Microanal 2003;9:190–201.
[42] Wilding M, Dale B, Marino M, di Matteo L, Alviggi C, Pisaturo ML, et al. Mitochondrial aggregation patterns and activity in human oocytes and preimplantation embryos. Hum Reprod 2001;16:909–17.
[43] Lonergan T, Brenner C, Bavister B. Differentiation-related changes in mitochondrial properties as indicators of stem cell competence. J Cell Physiol 2006;208:149–53.

[44] Varum S, Rodrigues AS, Moura MB, Momcilovic O, Easley 4th CA, Ramalho-Santos J, et al. Energy metabolism in human pluripotent stem cells and their differentiated counterparts. PLoS One 2011;6:e20914.
[45] Prigione A, Adjaye J. Modulation of mitochondrial biogenesis and bioenergetic metabolism upon in vitro and in vivo differentiation of human ES and iPS cells. Int J Dev Biol 2010;54:1729–41.
[46] Armstrong L, Tilgner K, Saretzki G, Atkinson SP, Stojkovic M, Moreno R, et al. Stem Cells 2010;28:661–73.
[47] Funes JM, Quintero M, Henderson S, Martinez D, Qureshi U, Westwood C, et al. Transformation of human mesenchymal stem cells increases their dependency on oxidative phosphorylation for energy production. Proc Natl Acad Sci USA 2007;104:6223–8.
[48] Folmes CD, Nelson TJ, Martinez-Fernandez A, Arrell DK, Lindor JZ, Dzeja PP, et al. Somatic oxidative bioenergetics transitions into pluripotency-dependent glycolysis to facilitate nuclear reprogramming. Cell Metab 2011;14:264–71.
[49] Chung S, Dzeja PP, Faustino RS, Perez-Terzic C, Behfar A, Terzic A. Mitochondrial oxidative metabolism is required for the cardiac differentiation of stem cells. Nat Clin Pract Cardiovasc Med 2007;4(Suppl. 1):S60–7.
[50] Birket MJ, Orr AL, Gerencser AA, Madden DT, Vitelli C, Swistowski A, et al. A reduction in ATP demand and mitochondrial activity with neural differentiation of human embryonic stem cells. J Cell Sci 2011;124(Pt 3):348–58.
[51] Collins TJ, Berridge MJ, Lipp P, Bootman MD. Mitochondria are morphologically and functionally heterogeneous within cells. EMBO J 2002;21:1616–27.
[52] Freyer JP. Decreased mitochondrial function in quiescent cells isolated from multicellular tumor spheroids. J Cell Physiol 1998;176:138–49.
[53] Radley JM, Ellis S, Palatsides M, Williams B, Bertoncello I. Ultrastructure of primitive hematopoietic stem cells isolated using probes of functional status. Exp Hematol 1999;27:365–9.
[54] Piccoli C, Ria R, Scrima R, Cela O, D'Aprile A, Boffoli D, et al. Characterization of mitochondrial and extra-mitochondrial oxygen consuming reactions in human hematopoietic stem cells. Novel evidence of the occurrence of NAD(P)H oxidase activity. J Biol Chem 2005;280:26467–76.
[55] Garland JM, Katz F. Relationship of factor-induced proliferation to respiratory status in marrow progenitor cells. Leukemia 1987;1:558–64.
[56] Ploemacher RE, Brons NH. In vivo proliferative and differential properties of murine bone marrow cells separated on the basis of rhodamine-123 retention. Exp Hematol 1988;16:903–7.
[57] Srour EF, Leemhuis T, Brandt JE, vanBesien K, Hoffman R. Simultaneous use of rhodamine 123, phycoerythrin, Texas red, and allophycocyanin for the isolation of human hematopoietic progenitor cells. Cytometry 1991;12:179–83.
[58] Bertoncello I, Bradley TR, Hodgson GS, Dunlop JM. The resolution, enrichment, and organization of normal bone marrow high proliferative potential colony-forming cell subsets on the basis of rhodamine-123 fluorescence. Exp Hematol 1991;19:174–8.
[59] Udomsakdi C, Eaves CJ, Sutherland HJ, Lansdorp PM. Separation of functionally distinct subpopulations of primitive human hematopoietic cells using rhodamine-123. Exp Hematol 1991;19:338–42.
[60] Kim M, Cooper DD, Hayes SF, Spangrude GJ. Rhodamine-123 staining in hematopoietic stem cells of young mice indicates mitochondrial activation rather than dye efflux. Blood 1998;91:4106–17.

[61] McKenzie JL, Takenaka K, Gan OI, Doedens M, Dick JE. Low rhodamine 123 retention identifies long-term human hematopoietic stem cells within the $Lin^{-}CD34^{+}CD38^{-}$ population. Blood 2007;109:543–5.
[62] Piccoli C, Agriesti F, Scrima R, Falzetti F, Di Ianni M, Capitanio N. To breathe or not to breathe: the haematopoietic stem/progenitor cells dilemma. Br J Pharmacol 2013;169:1652–71.
[63] Romero-Moya D, Bueno C, Montes R, Navarro-Montero O, Iborra FJ, López LC, et al. Cord blood-derived $CD34^{+}$ hematopoietic cells with low mitochondrial mass are enriched in hematopoietic repopulating stem cell function. Haematologica 2013;98:1022–9.
[64] Piro D, Piccoli C, Guerra L, Sassone F, D'Aprile A, Favia M, et al. Hematopoietic stem/progenitor cells express functional mitochondrial energy-dependent cystic fibrosis transmembrane conductance regulator. Stem Cells Dev March 1, 2012;21(4):634–46.
[65] Blank U, Karlsson G, Karlsson S. Signaling pathways governing stem-cell fate. Blood 2008;111:492–503.
[66] Mantel C, Messina-Graham S, Broxmeyer HE. Upregulation of nascent mitochondrial biogenesis in mouse hematopoietic stem cells parallels upregulation of CD34 and loss of pluripotency: a potential strategy for reducing oxidative risk in stem cells. Cell Cycle 2010;9:2008–17.
[67] Mortensen M, Soilleux EJ, Djordjevic G, Tripp R, Lutteropp M, Sadighi-Akha E, et al. The autophagy protein Atg7 is essential for hematopoietic stem cell maintenance. J Exp Med 2011;208:455–67.
[68] Chen C, Liu Y, Liu R, Ikenoue T, Guan KL, Liu Y, et al. TSC-mTOR maintains quiescence and function of hematopoietic stem cells by repressing mitochondrial biogenesis and reactive oxygen species. J Exp Med 2008;205:2397–408.
[69] Norddahl GL, Pronk CJ, Wahlestedt M, Sten G, Nygren JM, Ugale A, et al. Accumulating mitochondrial DNA mutations drive premature hematopoietic aging phenotypes distinct from physiological stem cell aging. Cell Stem Cell 2011;8:499–510.
[70] Simsek T, Kocabas F, Zheng J, Deberardinis RJ, Mahmoud AI, Olson EN, et al. The distinct metabolic profile of hematopoietic stem cells reflects their location in a hypoxic niche. Cell Stem Cell 2010;7:380–90.
[71] Mylotte LA, Duffy AM, Murphy M, O'Brien T, Samali A, Barry F, et al. Metabolic flexibility permits mesenchymal stem cell survival in an ischemic environment. Stem Cells 2008;26:1325–36.
[72] Katajisto P, Döhla J, Chaffer CL, Pentinmikko N, Marjanovic N, Iqbal S, et al. Stem cells. Asymmetric apportioning of aged mitochondria between daughter cells is required for stemness. Science 2015;348:340–3.
[73] Staiber W. Asymmetric distribution of mitochondria and of spindle microtubules in opposite directions in differential mitosis of germ line cells in *Acricotopus*. Cell Tissue Res 2007;329:197–203.
[74] Parker GC, Acsadi G, Brenner CA. Mitochondria: determinants of stem cell fate? Stem Cells Dev 2009;18:803–6.
[75] Morita M, Best JB, Noel J. Electron microscopic studies of planarian regeneration. I. Fine structure of neoblasts in *Dugesia dorotocephala*. J Ultrastruct Res 1969;27:7–23.
[76] Gschwentner R, Ladurner P, Nimeth K, Rieger R. Stem cells in a basal bilaterian. S-phase and mitotic cells in *Convolutriloba longifissura* (Acoela, Platyhelminthes). Cell Tissue Res 2001;304:401–8.
[77] Lopes KA, DE Campos Velho NM, Pacheco-Soares C. Method of isolation and characterization of *Girardia tigrina* stem cells. Biomed Rep March 2015;3(2):163–6.

[78] Boxma B, de Graaf RM, van der Staay GW, van Alen TA, Ricard G, Gabaldón T, et al. An anaerobic mitochondrion that produces hydrogen. Nature 2005;434:74–9.
[79] Sakai C, Tomitsuka E, Esumi H, Harada S, Kita K. Mitochondrial fumarate reductase as a target of chemotherapy: from parasites to cancer cells. Biochim Biophys Acta 2012;1820:643–51.
[80] Weinberg JM, Venkatachalam MA, Roeser NF, Nissim I. Mitochondrial dysfunction during hypoxia/reoxygenation and its correction by anaerobic metabolism of citric acid cycle intermediates. Proc Natl Acad Sci USA 2000;97:2826–31.
[81] Weinberg JM, Venkatachalam MA, Roeser NF, Saikumar P, Dong Z, Senter RA, et al. Anaerobic and aerobic pathways for salvage of proximal tubules from hypoxia-induced mitochondrial injury. Am J Physiol Renal Physiol 2000;279:F927–43.
[82] Chouchani ET, Pell VR, Gaude EAksentijević D, Sundier SY, Robb EL, Logan A, et al. Ischaemic accumulation of succinate controls reperfusion injury through mitochondrial ROS. Nature 2014;515:431–5.

Evolutionary Origins of Stemness: Relationship between Self-Renewal and Ancestral Eukaryote Biology; Conservation of Self-Renewal Principle in Parallel with Adaptation to O_2

Chapter Outline

The stem cell entity was discussed from different viewpoints in previous chapters (see Chapters 1–7). These considerations mainly concern mammalian stem cells (i.e., human and rodent ones). The mammals are, from an evolutionary viewpoint, the most complex organisms. Nevertheless, the phenomenon of stemness seems to be well conserved through the evolution, which will be elaborated in this chapter.

Anaerobiosis and Stemness. http://dx.doi.org/10.1016/B978-0-12-800540-8.00010-7

10.1 Stemness as Perceived on the Basis of Mammalian Studies

The typical stem cell properties perceived mainly on the basis of human and murine studies refer to (1) self-renewal and (2) differentiation capacity (totipotency, pluripotency, multipotency, oligopotency, and unipotency) phenomena. The first one ensures the maintenance of undifferentiated, stem cell populations, whereas the second one enables either development or regeneration of various tissues of the organism. This is elaborated by proliferation or cell divisions. Typically, self-renewal refers to division after which both or at least one daughter cell conserves the stem cell properties, while the commitment is related to activation of programs leading to realization of specific differentiation pathways, which is incompatible with the maintenance of stemness. At first glance, these two processes are incompatible but, in fact, this is not the case: the phenomena of self-renewal and commitment (preparation event for differentiation) should be considered at the cell population level. As a matter of fact, the two types of divisions with respect to self-renewal and commitment are possible: the symmetric and asymmetric ones. The symmetric scenario provides either two cells identical to the mother cell (i.e., maintaining completely stem cell characteristics that can be considered as "full" self-renewal) or two daughter cells in which the commitment events were engaged, that could be considered as "full commitment." The asymmetric scenario refers to a stem cell division after which one daughter cell maintains the stem cell properties and the second one undergoes commitment (see review [1]). The symmetric full-stem cell renewal exists during the embryonic life at the level of the first blastomers produced by zygote divisions, but they exist during some later stages as well as in adult somatic stem cell systems.

In mammals, totipotency and pluripotency seem to be the features restricted to some early embryonic cells since, so far, the pluripotent and totipotent adult somatic stem cells were not discovered. The asymmetric cell division model is visible in most embryonic and adult stem cell systems and it represents the major mechanism of regulation of stem cell and committed progenitor cell population maintenance, regulation, and tuning in steady and pathophysiological states. Also, the symmetric full commitment (differentiation process) is widely present in the compartments of committed progenitors, ensuring an efficient and rapid amplification of these cell compartments. While asymmetric division can ensure the maintenance of the stem cell pool, it cannot efficiently amplify the stem cell compartment, which is necessary in certain situations. In mammals, the symmetrical divisions of stem cells were documented during expansion of the stem cell pool in the course of embryonic and early fetal development. A good example is amplification of hematopoietic stem cells (HSCs) in mouse during mid-gestation, when their number doubles every day [2]. It is considered that the homeostasis of stem cell systems in adult mammals operates predominantly by asymmetric divisions. However, since in mammals it is not possible to detect the morphological difference between a stem cell and a committed progenitor cell, the other option, balance between symmetrical divisions of stem cells and committed progenitors, cannot be excluded (see review [1]). Even if asymmetric stem cell divisions seem to be predominant in steady state, a number of studies suggest that the symmetric divisions take place in adult mammalian tissues especially during regeneration (or tentative regeneration). For example, in rats, the neural stem cells in the subventricular zone shift from asymmetrical to symmetrical division after a stroke [3]. Muscle "satellite cells" [4], retinal stem cells [5], skin stem cells [6], and such also exhibit a capacity of symmetrical divisions.

The third typical property of a stem cell is so-called "high proliferative capacity." In fact, the high proliferative capacity results from the proportion of self-renewing divisions in the proliferative history of a stem cell clone. This property refers to the final number of cells that can be produced by a stem cell after all self-renewing, commitment, and differentiation divisions until final exhaustion of the stem cell potential in the clone issued from a stem cell. In that respect, the analytic modeling of stem cell studies, especially in the field of HSCs, revealed the notions as "differentiation pressure" and "self-renewal pressure" [7] as opposite trends influencing a stem cell stock. The mammalian organism is equipped with the stocks of multipotent and tissue-specific stem cells, which would be enough for several life spans of organisms. Thus, in a normal situation, the stem cells are not being exhausted during a life span. However, this can change in pathological conditions.

There is another hallmark of stem cell biology: the phenomenon of quiescence; that is, residing of cells in G_0 phase of cell cycle. The relatively higher proportions of primitive stem cells are quiescent in physiological conditions, comparing to the less primitive stem cell populations and committed progenitors. This, however, does not mean that all stem cells are quiescent (see Chapter 5). Neither does it mean that the phenomenon of cell quiescence is specific for stem cells since, in some conditions, committed progenitors, as well as more mature cell populations, can enter and reside in G_0 (i.e., be quiescent).

On the basis of all available data, it can be concluded that the three specific cell properties can be related to the notion of stem cell (1) capacity of self-renewal (self-renewal can be considered as a cell division without activation of commitment-differentiation events, hence providing daughter cells identical to the mother stem cell one), (2) differentiation potential allowing to give after division, at least one cell in which the commitment-differentiation events will be triggered in order to produce a cell different from the mother cell, and (3) capacity of quiescence (dormancy) of prolonged survival in the G_0 phase of cell cycle.

To get closer to the origin of stemness, we are analyzing some "model organisms"—those that are believed to provide the data about the first somatic full-established stem cell systems. Thus, in order to illustrate the phenomenon of stemness in the early phases of evolution we will analyze the most primitive multicellular organisms and some elements of protozoa life cycle emerging as a new and original vision of routs of stemness.

10.2 Stem Cells in Bilateria

10.2.1 *Platyhelminthes (Flatworms) and Nematodes (Roundworms)*

10.2.1.1 *Flatworms: Stem Cells in Planarians*

The first written texts describing a planarian organism with the notions of its regenerative capacities was published in China (Taiwan) 860 years BC. The data and descriptions with drawings of some planarian species can be found in a Japanese manuscript from the seventeenth century. One century later, the number of reports concerning planarians increased. Müller published a work in 1773 describing a number of planarian species erroneously classed in the trematode genus *Fasciola* (reviewed in [8]). In spite of all these reports, the beginning of planarian studies is usually associated with the work of Pallas published in 1774, reporting the data obtained during the exploration of the Ural Mountains and observing that Planarians regenerate missing body parts [9]. These animals attracted the attention of biologists in the nineteenth and twentieth centuries. In the late nineteenth century, the light microscope allowed scientists the cellular level approach to the analysis of planarians. This way, a special population of parenchymal (mesenchymal) cells actively dividing was evidenced on the basis of mitotic features. These are the cells with a high nucleo-cytoplasmic ratio (a large nucleus and scarce basophilic cytoplasm) [10–12]. Apart from free ribosomes and **few mitochondria**, the thin cytoplasm surrounding the nucleus does not contain the cellular organelle [13]. Different scientists named these cells by different terms (Table 10.1). Of note, 120 years ago, Keller [11] called them *Stamzellen* (stem cells). Finally, the term "neoblast" was adopted by Buchanan [14] and Wolf and Dubois [15]. These scientists used the term already applied by Randolf [16] to describe the cells of similar morphology and function in the earthworm, *Lumbriculus* [16].

By analogy to some other primitive animal models, it was proposed that these cells are the result of either of the differentiation of mature cells or an embryonic "stock of blastomere-related cells" that persisted into adulthood. The substantial advance in the neoblasts study allowed the ionizing radiation approach. First, it was shown that the whole-body planarian irradiation with a dose sufficient to kill neoblasts (which are proliferating cells) without killing the animal induces a loss of its ability to regenerate [17]. In further experiments, Dubois [18] demonstrated that the neoblasts are migrating in order to regenerate cut parts. In fact, they are able to go from intact parts through the irradiated parts of the body to reach the part that should be regenerated. Indeed, the start of the regenerating effect was proportional to the length of irradiated part. These experiments are very similar to those done decades later, in which the migration of the first hematopoietic stem/progenitor cell class discovered, CFU-S, was studied [18–21]. Altogether, these data strongly suggest that the viable neoblast

Table 10.1 Terms Used for the Planarian Cells Today Known as Neoblasts

Year	Original Term	English Translation	References
1884	*Verästelten bindengewebszellen*	Branching connective tissue cells	[177]
1890	*Bildungszellen*	Forming cells	[10]
1894	*Stamzellen*	Stem cells	[11]
1891	*Stoffträger*	Support material	[178]
1898	*Ersatzzellen*	Replacement cells	[12]
1922	*Cellules libre de parenchyme*	Free parenchyma cells	[179]
1923	*Regerationszellen*	Regeneration cells	[180]
1925	*Wanderzellen*	Migratory cells	[181]

migrated to the wound to regenerate the missed body part (i.e., the regeneration is not coming from the differentiation of mature cells, as proposed by the "dedifferentiation" hypothesis). This point was strengthened in experiments in which the cell fractions enriched for either neoblast or differentiated cells were isolated [22]. Only the neoblast-enriched fraction rescued irradiated animals [22].

In line with these experimental data are those obtained by another approach: grafting of irradiated tissue parts into irradiated hosts. Using this approach, it was evidenced that the dividing viable neoblasts migrate from grafted unirradiated tissue into the tissue of an irradiated host. Thus, in any case, the regeneration capacity depends on the viable proliferating neoblasts. On the basis of these data, collectively, the neoblasts were considered as a totipotent stem cell population, as long as they were studied at the level of a population [23–28]. This approach did not allow to conclude if the all neoblasts are totipotent stem cells or if the population morphologically defined as neoblasts was heterogeneous and contained different subpopulations varying in their potency and proliferative capacity, similar to the mammalian somatic stem cell systems (i.e. HSC system). Applying the single-cell transplantation approach combined with ionizing radiation, Wagner et al. [29] demonstrated that only a subpopulation of neoblasts can form large descendant cell colonies in vivo, in the host tissue. These neoblasts, endowed by in vivo clonogenic capacity (similar by analogy to CFU-S in hematopoietic system [30]; see Chapter 3), produce cells that differentiate into neuronal, intestinal, and other postmitotic cell types in planarians. This neoblast cell population was termed "c-neoblasts." Furthermore, the authors showed that a single c-neoblast transplanted is able to restore the regeneration capacity in a lethally irradiated host animal [29]. The other studies demonstrating heterogeneity of a neoblast population are in line with these results. They revealed heterogeneity of a neoblast cell population at the level of gene-expression signature both during the regeneration [31–34] and in steady state [35,36].

We consider very interesting the study of Hayashi et al. (2006) in which the neoblasts were found to belong to two radio-sensitive populations ("X1" and "X2" populations) distinguishable on the basis of Hoechst33342/Calcein AM flow/cytometry profiles. Furthermore, combining the Hoechst blue and Hoechst red staining, the authors revealed a real "side population"—the cells with intensive Hoechst stain efflux (see Chapters 1 and 12). These cells are radio-resistant, but their functional properties remain to be evaluated [37]. The same group further characterized the two populations of radio-sensitive cells by means of electron microscopy and cell cycle analysis. They evidenced two major phenotypes of neoblasts: cells whose nucleus appeared to contain a large amount of euchromatin (A type) or heterochoromatin (B type). The "A type" stem cells were enriched in X1 fraction, which contains actively proliferating cells. In contrast, "B type" cells were only scarcely detected in X1 fraction and concentrated in X2 fraction. The authors present the arguments suggesting the existence of committed progenitor cells that can maintain the high proliferative potential [38]. This is in line with the data of Moritz et al. [36], who demonstrated that about two-thirds of the cycling planarian adult stem cells showed a specific membrane signature coupled with the expression of markers hitherto considered to be restricted to differentiating, postmitotic progeny. The analysis of two of the clones generated revealed that a subset of cells of the X1 population expresses early and late progeny markers, which might indicate that these cells are committed while still proliferating [36]. The "B" cells, which represent, in fact, a new class of neoblasts proven to exist in steady state planarian tissue, are small in size with few chromatoid bodies and a heterochromatin-rich nucleus. Interestingly, they were concentrated in the X2 fraction, containing G_0/G_1 phase cells [38] (no discrimination between G_0 and G_1 cells presented in this study). However, Verdoodt et al. [39] demonstrated, using the planarian *Macrostomum lignano*, that the neoblasts can be divided in at least two distinct subpopulations: quiescent and active neoblasts. Using the BrdU pulse-chase approach combined with "Vasa" labeling they showed that, in addition to differentiated cells, some neoblasts behave as "label-retaining cells" (LRCs) (for LRC, see Chapter 5), a typical characteristic of adult mammalian stem cells.

The real breakthrough, however, was provided by the study of van Wolfswinkel et al. [40]. Combining high-dimensional single-cell profiling with the functional irradiation/regenerating assays, the authors provided the compelling evidence on the existence of two major classes of neoblasts: the first class called "sigma neoblasts," which proliferates in response to injury exhibits a broad differentiation potential and can give rise to the second class of neoblast called "zeta neoblasts," which, in turn, encompasses specified cells that give rise to an abundant postmitotic lineage including epidermal cells. However, neither the sigma class, nor the zeta class can be a homogeneous cell population, which is evident on the basis of their gene expression profile [40]. Within the sigma class neoblasts, the authors suggest the existence of another subpopulation called "gamma class," an even more complex expression/functional profile of sigma neoblasts containing the other subpopulations.

Concerning the zeta neoblasts representing, in fact, some kind of committed progenitors, the authors hypothesized the existence of several subclasses of lineage-committed and precursor cells [40]. Altogether, the neoblasts cell system is similar to a classical mammalian stem cell system typically composed of non-differentiated cells with high nucleus cytoplasmic ratio exhibiting a complex functional and transcriptional heterogeneity, completely corresponding to the general frame of generation-age hypothesis [41]. In addition, the neoblast stem cell system exhibits a subpopulation in quiescence and/or in slow-cycling state completely corresponding to the "strategic reserve" stem cells in mammalian tissues. In view of its evolutionary position, the neoblast stem cell system, having all characteristics of one-tissue, two-tissue, or more stem cell systems in the evolutionary most complex organisms might be considered a prototype of a "complete" stem cell system. The planarians are the most studied flatworm model but the neoblast-like cells were evidenced in other flatworms including *Schistosomoe* (parasitic trematodes). Transmission of these parasites relies on a stem-cell-driven, clonal expansion of larvae inside a molluscan intermediate host. This proliferative larval cell population (germinal cells) shares some molecular signatures with stem cells from diverse organisms, in particular neoblasts of planarians (free-living relatives of schistosomes) [42]. In *Schistosoma mansoni* the somatic cells corresponding to the planarian neoblasts were evidenced on the basis of phenotypic, functional, and molecular criteria [43].

10.2.1.2 Roundworms and Other Worms

A quite similar neoblast stem cell system was documented in free-living nematodes; for example, in earthworms: *Enchytraeus japonensis* (*Oligochaeta, Annelida*) [44,45], *Lumbriculus variegatus* (*Oligochaeta, Annelida*) [46]. However, some species do not possess the neoblasts, which results in less universal regeneration capacities, the appearance of antero-posterior gradient of regeneration ability along the body axis [47]. The same author suggests that the annelid neoblasts are more essential for efficient asexual reproduction than for the regeneration of missing body parts. The neoblasts were also observed in tubeworms (*Spirorbis vitreus* and *Spirorbis borealis*) [48]. This stem cell population in roundworms and other worms was not studied in details as in Planarians, so the data concerning the individual neoblast functional capacities are lacking.

10.3 Stemness in Basal Metazoans

10.3.1 Hydra (Cnidaria, Hydrozoa)

Two-hundred and forty years ago, Trembley [49] published *Mémoires pour servir à l'histoire d'un genre de polypes d'eau douce, à bras en forme de cornes*. Trembley cut a "polyp" into pieces and observed that, in the next days, each piece gave a new, fully operational organism. On page 234 (*Mémoire* 4) he wrote, "wherever the polyp is cut … the portions become the complete polyps." On the basis of this regeneration capacity remembering one of a creature from the ancient myths, this small organism is going to be named "Hydra." In actual classification, Hydra represents a genus of small, freshwater cnidarians (coelenterates), which belong to the class Hydrozoa. Cnidarians are found in Cambrian fossil record: Actinaria-related fossils from Lower Cambrian Chengjiang biota of Yunnan, China [50] and the jellyfish fossils from the Middle Cambrian (approximately 505 million years old) [51]. The origin of cnidarians might extend back to ~700 million years ago as suggested by "molecular clock" studies [52].

It is known that a small piece of Hydra tissue (about 2 mm or 300 cells) represents the "minimal unit" allowing full organism regeneration. This "critical mass" of tissue, in fact, is necessary to capture the three self-renewing cell lines necessary to complete hydra regeneration [53]. These three lines are coming from three different cell types, considered by many authors as stem cells. The transition of a cell from one somatic lineage to another has never been observed in Hydra. The lineages that form ecto- and endoderm of Hydra polyp (i.e., exterior and interior unicellular sheets) are composed of cells of epitheliomuscular nature, usually termed "epithelial cells" in the literature. Positioned between two epitheliomuscular layers are interstitial cells ("I-cells").

10.3.1.1 Ectodermal Epithelial Cells

In steady state growth, the cells of first-line, ectodermal epithelial cells give rise to three types of terminally differentiated cells: tentacle-specific battery cells, hypostome-specific cells, and basal-disc-specific secretory cells. The ectodermal epithelial cells exhibit the characteristics of a true epithelium as apical-basal polarity, apical junctional complexes, and hemidesmosome-like junctions (to attach to mesoglea). The ectodermal epithelial cells maintain their basic epithelial sheet organization during regeneration. Dubel et al. [54] demonstrated that, in steady state, the ectodermal epithelial cells of the foot, the foot mucous cells, and the ectodermal epithelial cells of the tentacles, the battery cells, differentiate from gastric ectodermal epithelial stem cells. They can differentiate from the "stem cell" to a terminally differentiated cell after only one complete cell cycle to reach a G_2 definitive arrest. Gastric ectodermal epithelial cells may also differentiate into the hypostomes (a part of the head structure), but only during the head regeneration and budding. However, in this case, the differentiated hypostome cells maintain the capacity of proliferation, which allows them to self-renew in steady state without the need for a repopulation from gastric stem cell pool [54]. So, the local regulation of cell proliferation upon injury may completely change the properties of differentiated cells issued from the mid-gastric ectodermal epithelial cells [55]. The ectodermal endothelial cells probably, during the sexual reproduction, originate from "female egg cup," an egg holding structure placed in mid-body column as well as from testis localized closer to the head.

10.3.1.2 Endodermal Epithelial Cells

In contrast to ectodermal epithelial cells, the endodermal epithelial cells do not participate in sexual reproduction and do not present the junctions to attach mesoglia. Maybe due to this feature, the endodermal endothelial cells exhibit greater morphogenic flexibility: within a few hours after tissue recombination, the endoderm lost its epithelial sheet organization and turned into a mass of irregularly shaped cells without the apical-basal cell polarity initially present. The endodermal epithelial cells are very mobile during regeneration, when a single endodermal epithelial cell placed in contact with an ectodermal aggregate actively extends pseudopod-like structures and migrates toward the center of the ectodermal aggregate [56], as well as in terms of tissue recruitment into the newly formed bud [57]. If single transgenic endodermal epithelial cell is transplanted into wild-type host animal, it proliferates and is ultimately able to regenerate polyps in which the entire endothelium originates from that transgenic cell [57]. The endodermal epithelial cells can differentiate into four cell types: digestive cell, foot endodermal cells, tentacle endodermal cells, and hypostome endodermal cells. Endoderm epithelial and ectoderm epithelial cells exhibit some particularities: a very short or even absent G_1-phase in cycling cells and so-called "G_2 posing" [58].

10.3.1.3 Interstitial Stem Cells ("I-cells")

Interstitial cells, positioned between two epitheliomuscular layers, comprise about 80% of total Hydra cells. They are described as "small cells with an interphase nucleus and a ribosome-rich cytoplasm" [59]. These cells can differentiate into the somatic cells (nematocytes, ganglion neurons, basal neurons, and gland cells) and produce the gametes (sperm and oocyte progenitors; schematic representation in [58]). It is important to stress that the interstitial stem cells undergo several cell cycles to produce a differentiated cell. For example, to give the nematocytes, they go through four cell cycles to produce first the clusters (16 cells) that are going to amplify exponentially through a compartment of nematoblasts to finally differentiate and maturate into nematocytes (presented in Figure 5 of [60]). They also showed that the nerve cell production starts at the stage of two interstitial-cell containing clusters but not later. This stage is followed either by a short amplification through a nerve committed progenitor and precursor compartment before producing the terminal cell differentiation (schematic presentation given in Figure 1 in [61]) or by direct differentiation of precursors descending from stem cells into mature neurons [62]. The short precursor cell amplification before terminal differentiation, starting directly from committed cells, was proposed also for secretory cells (schematic presentation given in Figure1 in [61]). Note that the principle of amplification through the committed progenitor compartments followed by the differentiation of precursors is typical for the stem cell systems in further evolution and is conserved in all "higher" classes including the mammals and well represented by the model of HSC commitment and differentiation, and completely "enters" in concept "generation-age hypothesis" of Rosendaal [41]. Furthermore, on the basis of kinetics data, four decades ago, David and Gierer concluded, "The continuous differentiation of nerve cells and nematocytes from interstitial cells implies that, among interstitial cells, there is a sub-population of 'stem' cells which are capable of producing both new stem cells, as well as determined cells" [60], clearly pointing to the phenomenon of stem cell self-renewal in its most "orthodox" meaning, but also to the heterogeneity in capacities of individual interstitial cells (the principle still present in mammal stem cell considerations; see Chapter 1).

It is particularly interesting to note that Buzgariu et al. [58] found interstitial cells with a shorter G_2 that "pause" in G_1/G_0. Furthermore, the undifferentiated cells of all three lineages in Hydra show heterogeneity in their proliferative potential and cell cycle distribution and a significant proportion of these cells does not proliferate for quite long periods (8–10 cell cycle times), as deduced on the basis of a thymidine analogue, 5-ethynil-2′-deoxyuridine (EdU) retention [63]. This phenomenon of slow-cycling cells in Hydra, as demonstrated by EdU retention, seems to be very similar to one of "BrdU label-retaining cells" in mammalian HSC systems (see Chapter 5). While in the cell cycle studies of Hydra cells, only "G_0/G_1 pick" was presented on the basis of DNA ploidy analysis, it would be of particular interest to discriminate the phase G_0 from G_1 one. Indeed, the possibility of G_0 phase quiescence (especially of Hydra interstitial cells, which seem to approximate better a classic stem cell system) should be explored to evaluate the functional analogies between this early evolutionary stem cell system and other stem cell systems.

Removing Hydra interstitial cells by either chemical (colchicine) or physical (low temperature) agents does not have a lethal effect but results in so-called "I cell-free Hydra" lacking nematocytes, gametes, neurons, secretory, and sensory cells. Furthermore, if appropriately fed, this Hydra composed of epithelial cells only will continue with asexual reproduction (budding), its epithelial cells self-renew, and it can regenerate a lost head [64,65]. This is very interesting since the ectodermal and endodermal cell lineages correspond to outer (epidermal) and inner (gastrodermal) cell layers, whose evolutionary origins can be traced back to sponges (see below). Furthermore, apart hydrozoans, the other cnidarian classes do not have the interstitial cells with stem cell potential [66]. The most ancient cnidarians groups, anthozoans, are composed of only ectodermal and endodermal epithelium. Having in mind these facts, we can consider that the result of artificial removing of Hydra interstitial cells was the creation of an organism that mimics the preceding stages in cnidaria evolution (or mimics just a spontaneous regression operated by Nature in the case that anthozoans represent the results of a regressing process from a hydra-like organism?). This also acts in favor of assumption that the interstitial cells are, in fact, the hydrozoa evolutionary innovation. However, the nonhydrozoan cnidaria as anthozoan sea anemone *Nematostella* [67] and the jellyfish *Aurelia* appear to have another undifferentiated cell type—amebocytes (amoebocytes) (reviewed in [52]). Until now, there is no data proving that the amebocytes can act as stem cells, although it cannot be excluded that the "I-cells" in hydrosomes were evolutionary derived from amebocytes (reviewed in [53]).

Unlike in Hydra, where only through the germinative cells in sexual reproduction the "I-cells" reveals their totipotency, in close Hydra relatives, the colonial hydroids Hydractinia, the somatic interstitial cells are totipotent [66,68]. Muller et al. [66] performed experiments in which the interstitial cells were eliminated from subcloned wild-type Hydractinia animals and, subsequently, the interstitial cells from mutant clones introduced. This resulted with the time in a complete conversion of recipient into the phenotype and genotype of the donor. Thus, not only interstitial cells, but also epithelial ones were produced from interstitial donor cells. Due to the fact that Hydra is a derived solitary hydroid, both groups consider that the Hydra interstitial cells may have lost their totipotency since extensive loss of tissue in a solitary Hydra polyp cannot be compensated by migrating interstitial cells from other polyps of colony due to the lack of tissue continuity among clonemates [66]. Thus, the loss of "stem cell supply system" may have accompanied the evolution of autonomous epithelial stem cell lineages in Hydra [66].

10.3.2 Sponges (Porifera)

Even if it is still a controversial issue [69], sponges are thought to be the earliest branching multicellular lineage of extant animals [70]. Sponges are divided into four classes: hexatinellidis (glass sponges), desmosponges [70] (considered to include 95% of Porifera species), calcareous sponges, and homoscleromorphs. Some authors believe that the Porifera represent the earliest branches on phylogenetic trees of all animals [71] and approximate the last common ancestor of metazoans—the Ur-metazoan, so they can yield insights into the biology of earliest animals [72]. In fact, among a dozen types of cells identified by morphological features in desmoponges [73], there are the flagellated feeding cells, choanocyes, which morphologically resemble choanoflagellates, single-celled eukaryotes considered to be the closest living relatives to all animals [74,75]. It seems that a consensus exists that choanoflagellates are not derived from metazoans, but instead, represent a distinct lineage that evolved before the origin and diversification of metazoans [74].

Sponges exhibit an extensive regenerative capacity, which is due to the presence in the inner space of their body ("mesohyl") of a special type of cells from which all other cells differentiate—archeocytes.

Archeocytes are considered as a totipotent (by up-to-date terminology) stem cell [76–78]. Indeed, after dissociation of desmosponge cells, some cells die and other migrate and adhere to each other to form cell aggregates. If archeocytes were eliminated, these aggregates do not attach to the substratum and do not reconstruct the sponge body, which, in contrast, occurs with the archeocyte-rich fraction (reviewed [73]). The archeocytes are in origin of so-called "gemmule hatching," which is, in fact, one of the variants of asexual reproduction systems of sponges. For sexual reproduction, the desmosponges do not possess the "prefabricated" germ line stem cells—their gametes are directly produced either by archeocytes (oocytes; [79]) or choanocytes (spermatozoids; [80]).

Archeocytes are from a morphological viewpoint large ameboid cells containing a large nucleus with a single large nucleolus, mitotically active, motile, and containing numerous vitelline platelets during the early developmental stages of gemmule hatching. They have phagocytic activity [72,73,81] defined, using the freshwater sponge (*Ephydatia fluviatilis*), the molecular markers *EflPiwiA* and *EflPiwiB*, which mRNA is expressed specifically in archeocytes and choanocytes. The same group evidenced an archeocyte-specific molecular marker: mRNA expression of EflMusahiA (*EflMsiA*) [82]. These molecular markers allowed the authors to conclude that archeocytes divide symmetrically for both self-renewal and differentiation. The maintenance of a high level both of *EflMsiA* mRNA and protein would condition a self-renewing division, while their decrease induces some kind of commitment leading to differentiating and finally differentiated cells in which neither mRNA nor protein are expressed. Archeocyte may be in a "resting" state, a special form called "thesocytes." Another transcription factor specifically expressed by archeocytes is *EmH-3*, which belongs to the *Tlx* homeobox gene family. Similarly to the impact of *Tlx* homeobox gene expression in murine and human somatic cells (abrogation of differentiation progress or even reprogramming the mature toward primitive cells), expression of these genes in archeocytes is associated with the mitoses of immature cell phenotype and a delay or abrogation of the differentiation [83].

Choanocytes exhibit morphology of fully differentiated cells with specialized function composed of a single flagellum surrounded by microvilli interconnected laterally to form a cylindrical "collar." Packed choancytes form a network of interconnected "chambers," generate water currents, take up the nutrients from the water, and transfer them to other cells via vesicles (reviewed in [73]). As mentioned earlier, choanocytes exhibit morphology almost identical to choanoflagellates [84], protist closest to metazoans in molecular phylogenetic trees [85]. The choanocytes proliferate to maintain the choanocyte chambers (i.e., self-renew, which is a parallel to the nonsexual reproduction of choanoflagellates). In addition to the capacity to directly transform in situ into spermatozoids, the classic microscopic observation studies suggested ability of choanocytes to transform into archeocytes. In fact, several publications dating from the 1970s evidenced such a transformation both during gemmule formation (when choanocytes deepithelialize and move into the mesohyl to transform into archeocytes) [86] and upon disruption of sponge tissue and subsequent remodeling (Diaz 1979, cited in [73]). These very interesting data, however, remain to be confirmed by up-to-date approaches. Nevertheless, Funayama [73] proposed existence of a two-stem-cell-based system in desmosponges consisting of archeocytes and choanocytes, both cell types characterized by expression of *EflPiwiA*, which could mutually *trans*-differentiate. This scenario considers the archeocytes as "active stem cells" and choanocytes as a "storage state" of archeocytes. The authors favor a hypothesis that the original stem cell in Ur-metazoan was similar to choanocytes, and, hence, that the metazoan stem cells evolved from choanoflagellates-like protist. That would be in line with the King's concept [87,88], proposing that choanoflagellates or their ancestors acquired genes that allow the cohesion of cells, essential for multicellularity, first as colonies, then as more complex sponges.

However, the structural similarity between choanoflagellates and sponge choanocyte does not necessarily mean that choanoflagellate-like organisms were ancestors of the extant sponges and the remaining metazoans. An alternative hypothesis regards choanoflagellates as derived from evolutionary simplification of sponges [89]. A 2008 molecular phylogeny analysis [90] strongly rejects the hypotheses proposing that either group (choanoflagellates and sponges) is derived from the other. As concluded by Carr et al. [90], "it appears instead that both groups are descendent from a common marine protistan ancestor. From this ancestor, the Metazoa evolved into truly multicellular organisms, whereas the choanoflagellates have maintained a predominantly solitary existence, albeit with a widespread ability to form colonial stages in their life cycle." Coutinho and Maia [83] point to "mesenchymal cells" in ancestral spongiomorph Ur-metazoa and relate rather archeocytes (capable to "generate" choanocytes) to the ancestral stem cell lineage [83]. They stress that the property of generating all sponge cell types is unique to ameboid mesenchymal stem cells only (i.e., archeocytes). Indeed, the simplest example of sponges' development is that of Tetillidae (*Spirophorida*), in which the substrate-adherent egg generates, by symmetric divisions, an ameboid cell mass, which differentiates directly into an adult sponge [91].

10.3.3 Placozoa

The strange macroscopic (a couple of millimeters big) giant amoeba-looking organism, covered by cilia and adhered to the glass, was discovered in 1883 in a seawater aquarium of Institute of Zoology in Graz (Austria) by Franz Eilhard Schulze, to become the most controversial organism with respect to its phylogenetic position and to the place in evolution of animals (reviewed in [92]).

This animal, named *Trichoplax adhaerens*, is the only extant representative of the phylum *Placozoa*. Due to misinterpretation of the life cycle and some more recent molecular phylogenetic analyses, *Trichoplax* was considered a derived species within the *Cnidaria*, but the molecular morphological characters provide compelling evidence that this is not the case [93]. This was confirmed by a comparative mitochondrial genome analysis that firmly suggested that the *Placozoa* represents the basal lower metazoan phylum, and implies that they would have arisen relatively soon after the evolutionary transition from unicellular to multicellular forms. Thus, *T. adhaerens*, considered to be close to the last common metazoan ancestor, would be an intriguing model for "Ur-metazoan" hypothesis. It should be stressed, however, that the relationships between the four nonbilaterian metazoan phyla (Porifera, Placozoa, Ctenophora, and Cnidaria) remain a controversial issue [94–97].

Trichoplax lacks any organs and body symmetry and is composed of only six cell types (of note, the sponges contain at least 12 cell types). In fact, we can distinguish the top and bottom epytheloid layers (dorsal and ventral epithelial cells, each one presenting one cilium) and fluid-filled cavity with the mesenchymal-like cells known also as "ameboid cells" or "syncytium fiber cells" [98]. The fourth type is the ventral gland cell type. To these four, two newly identified cell types were added recently: lypophile and crystal cells [99]. This latest study also suggests that gland cells are neurosecretory cells and could control locomotion and feeding behavior. Usually, *Trichoplax* reproduces asexually (dividing and budding) but may also exhibit the sexual reproductive cycle.

Totipotent or pluripotent cells, or any cells that can differentiate into other cell types, have not yet been demonstrated unambiguously in *T. adhaerens*, in contrast to the case of the *Eumetazoa*. However, the expression of Trox-2, which can be considered a candidate for a proto-Hox gene, from which the other genes in this important family could have arisen through gene duplication and variation, is evidenced only in small cells forming a ring around the periphery of *Trichoplax*, located between the outer margins of the upper and lower epithelial cell layers [100]. Importantly, inhibition of Trox-2 function, either by uptake of morpholino antisense oligonucleotides or by RNA interference, causes complete cessation of growth and binary fission of *Trichoplax* [100]. Logically, the authors suggest that inhibition of the Trox-2 function is blocking the supply of differentiated somatic cells derived from Trox-2-expressing multi- or omnipotent stem cells at the periphery of the animal. The existence of these small marginal cells was confirmed by Guidi et al. [101]. These authors consider, in fact, that smaller size and arrangement to form a thick cord around the animal body of small marginal cells are unique characteristics and propose that they represent a new cell type in *Trichoplax*.

10.4 Stemness Features in Protists

Although traditionally, the terms "self-renewal," "differentiation," and "quiescence" are associated with metazoans and the stem cell concept in multicellular context, and the pluripotent stem cell systems supporting asexual reproduction in sponges was traditionally considered [102] as the origin form of stemness, to reach the roots of stemness it is necessary to analyze the protists, their life cycle, and their relationship with the ecological niche. In fact, if nobody questions the fact that all multicellular organisms depend on stem cells for their survival and perpetuation [103], the idea that somatic stem cells may be ancestral, with germ line stem cells being derived later in the evolution of multicellular organisms, is not yet generally accepted. In most cases the appearance of stemness is associated with the appearance of multicellular organisms. Indeed, as discussed above, the stem cells are evidenced in all multicellar organisms from sponges to primates, and their existence is probable in Placozoa. But, are the main features endowing the stemness really appearing first in metazoans?

During the last years, the direct analogies between the clonal aging in protozoa and the decline of "immortality" of somatic totipotent/pluripotent stem cells were established [104]. Also, several interesting papers of Vladimir Niculescu appeared, providing a logical dissection of the life cycle of protists with respect to the basic features of stemness: self-renewal (undifferentiated proliferation "*à l'identique*") and differentiation as well as quiescence. The analysis of the life cycle with respect to the environmental conditions (oxygen content), Niculescu [105] showed that *Entamoeba invadens* possesses all basal mechanisms of stemness observed in higher eukaryotes: in culture, the hatched amoebulae (ex-cysted A-cells) give rise to the primary stem cells, which are capable of asymmetric cell divisions producing two cell types: (1) cells that continue to proliferate asymmetrically and (2) cells that exit the cell cycle assuming a G_1/G_0 phenotype (they can reenter in active cell cycle in secondary cultures). The primary multipotent population contains a single stem cell line consisting of undifferentiated cells (self-renewal) and quiescent cells (mitotic arrest). Cells definitively exiting mitotic cycle (differentiation precursors) differentiate terminally to cysts (Niculescu, personal communication). The primary population (cyst-free during growth) remains undifferentiated [105]; at 96 h of initiation, the primary culture described by Niculescu [105] contains 86% quiescent and 14% proliferative, self-renewing cells. Oxygenation converts the multipotent proliferative cells (primary cells) into oligopotent proliferative cells (secondary cells) that generate precursors for oxygenic differentiation. Less oxygenic conditions convert oligopotent cells to tertiary unipotent subpopulations lacking precursor cells. Tertiary proliferative cells may come back to the primitive symmetrical division in "severely hypoxic conditions" (our estimation: anoxic or nanooxic conditions), revealing first **a slow self-renewing proliferation (full/complete self/renewal**) and finally, the proliferation stop.

Entamoeba cell forms can be classified into three categories with respect to their differentiation capacities and oxygen requirements: the strictly hypoxic, the moderate hypoxic, and the most oxygenic cell forms (the "strictly hypoxic" one according to author's terminology corresponds to the anoxic or nanooxic full self-renewing form—"stem cells"). Two distinct differentiation patterns start terminal differentiation (encystment) in response to an increase in environmental O_2 concentrations: (1) autonomous encystment of secondary mitotic arrested cells (in growing secondary cell populations) and (2) induced encystment of tertiary populations transferred in hypo-osmotic nutrient-free environment ([105]; Niculescu, personal communication). Thus, the life cycle of *E. invadens*, a single-celled eukaryote of anaerobe metabolism, displays all basal mechanisms of stemness and cell differentiation observed in higher eukaryotes, although the differentiation options at its evolutionary level are rather restricted. Niculescu resumes, "The ancient stem cells of *E. invadens* (AnSC) have all standards of the stem cell definition [self-renewing and quiescent cells, cells withdrawing mitotic cell cycle and differentiation precursors (committed progenitors), glycolytic metabolism, stem cell plasticity, reprogramming and induced totipotency recovery, and both induced and autonomous terminal differentiation pathways]" [106]. If we consider the ameboid form of *E. invadens* as a "stem cell entity," then the analogy with the stem cells in sponges—archeocytes (see above) according to Niculescu [105]—is evident: the archeocytes are large ameboid cells exhibiting a high phagocytic activity and proliferative capacity and possessing the capacity of totipotency (of course, this potential includes the choanocytes); in contrast, choanocytes can only either convert back to archeocytes or produce the gametes. This analogy can be fully extrapolated to the Hydroid "I-Cells" as well as to Hydra "I-Cells," which are, however, "only" multipotent in somatic/mitotic terms but reveal their totipotency through the germinative cells in sexual reproduction (see above).

Reinterpreting from the same viewpoint, the literature data for the life cycle of *Giardia lamblia*, especially in relation to the "ecological niche" and the low oxygenation as its major component (which is, in the multicellular organism terminology "microenvironmental stem cell niche"), the author reveals the same principle: the life cycle of single-celled protists provides the basic features of what will be called "stemness" in multicellular organisms [106]. The similar analysis of the life cycle of another protist, *Colpoda cucullus*, reveals that this ciliate exhibits a cell-renewing form that proliferates by asymmetric division, giving rise to cell mitotically arrested cells in addition to proliferative ones. Proliferation and terminal differentiation (encystment) are hypoxia-/oxygen-dependent.

Based on the above, protists as *E. invadens* exhibiting two proliferative cycles (the *mitotic cycle* giving rise to omnipotent/unipotent daughter cells (trophozoites) and the *endopolypoid cell cycle* leading to terminal differentiation and multiple totipotent daughter cells (amoebulae, A-cells)) (Niculescu, personal communication), exhibiting at least two life-forms (vegetative one ("trophozoites") and dormant/resistant form (cyst)) are capable of differentiation but also of self-renewal, presenting "hidden stem cell lineages" in their life cycle [107]. Prominent commensals, pathogens, and also many free-living protists have the described life cycle [108]. It is interesting to mention that, one century ago, it was observed that the large, complex ciliate *Bursaria* adult form dedifferentiate before asexual symmetric division after which the daughter cells again differentiate, reconstructing the "adult" structures [109]. This example seems to be an instructive prototype of the differentiation/dedifferentiation principle.

Since the asymmetrical division, pointed out as the first event in the origin of stem cells [110] (rather of differentiation in our opinion), is widely spread in extant protists, it is highly probable that these features existed in the last eukaryotic common ancestor (LECA), a hypothetical lineage that gave rise to all modern eukaryote supergroups.

10.5 The Oxygen and Stem Cell Entity

Previous chapters discussed the anaerobic/aerophilic nature of mammalian stem cells through the different stem cell systems and discussing in vivo and in vitro data. The various classes of stem cells, are, in fact, "hidden" in low-O_2 niches where they can either divide slowly or stay in quiescent state. This slow proliferation, associated with an anaerobic metabolic profile, seems to be related to the self-renewal. It is highly probable that the "place of cursor" on the energetic metabolism scale, varying from anaerobiosis through the nanoaerophilia and microaerophilia until aerobiosis (see Chapter 8), determines, in fact, if these divisions will be symmetric or asymmetric and which proportion of these two modalities will be established.

In the ancestral stem cell "prototype" systems, this feature remains more or less evident: the bilaterians having the neoblast stem cell system may be aerobes, "facultative respirators," or anaerobes (for review see the comprehensive work of Muller et al. [111]). Of note, most anaerobic organisms do not rely only on glycolysis, but perform also mitochondrial anaerobic respiration using as electron acceptors molecules other than O_2, mainly fumarate, and producing succinate (which in certain organisms can be metabolized to propionate). In those that do not possess an independent circulatory system, O_2 is available only by diffusion though the tissue. Whether they exhibit a parasitic/commensal or free life style, they are not exposed to the high O_2 concentrations in their typical ecological niche. A long time ago, it was noticed in some worms [112] and planarians [113,114] that during mobilization and proliferation of neoblasts, a phase preceding regeneration, the expected respiratory increase lacked. Also, under anaerobic conditions, planarian *Polycelis nigra* produces lactic acid but the production of succinate, acetate, or propionate (usually produced by parasitic platihelmints) was not evidenced [115].

Of note, intact and adult planarians present a predominantly anaerobic metabolism [116]. Furthermore, during starvation, when the O_2 consumption, depending on duration, either decreases or remains stable (did not increase with respect to steady state) [117], which is followed by a growth decrease, the pool of primitive stem cells (totipotent neoblasts) is maintained as well as their proliferation (self-renewal) rate, while committed neoblast progenitor population decreases [118]. On the basis of these indirect data and the fact that, similarly to mammalian adult stem cells, they possess only few mitochondria [13,119,120], it might be concluded that neoblasts (at least the most primitive subpopulation) probably exhibit an anaerobic/microaerophilic metabolic character. However, the up-to-date technology-based energetic metabolism analyses in different neoblast subpopulations (see above) are still lacking.

In cnidarians, which also lack a real circulatory system (the tiny tissue enables the minimal diffusion distances), the gas exchange occurs across the internal and external body surfaces. The respiration can be either aerobic or facultative anaerobic (especially in anemones, which are buried in soft sediments). In steady state, the Hydra respires at 34 μl O_2/mg protein nitrogen/h, which is comparable with the respiration rates of higher animals [121]. This steady state respiration rate is independent of (1) pH between 4 and 10, (2) osmotic pressure below 350 mm Hg., and (3) gas-phase oxygen concentration between 10% and 100% with respect to one in atmosphere air., 2.1% and 21% O_2 (44.1–160 mm Hg) [121]. Concerning an individual polyp, the proximal-distal gradients in hydra were examined, and these gradients suggested a switch to a more anaerobic type of metabolism and an elevation of the pentose phosphate pathway as the basal region was approached [122]. It is particularly interesting to mention that, more than seven decades ago, Needham [123] published data showing that in Hydra (*H. littoralis*) the regeneration process is preceded by a decreased O_2 consumption. The respiration rate measurement in some Hydra species is complicated due to the presence of the symbiotic photosynthetically active green algae in their digestive cells [124]. Unfortunately, there is no current data concerning the energetic metabolism in Hydra and particularly in its stem cells (especially in "I-cells"). The tissue of Hydra contains the antioxidative enzymes [125,126], but there are no data on their activity specifically in "I-cells."

Concerning the hydroid polyps, an old study suggested that the metabolic processes seem to be the consequences of the morphogenetic processes but not vice versa [127]. However, the modern studies stress that redox state and reactive oxygen species (ROS) are very important factors in colony hydroids, obviously acting as the adaptative mechanisms during the feeding and environmental changes, but also in the steady state growth and development [128,129]. Treatments of colonial hydroids with azide (an inhibitor of complex IV) lead to relatively reduced redox states and do not arrest the growth but rather change the shape of hydroids (results in "runner-like growth"); in contrast the treatment with dinitrophenol (an uncoupler of oxidative phosphorylation) leads to relative oxidation and sheet-like growth [130]. While the treatment with some antioxidants (vitamin C) has been shown to trigger stolon regression and sheet-like morphology [131,132], the other antioxidants (glutathione and reduced glutathione ethyl ester) treatment results, paradoxically, in higher levels of ROS and, consequently, in a runner-like growth [133]. In fact, whereas moderate levels of ROS may lead to rapid colony growth and a runner-like form, high levels of ROS are involved in stolon regression, which, in turn, may lead to more sheet-like growth [134–136]. While all these data imply that the energetic metabolism, ROS, and NOS aspects influence differentiation, commitment, and self renewal of hydroid stem cells, the studies dealing with this specific issue, as well as those dealing with the respiration type of "I-cells," the most ancestral full-stem cell prototype model characterized until today, are lacking.

Sponges are adapted to aerobic conditions, but display a remarkable heterogeneity in the metabolic intensity between the species. Curiously, a *Petrosaspongia mycofijiensis* marine sponge contains mycothiazole, a molecule that inhibits hypoxia-inducible transcription factor-1 (HIF-1) [137], implying a particular way of the energetic metabolism regulation (HIF-1 was evidenced in sponges in 2009 [138]). Under 1.5–2.0 ppm dissolved oxygen concentration (45.6–69.8 μM), desmosponge *Haliclona pigmentifera* show adherent growth for 42 ± 3 days [138]. However, in nature, sponges were observed in nano- and microaerobic environments (oxic–anoxic interface of Saanich Inlet salt water on Vancouver Island) [139] of microaerophylic environments (4.5–7.2 μM dissolved O_2 concentration) [140]. A recent study [141] demonstrated that the modern desmosponge *Halichondria panice* can survive under low-oxygen conditions corresponding to 0.5–4.0% of present atmospheric levels (PAL); that is, 0.1–0.84% O_2 in gas phase (at 20 °C; standard atmospheric pressure the dissolved O_2 concentration would be 1.35–11.31 μM). Furthermore, in the sponge *Geodia barretti*, internal milieu environment (in which reside the anaerobic bacteria including the sulfate-reducing ones) represents an anoxic microecosystem [142]. Also, the Mediterranean sponge *Aplysina aerophoba* can stop pumping for several hours or even days, developing an anoxic internal environment [143]. Although the technique for isolation and ex vivo cultivation of archeocytes was published [144], the direct data revealing the metabolic properties of these cells are still lacking.

Concerning the Placozoa, it was shown that incubation of *T. adhaerens* in hypoxic seawater (5% atmospheric O_2) did not affect viability for 42 days; at 3%,

2%, and 1% atmospheric O_2, *T. adhaerens* survived in seawater for ~1.5 days, >16 and ~5 h, respectively [145], showing that the modern *T. adhaerens* is an aerobic organism. In this simplest animal, a complete oxygen sensing/effecting system was found: the oxygen sensing protein proline hydroxylase domain (PHD) enzyme, the hypoxia response protein called hypoxia-inducible transcription factor (HIF) that can be switched on or off by PHD, and a "trash-tagging" protein called von Hippel Lindau protein [145]. HIF induces the expression of more than 100 genes that are required to increase oxygen delivery and to reduce oxygen consumption, but also to regulate the energetic metabolism. HIF-1α plays a critical role in the growth factor-dependent regulation of both anaerobic glycolysis and oxidative phosphorylation. In fact, when the O_2 is low, HIFα is fully expressed. It orients the metabolism toward anaerobic glycolysis, a basic energy-producing way inherited from the ancestral predecessors. Thus, it strongly suggests that the degradation of HIFα upon O_2 sensing, from an evolutionary viewpoint, represents the adaptation to oxygen. This perception was reversed considering modern atmospheric O_2 concentration as "normoxia" and taking the highly oxygen-adapted cells (those from mammals) as a start point (discussed in Chapter 2). When the PHD from *T. adhaerens* (taPHD) were expressed ectopically in human cells, all the constructs reduced HIF-1α levels to a similar extent in response to low O_2 concentration, in which they are reduced by endogenous "human" PHD, in spite of the fact that *T. adhaerens* (taHIFα) oxygen-dependent degradation domain differs substantially from human HIF-1α) [145]. This conservation of function for at least 580 million years of evolution witnesses the importance of the O_2 sensing/responding system for multi-cellular organisms. So far, the data concerning the respiration and metabolism of hypothetical stem cells in the "peripheral ring" of *Trichoplax* were not published.

A very interesting finding [146] connects the Placozoa with the other eight phyla of basal animals (Porifera, Cnidaria, Mollusca, Annelida, Nematoda, Echinodermata, Hemichordata, and Chordata): an ubiquinol oxidase named "alternative oxydase." This enzyme introduces a branch point into the respiratory electron transport chain, bypassing complexes III and IV and resulting in cyanide-resistant respiration [147]. The presence of this enzyme in these animals, obviously an ancestral feature lost in vertebrates, might witness another ancestral O_2-related adaptation mechanism.

Finally, the relationship between anaerobiosis/nanoaerobiosis/microbiosis and "stemness" seems to be obvious in protozoa conserving the features of "hidden stemness": the primitive nondifferentiated forms capable of full/complete self-renewal appeared only in anoxic/nanoaerobic environmental conditions. Advanced oxygen depletion, for example, switch hypoxic forms of *E. invadens* from asymmetric to symmetric divisions (full/complete self-renewal) and finally to proliferation stop (anoxic/nanooxic cell type) [108]. Slight O_2 increase reverses this conversion process. Most cell types of *E. invadens* prefer higher oxygenation contents, leading to fast cycling (i.e., partial self-renewal) and autonomous differentiation of the cell type exiting definitively the mitotic cell cycle [108]. The symmetrical self-renewing divisions in nanooxic conditions are of paramount interest for comprehension of the self-renewal phenomenon conserved from protist to mammals.

10.6 The Oxygen Evolutionary Paradigm and Stemness

Studying the respiration of water-breathers, Massabuau concluded that the physiological mechanisms in all aquatic animals have the goal to maintain, independent of environmental changes, the constant tissue O_2 levels in a low, narrow range of 1–3 kPa (7.5–21 mm Hg or 1–3%) [148]. Of note, in some mammalian tissues, the similar low O_2 concentration range was measured (see Chapter 2). As this low O_2 concentration range was found in tissues of very primitive and archaic organisms (e.g., it did not change in *Cilindroleberididae*) since Paleozoic [149], it seems that it is a consequence of an early adaptation strategy [148]. Analyzing the appearance of specialized enzymes ensuring the antioxidative activity, as well as other factors, he concludes that the cell oxygenation status represents a primitive feature that did not change from the appearance of the water-breathing organism until today [150]. This "low O_2 tissue strategy" [151] enabled primitive and modern water-breathers to maintain pO_2 in their internal environment at levels similar to those when their evolution started (>3%) [148,150,152] (see Chapter 8). This strategy is elaborated primarily by the ventilatory regulation, allowing to maintain a constant range of pO_2 in arterial blood. The more primitive species, lacking any mechanism that can influence ventilation, use behavioral rather than physiologic strategy, avoiding the O_2-rich areas or migrating through the O_2 gradient to less oxygenated sediment layers [149,152].

In mammalian organs, the intracellular O_2 concentration is in the range of 5–25 μM (4–20 mm Hg or 0.5–3.3%) and the critical concentration below which respiration becomes anaerobic is in the range of 2–6 μM (1.6–4.8 mm Hg or 0.2–0.63%) (reviewed in Boveris et al. [153]). (Note that these values are practically in range of the "Pasteur point"; see Chapter 8.) The "low O_2 strategy" (i.e., maintenance of the physiologically low tissue pO_2 range in mammals) is extremely elaborated: it is performed not only by the regulation of ventilation, but also by the control of central and local blood flow and density adjustment of the peripheral microartery bed [152]. Doing so, O_2 tissue flux may vary by a factor of 20 with the same arterial pO_2 and nearly constant O_2 tissue concentration (which is in function of O_2 consumption and supply) (reviewed in [154]). These considerations are in line with those of Massabuau cited below, derived from the physiology of water-breathers: in contrast to the traditional idea about O_2 transport, which is focused only on the adequacy of supply, the oversupply should not be ignored [154]. This notion stresses at which point the ancestral, Paleozoic "standard," established 400–450 million years ago, is important, as well as why it is maintained until now. The ventilation and circulation system serves, in fact, to maintain this low O_2 range at the cellular level, to enable the "low O_2 strategy" rather than to supply as much O_2 as efficiently possible to the tissues.

If the tissues (most of the cells making a tissue are mature cells) are characterized by aerobic metabolism in at least microoxic conditions (above Pasteur point), then the stem cells, which reside in the niches, are characterized with particularly low O_2 concentrations (see Chapter 3) and exhibiting the anaerobic metabolic properties (see Chapter 6) have to be exposed to nanooxic or even anoxic conditions. If the overall tissue O_2 homeostasis was conserved from Paleozoic era, then the O_2 homeostasis of stem cells should be conserved from Proterozoic, at least during the last 2 billion years (see Chapter 8, and specifically Figure 8.1). Facing these conditions, the stem cells have to deal with the O_2 concentrations below the Pasteur point, as was the case of the first eukaryotes. This point again connects the phenomenon of stemness to Proterosoic and to the ancestral eukaryotes.

On the basis of anaerobic and nanoaerophilic "connection" with the maintenance of human stem cells, we hypothesized in 2009 [155] that the self-renewing division is nothing other than a simple division *à l'identique*, a replication based on a low-energy (slow) proliferation reminiscent of first single-celled eukaryotic ancestors that were not completely adapted to atmospheric oxygen and still did not accumulate, through the evolution, the molecular extensions enabling the functional and morphological changes called "differentiation" [155–157]. In other words, in a human cell showing the "pluripotency," which exhibits a full differentiation potential accumulated during evolution from the first eukaryotic common ancestor (FECA) until humans, the expression of all genes that allow activation of differentiation programs should be inhibited while only the "minimal essential genome" enabling the elementary life functions and proliferation (replication) is expressed and activated. Thus, our "oxygen stem cell paradigm" related the self-renewing stem cell to the simple asexual (mitotic) reproduction of the earliest eukaryotes [155]. Following this logic, we should examine the up-to-date knowledge concerning these first eukaryotes, notably the LECA and FECA.

LECA most probably possessed a mitochondrion and a modern nucleus and a complex membrane-trafficking system with a near modern array of organelles and representatives of almost all the major protein families, a cytoskeleton, and even a flagellum. It was able to undergo meiosis (sexual reproduction) to perform the phagocytosis and had an almost modern eukaryotic cell cycle regulation system [158]. So, the LECA probably had a Choanoflagellata-like—already a differentiated—phenotype. Thus, it probably had a life cycle composed of two to three forms allowing the adaptation with respect to the oxygen (anoxic, nanooxic, and maybe microoxic, ecologic niches) and availability of the nutrients, just as it is the case of the existent protozoa (of note, the modern Choanoflagellata may undergo the encystment [159] and possess the meiosis-specific genes [160]). By analogy, one of these forms of LECA life cycle had a nondifferentiated (ameboid?) aspect and had been slow dividing and anaerobic; that is, self-renewing form. As the similar forms in modern day protists (as *E. invadens* for instance [108]), this nondifferentiated form was capable of symmetric division in anoxic or nanooxic conditions (see Table 10.1).

It is interesting to mention in this context, the 570-million-year-old fossils previously interpreted as "embryos of some of the earliest animals" (i.e., as cnidarian and bilaterian gastrulae) [161]. Later, a very consistent interpretation [162] suggested that these fossils represent a developmental pattern comparable with nonmetazoan holozoans and belong either to total-group Holozoa (the clade that includes Metazoa, Choanoflagellata, and Mesomycetozoea) or even to more distant branches in the eukaryote tree. These ancestral protists exhibited a life cycle comprising the hypertrophy of a single-cell free-living organism followed by encystment and consequent formation of a cell cluster by palintomic divisions (cell *division* without a preceding compensatory increase in size) inside an enclosed multilayered envelope with tuberculate surface texture (Megasphaera). This internal envelope was itself enveloped in a process-bearing cyst wall [162]. Eventually, when the symmetrical divisions result in hundreds of thousands of tightly packed cells, the outer envelope wall is ruptured, and the more pliable inner wall with its content emerged in fingerlike protrusions. The cells resulting from the cleavage process escaped as propagules, through dissolution of the inner wall [162]. Thus palintomic cleavage (although precluding the embryonic development of animals) results in dissemination of single-cell eukaryotes (in metazoans, these cells would remain together to form a multicellular organism). It is obvious that these divisions, although without proportional cytoplasmic volume increase, represent, in fact, the self-renewing divisions producing the new single-cell eukaryotes in a form of undifferentiated free-living cell forms, which are, in turn, going to differentiate (hypetrophy and encystment), probably in response to the ecological niche factors.

Since the geological period in question is characterized with the increase in microoxic ecological niches (we believe that the life cycle described resulted as adaptation to increased oxygenation of environment; Figure 10.1). In fact, the encystment enabled

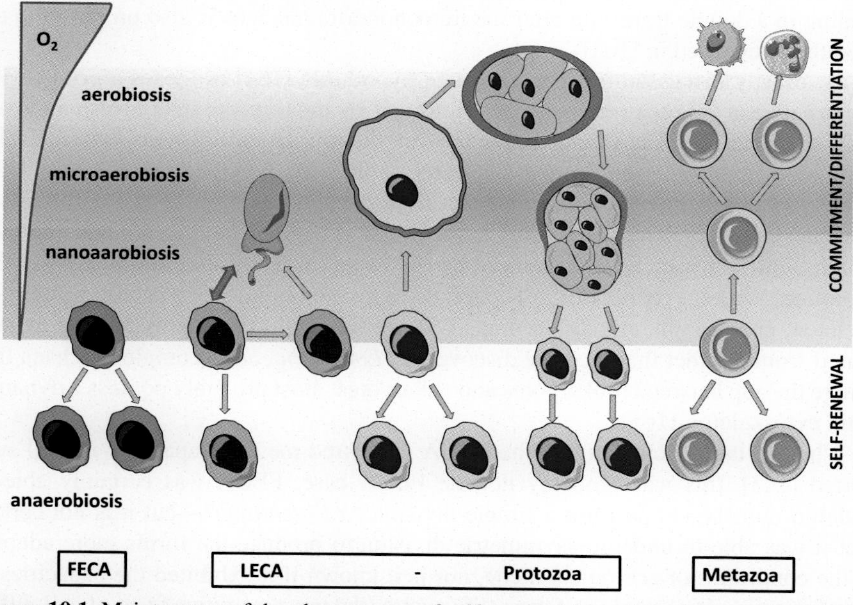

Figure 10.1 Maintenance of the phenomenon of self-renewal through evolution and analogies between the life cycle of protists with the commitment and differentiation in metazoans. The undifferentiated cell symmetric, self-renewing divisions are related to anaerobic metabolism and anoxic/nanooxic environment throughout the whole evolution period; the asymmetric divisions are related to the availability of microoxic niches to the daughter cells. To preserve self-renewing divisions in the life cycle of some protists appeared encystment, allowing isolation from the outside environment and maintaining an anoxic/nanooxic microenvironment necessary to self-renewing divisions but with diminution in cell size (palintomic cleavage). This "differentiation" is related to increasing O_2 availability, and a big number of molecular/functional extensions appeared and accumulated in the course of evolution, representing the differentiation potential of the modern stem cells. The paleo-prototypes of self-renewal and differentiation are still visible in the highest organisms, mammalians. FECA, First Eukaryotic Common Ancestor; LECA, Last Eukaryotic Common Ancestor.

an anoxic (or may be nanooxic) environment, providing the conditions for replication of initial single-cell living eukaryote in form of a nondifferentiated eukaryotic cell, enabling the basic mitotic division of a most ancestral eukaryote. We consider that this paleo-proliferation pattern of the low-energy, slow cell division resulting in two identical daughter cells might be inherited from another hypothetic eukaryotic ancestor: FECA. This hypothetical organism was the first true eukaryote (i.e., nucleated cell) and was a direct descendent of a prokaryote and predated LECA. One of the theories considers that the FECA acquired not only nucleus, but also mitochondria (which was a facultatively anaerobic α-proteobacterium, see Chapter 9) and possessed the internal membrane system (reviewed in [163]). The two major models, the syntrophic and phagotrophic, suggest that the mitochondrion was the first event or that, as the latter model suggests, evolution of the nucleus and more complex intracellular structures occurred prior to phagocytosis of the mitochondrial ancestor. A third complex path that incorporates additional evolvable systems like a sophisticated cytoskeleton leading to a double transition after the mitochondrion/nucleus is also proposed in the literature (reviewed in [163]).

As already discussed in Chapter 8, one hypothesis [164] proposes a prior fusion event between archaea (seems to be confirmed by most recent data [165]) and bacteria, to produce a FECA, which then took on the mitochondrial endosymbiont. This hypothesis represents, in fact, a synthesis of the two previous models—the third model. Thus, all models agree that the earliest eukaryotes possessed a functional mitochondrion, a nucleus, and other intracellular structures, but there is no consensus which of these features were acquired by FECA and which during the FECA–LECA evolution. We believe that FECA had a fluid outer membrane and a prototype of cytoskeleton, enabling the nondifferentiated "ameboid phenotype." This may be extrapolated from the fact that recently discovered "*Lokiarchaeota*," complex archaea that bridge the gap between prokaryotes and eukaryotes, most probably possess a dynamic actin cytoskeleton [165].

Whereas the conclusion that the LECA possessed meiotic capability is well supported [163], this was probably not the FECA case. FECA was certainly able to undergo mitosis—to perform a simple division "*à l'identique*"—but it is not certain that it was able to undergo asymmetric division to produce the forms more adapted to the conditions of ecological niche, nor is it known if it exhibited the capacities of transformation in several life forms (life cycle) (the most primitive form of cell differentiation). Regardless of the case that was taking place (the simple proliferation; i.e., symmetrical mitotic division, asymmetric division providing one identical daughter cell and another that was different (primitive prototype of differentiation)), FECA still provides a prototype of the self-renewal principle. It is also probable that the FECA was, due to its lower level of complexity (exhibiting aspect of a nondifferentiated cell), more dependent on the ecological niche conditions. We believe that the results obtained with the ameboid form of *E. invadens* that self-renews (undergoes symmetrical division) in anaerobic conditions [105] approximate the hypothetical FECA entity. Furthermore, this simple proliferation without activating the extensions providing disposal for differentiation (commitment) represent the prototype of self-renewal maintained in stem cell systems through the evolution.

Therefore, by pointing out to the FECA, LECA and ancestral eukaryotes in the quest for the roots of stemness (self-renewal and differentiation), a half-century later, we are revisiting the basic idea of *E. Haeckel* that the ancestral single-celled organisms represents the *Stammzellen* (stem cells), the departure point of the evolution of multicellular organisms [166].

10.7 Integrative Model of Stemness

The morphofunctional organization of somatic stem cells during the animal evolution, starting from the most primitives (as reviewed in this chapter for sponges, cnidarias, and worms and confirmed by a Russian group who also extended their analyses to Arthropoda and Chordata [167]) is based on the same principles: (1) ability of self-renewal and (2) differentiation capacity (mammalians HSC as an example, [168,169]). As proposed above, the self-renewal should represent nothing else but division without "unlocking" molecular extensions, which will allow the differentiation. The differentiation capacity, however, depends on the accumulation of the extensions for new functions during the evolution: in protozoa these are hypertrophy and encystment (see Figure 10.1), while in metazoans the number and complexity of these extensions increase exponentially, leading to the diversification of the new cells types as it was previously discussed for placozoa, sponges, cnidaria, worms, and mammals.

The complexity level of a stem-cell system at one evolutionary stage (e.g., totipotency of c-neoblasts in Planaria) [40] can correspond to the complexity of another tissue stem-cell system in the higher animals (e.g., HSC system in mammals) which is, in the context of a mammalian organism complexity, only multipotent. Indeed, these two systems have comparable differentiation potentials since both stem cells, HSC and c-neoblast ("sigma neoblast") [40], are capable of producing approximately the same number of different cell types; they also respond in a similar way to insults, are capable of migration, and of giving the progeny (committed progenitors) to regenerate the totality of tissue (see above). However, whereas a planaria is an individual biological unit (organism), the hematopoietic tissue is not—it only represents one of physiological systems in a more complex organism. According to the actual knowledge, unlike during blastulation, the stem cells in adult organisms of highest animals do not exhibit the totipotent potential (capacity to generate the whole organism), and very rarely the pluripotent one (capacity to generate all tissues of an organism). Since all cells of an organism have the same genome, all cells must have the totipotent potential, which is not, however, exhibited in the physiological conditions. This is confirmed by the reproduction of mammals by nuclear transfer ("cloning") in which the nucleus from mature cells are inserted in an oocyte being reprogrammed until the totipotency [170]. Also, the reprogramming by gene transfer results in transformation of mature cells to pluripotent stem cells (induced pluripotent stem cells, iPSCs) [56,171]. Thus, to enable self-renewal, the differentiation potential in a stem cell has to be "locked" and "silenced" ("…suppressing expression of genes that confer lineage commitment and/or tissue specificity" [172]) to allow the simple mitotic division of primitive, ancestral type.

During the divisions resulting in totipotency or pluripotency, practically all extensions gained during 2 million years of evolution should be locked and silenced to allow a symmetrical self-renewing division. Some unlocked extensions, if they are still "inhibited," will produce the stem cells where differentiation potential is reduced (multipotent, oligopotent, and unipotent), which can explain the phenomenon of self-renewal of these stem cell categories, as shown for myeloid-restricted progenitors with long-term repopulating activity [173]. In physiological conditions, these cells (e.g., HSC) will be limited only to hematopoietic descendence since only the "hematopoietic" repertoire was unlocked in these cells. However, to execute a differentiation program, the additional regulation is needed, which is highly dependent on extrinsic factors (cytokines, enough O_2, cell–cell interactions, even mechanical factors [174], and so on). If these factors were not met, the HSC will self-renew at its level of primitiveness. Since "unlocking" and activating depends on the microenvironment, the different microenvironment niches into which the cells after division were directed can explain the asymmetric division and partial self renewal, but an "intrinsic" mechanism can also operate (reviewed in [175]). Following this reasoning, we can derive the conclusion that, to obtain the high proliferative capacity and, hence, a long duration of a stem cell lineage, the mechanisms aimed to "lock," "inhibit," and "brake" disposal for differentiation and differentiation itself have been developed through the evolution in parallel (or soon after) acquiring of these extensions (i.e., mechanisms leading to the differentiation). Since the self-renewal is documented through the evolution of eukaryotes we believe that these mechanisms represent the features of the basic stem cell life and operate in the basic primitive conditions (i.e., maintaining stemness in these conditions). In that context acts the oxygen concentration: the absence or weak presence of O_2 establishes the conditions reminiscent of a paleo-ecological niche, induces the mechanisms locking and inhibiting the processes leading to differentiation and/or, simply, prevents the energy-consuming processes leading to differentiation simply by unavailability of energy. We believe that this is the frame allowing us to understand the multiple, complex, and redundant molecular mechanism related so far to the "pluripotency."

As already suggested in a previous discussion, the stem cell commitment and differentiation in a complex organism somehow (although not literary) mimics the acquisition of new cell functionalities (hence, new molecular extensions) during the evolution. The point of passage from self-renewal to commitment and consequent differentiation is reminiscent of evolution from FECA (simple mitotic divisions and reproduction "*à l'identique*"; i.e., self-renewal) to LECA and maybe to divergence of eukaryotes and all other forthcoming acquisitions of molecular extensions during the colonial unicellular evolution and early evolution of metazoans. For each of these extensions, a molecular mechanism had to develop to "lock" its execution in order to allow the self-renewing division and consequently, perpetuation of the lineage (which would be, without this mechanism, rapidly exhausted). According to this hypothesis, the commitment and differentiation are more or less spontaneous processes if energy is available; that is, they mainly depend on extrinsic factors (as was the acquisition of the new functions during the evolution, in function of adaptation to microenvironment) while the self-renewal cannot be realized without active "locking," "inhibiting," and "braking" of these molecular extensions leading to the differentiation.

As previously reviewed, the degree of the maintenance of "primitiveness" (referring to a combination of self-renewal ability and differentiation capacity) of a stem cell is associated with anaerobiosis, nanoaerobiosis, and microaerobiosis (as it is the evolution FECA-LECA-Metazoa), the mechanisms controlling "locking/unlocking" of a perspective for differentiation are obviously related to metabolic/energetic type and the oxygen-sensing [176]. Each of these mechanisms is related to the point in which it appeared in evolution and the ecological-niche conditions that allowed/favored its appearance. These conditions have to be, in general, conserved until today. Concerning the self-renewing division, once enabled, it should rely on expression of those genes that are strictly necessary for simple anaerobic (low-energy) cell cycle regulation and survival. This part of the genome we consider "minimal essential eukaryote genome," and it should not be much changed from LECA to human. Of course, the simultaneous expression of the "lockers," "inhibitors," and "brakers" will mask this minimal essential genome, demanding the evolutionary follow-up for each mechanism to reveal the most ancestral ones. For example, the transcription factors as Fox, Hox, Sox, POU family (includes Oct), PAS, and so on, as well as the parts of signaling pathways as Wnt (β-catenin), Notch (some corposants), TGfβ (some signaling composants), MAP kinases, and so on, appeared in protists and still serve in mammalian cells and are related to the stemness. We hypothesize that these factors appeared in relation to the diversification of life cycle in protist (presenting the prototypes of self-renewal, differentiation, and dedifferentiation) and may belong to the "lockers," "inhibitors," and "brakers," but some of them may also be part of the minimal essential genome.

References

[1] Morrison SJ, Kimble J. Asymmetric and symmetric stem-cell divisions in development and cancer. Nature 2006;441:1068–74.

[2] Morrison SJ, Hemmati HD, Wandycz AM, Weissman IL. The purification and characterization of fetal liver hematopoietic stem cells. Proc Natl Acad Sci USA 1995;92:10302–6.

[3] Zhang R, Zhang Z, Zhang C, Zhang L, Robin A, Wang Y, et al. Stroke transiently increases subventricular zone cell division from asymmetric to symmetric and increases neuronal differentiation in the adult rat. J Neurosci 2004;24:5810–5.

[4] Dumont NA, Wang YX, Rudnicki MA. Intrinsic and extrinsic mechanisms regulating satellite cell function. Development 2015;142:1572–81.

[5] Balenci L, van der Kooy D. Notch signaling induces retinal stem-like properties in perinatal neural retina progenitors and promotes symmetric divisions in adult retinal stem cells. Stem Cells Dev 2014;23:230–44.

[6] Dahl MV. Stem cells and the skin. J Cosmet Dermatol 2012;11:297–306.

[7] Zucali JR. Self-renewal and differentiation capacity of bone marrow and fetal liver stem cells. Br J Haematol 1982;52:295–306.

[8] Elliott SA, Sánchez Alvarado A. The history and enduring contributions of planarians to the study of animal regeneration. Wiley Interdiscip Rev Dev Biol 2013;2:301–26.

[9] Pallas PS. Spicilegia zoologica: quibus novae imprimis et obscurae animalium species iconibus,descriptionibus atque commentariis illustrantur. Berolini: Prostant apud Gottl. August. Lange; 1774.
[10] Wagner F. Zur Kenntnis der ungeschlechtlichen Fortpflanzung von Microstoma nebst allgemeinen Bemerkungen über Teilung und Knospung im Tierreich. Z Jahrb Abth F Anat u Ontog D Thiere 1890;4:349–423.
[11] Keller J. Die ungeschlechtliche Fortpflanzung der Südewasser-Tubellarien. Jen Zeits Naturw 1894;28:370–407.
[12] Flexner S. The regeneration of the nervous system of *Planaria torva* and the anatomy of the nervous system of double-headed forms. J Morphol 1898;14:337–46.
[13] Morita M, Best JB, Noel J. Electron microscopic studies of planarian regeneration. I. Fine structure of neoblasts in *Dugesia dorotocephala.* J Ultrastruct Res 1969;27:7–23.
[14] Buchanan JW. Regeneration in *Phagocata gracilis* (Leidy). Physiol Zool 1933;6:185–204.
[15] Wolff ET, Dubois F. Sur une methode d'irradiation localisee permettant de mettre en evidence la migration des cellules de regeneration chez les Planaires. C R Soc Biol Paris 1947;141:17–8.
[16] Randolph H. Regeneration of the tail in Lumbriculus. Zool Anz 1891;14:154–6.
[17] Dubois F, Wolff E. Sur une méthode d'irradiation localisée permettant de mettre en évidence la migration des cellules de régénération chez les planaires. Soc Biol Strasbg 1947;141:903–9.
[18] Dubois F. Contribution a l'etude de la migration des cellules de regeneration chez les planaires dulcicoles. Bull Biol 1949;83:213–83.
[19] Boggs DR, Chervenick PA. Migration of transplanted hematopoietic stem cells to the spleen of irradiated mice. Transplantation 1971;112:191–2.
[20] Petrov RV, Khaitov RM. Migration of stem cells from screened bone marrow following non-uniform irradiation. Radiobiologiia 1972;12:69–76.
[21] Croizat H, Frindel E, Tubiana M. The effect of partial body irradiation on haemopoietic stem cell migration. Cell Tissue Kinet 1980;13:319–25.
[22] Baguñà J, Saló E, Auladell MC. Regeneration and pattern formation in planarians. III. Evidence that neoblasts are totipotent stem-cells and the source of blastema cells. Development 1989;107:77–86.
[23] Reddien PW, Sánchez Alvarado A. Fundamentals of planarian regeneration. Annu Rev Cell Dev Biol 2004;20:725–57.
[24] Reddien PW, Oviedo NJ, Jennings JR, Jenkin JC, Sánchez Alvarado A. SMEDWI-2 is a PIWI-like protein that regulates planarian stem cells. Science 2005;310:1327–30.
[25] Eisenhoffer GT, Kang H, Sánchez Alvarado A. Molecular analysis of stem cells and their descendants during cell turnover and regeneration in the planarian Schmidtea mediterranea. Cell Stem Cell 2008;3:327–39.
[26] Newmark PA, Sánchez Alvarado A. Bromodeoxyuridine specifically labels the regenerative stem cells of planarians. Dev Biol 2000;220:142–53.
[27] Salvetti A, Rossi L, Deri P, Batistoni R. An MCM2-related gene is expressed in proliferating cells of intact and regenerating planarians. Dev Dyn 2000;218:603–14.
[28] Lange CS. A quantitative study of the number and distribution of neoblasts in *Dugesia lugubris* (Planaria) with reference to size and ploidy. J Embryol Exp Morphol 1967;18: 199–213.
[29] Watanabe Y. The development of two species of *Tetilla* (Demosponge). Nat Sci Rep Ochanomizu Univ 1978;29:71–106.
[30] Till JE, McCulloch EA. A direct measurement of the radiation sensitivity of normal mouse bone marrow cells. Radiat Res 1961;14:213–22.

[31] Lapan SW, Reddien PW. dlx and sp6-9 control optic cup regeneration in a prototypic eye. PLoS Genet 2011;7(8):e1002226.
[32] Scimone ML, Srivastava M, Bell GW, Reddien PW. A regulatory program for excretory system regeneration in planarians. Development 2011;138:4387–98.
[33] Cowles MW, Brown DD, Nisperos SV, Stanley BN, Pearson BJ, Zayas RM. Genome-wide analysis of the bHLH gene family in planarians identifies factors required for adult neurogenesis and neuronal regeneration. Development 2013;140:4691–702.
[34] Currie KW, Pearson BJ. Transcription factors lhx1/5-1 and pitx are required for the maintenance and regeneration of serotonergic neurons in planarians. Development 2013;140:3577–88.
[35] Hayashi T, Shibata N, Okumura R, Kudome T, Nishimura O, Tarui H, et al. Single-cell gene profiling of planarian stem cells using fluorescent activated cell sorting and its "index sorting" function for stem cell research. Dev Growth Differ 2010;52:131–44.
[36] Moritz S, Stöckle F, Ortmeier C, Schmitz H, Rodríguez-Esteban G, Key G, et al. Heterogeneity of planarian stem cells in the S/G2/M phase. Int J Dev Biol 2012;56:117–25.
[37] Hayashi T, Asami M, Higuchi S, Shibata N, Agata K. Isolation of planarian X-ray-sensitive stem cells by fluorescence-activated cell sorting. Dev Growth Differ 2006;48:371–80.
[38] Higuchi S, Hayashi T, Hori I, Shibata N, Sakamoto H, Agata K. Characterization and categorization of fluorescence activated cell sorted planarian stem cells by ultrastructural analysis. Dev Growth Differ 2007;49:571–81.
[39] Verdoodt F, Willems M, Mouton S, De Mulder K, Bert W, Houthoofd W, et al. Stem cells propagate their DNA by random segregation in the flatworm *Macrostomum lignano*. PLoS One 2012;7(1):e30227.
[40] van Wolfswinkel JC, Wagner DE, Reddien PW. Single-cell analysis reveals functionally distinct classes within the planarian stem cell compartment. Cell Stem Cell 2014;15:326–39.
[41] Rosendaal M, Hodgson GS, Bradley TR. Organization of haemopoietic stem cells: the generation-age hypothesis. Cell Tissue Kinet 1979;12:17–29.
[42] Wang B, Collins 3rd JJ, Newmark PA. Functional genomic characterization of neoblast-like stem cells in larval *Schistosoma mansoni*. Elife 2013;2:e00768.
[43] Collins 3rd JJ, Wang B, Lambrus BG, Tharp ME, Iyer H, Newmark PA. Adult somatic stem cells in the human parasite *Schistosoma mansoni*. Nature 2013;494:476–9.
[44] Yoshida-Noro C, Tochinai S. Stem cell system in asexual and sexual reproduction of *Enchytraeus japonensis* (Oligochaeta, Annelida). Dev Growth Differ 2010;52:43–55.
[45] Sugio M, Yoshida-Noro C, Ozawa K, Tochinai S. Stem cells in asexual reproduction of *Enchytraeus japonensis* (Oligochaeta, Annelid): proliferation and migration of neoblasts. Dev Growth Differ 2012;54:439–50.
[46] Stephan-Dubois F. Activation and migration of neoblasts in the caudal regeneration of *Lumbriculus variegatus* from healthy or irradiated regions. C R Seances Soc Biol Fil 1953;147:886–90.
[47] Myohara M. What role do annelid neoblasts play? A comparison of the regeneration patterns in a neoblast-bearing and a neoblast-lacking enchytraeid oligochaete. PLoS One 2012;7(5):e37319.
[48] Potswald HE. The relationship of early oocytes to putative neoblasts in the serpulid *Spirorbis borealis*. J Morphol 1972;137:215–27.
[49] Trembley A. Mémoires pour servir à l'histoire d'un genre de polypes d'eau douce, à bras en forme de cornes. Leiden: Jean and Herman Verbeek; 1774. Quatrième mémoire: 229–322.
[50] Hou XG, Stanley GD, Zhao J, Ma XY. Cambrian anemones with preserved soft tissue from the Chengjiang biota, China: Lethaia 38:193–203.

[51] Cartwright P, Halgedahl SL, Hendricks JR, Jarrard RD, Marques AC, Collins AG, et al. Exceptionally preserved jellyfishes from the Middle Cambrian. PLoS One 2007;2(10):e1121. 31.
[52] Erwin DH. Wonderful Ediacarans, wonderful cnidarians? Evol Dev 2008;10:263–4.
[53] Gold DA, Jacobs DK. Stem cell dynamics in Cnidaria: are there unifying principles? Dev Genes Evol 2013;223:53–66.
[54] Dübel S, Hoffmeister SA, Schaller HC. Differentiation pathways of ectodermal epithelial cells in hydra. Differentiation 1987;35:181–9.
[55] Galliot B. Injury-induced asymmetric cell death as a driving force for head regeneration in Hydra. Dev Genes Evol 2013;223:39–52.
[56] Takahashi K, Yamanaka S. Induction of pluripotent stem cells from mouse embryonic and adult fibroblast cultures by defined factors. Cell 2006;126:663–76.
[57] Wittlieb J, Khalturin K, Lohmann JU, Anton-Erxleben F, Bosch TC. Transgenic hydra allow in vivo tracking of individual stem cells during morphogenesis. Proc Natl Acad Sci USA 2006;103:6208–11.
[58] Buzgariu W, Crescenzi M, Galliot B. Robust G2 pausing of adult stem cells in hydra. Differentiation 2014;87:83–99.
[59] David CN. A quantitative method for maceration of hydra tissue. Wilhelm Roux' Archchiv Entwicklungsmechanik der Org 1973;171:259–68.
[60] David CN, Gierer A. Cell cycle kinetics and development of *Hydra attenuata*. III. Nerve and nematocyte differentiation. J Cell Sci 1974;16:359–75.
[61] Teragawa CK, Bode HR. Migrating interstitial cells differentiate into neurons in hydra. Dev Biol 1995;171:286–93.
[62] Hager G, David CN. Pattern of differentiated nerve cells in hydra is determined by precursor migration. Development 1997;124:569–76.
[63] Govindasamy N, Murthy S, Ghanekar Y. Slow-cycling stem cells in hydra contribute to head regeneration. Biol Open November 28, 2014;3(12):1236–44.
[64] Marcum BA, Campbell RD. Development of hydra lacking nerve and interstitial cells. J Cell Sci 1978;29:17–33.
[65] Holstein TW, Hobmayer E, David CN. Pattern of epithelial cell cycling in hydra. Dev Biol 1991;148:602–11.
[66] Müller WA, Teo R, Frank U. Totipotent migratory stem cells in a hydroid. Dev Biol 2004;275:215–24.
[67] Marlow HQ, Srivastava M, Matus DQ, Rokhsar D, Martindale MQ. Anatomy and development of the nervous system of *Nematostella vectensis*, an anthozoan cnidarian. Dev Neurobiol 2009;69:235–54.
[68] Künzel T, Heiermann R, Frank U, Müller W, Tilmann W, Bause M, et al. Migration and differentiation potential of stem cells in the cnidarian hydractinia analysed in eGFP-transgenic animals and chimeras. Dev Biol 2010;348:120–9.
[69] Halanych KM. The ctenophore lineage is older than sponges? That cannot be right! Or can it? J Exp Biol 2015;218:592–7.
[70] Adamska M, Degnan BM, Green K, Zwafink C. What sponges can tell us about the evolution of developmental processes. Zool Jena 201(114):1–10.
[71] Dunn CW, Hejnol A, Matus DQ, Pang K, Browne WE, Smith SA, et al. Broad phylogenomic sampling improves resolution of the animal tree of life. Nature 2008;452:745–9.
[72] Funayama N. The stem cell system in demosponges: insights into the origin of somatic stem cells. Dev Growth Differ 2010;52:1–14.
[73] Funayama N. The stem cell system in demosponges: suggested involvement of two types of cells: archeocytes (active stem cells) and choanocytes (food-entrapping flagellated cells). Dev Genes Evol 2013;223:23–38.

[74] King N, Westbrook MJ, Young SL, Kuo A, Abedin M, Chapman J, et al. The genome of the choanoflagellate *Monosiga brevicollis* and the origin of metazoans. Nature 2008;451: 783–8.
[75] Fairclough SR, Chen Z, Kramer E, Zeng Q, Young S, Robertson HM, et al. Premetazoan genome evolution and the regulation of cell differentiation in the choanoflagellate *Salpingoeca rosetta*. Genome Biol 2013;14:R15.
[76] Borojevic R. Etude expérimentale de la différentiation de cellules de l'Eponge au cours de son développement. Dev Biol 1966;14:130–53.
[77] Borojevic R. Différenciation cellulaire dans l'embryogenèse et la morphogenèse chez les Spongiaires. In: Fry WG, editor. The biology of the Porifera. London: Academic Press; 1970. p. 467–90.
[78] Müller WE. The stem cell concept in sponges (Porifera): metazoan traits. Semin Cell Dev Biol 2006;17:481–91.
[79] Saller U. Oogenesis and larval development of *Ephydatia fluviatilis* (Porifera, Spongillidae). Zoomorphology 1988;108:23–8.
[80] Gaino E, Burlando B, Zunino L, Pansini M, Buffa P. Origin of male gametes from choanocytes in *Spongia officinalis* (Porifera, Demospongiae). Int J Inv Rep Dev 1984;7:83–93.
[81] Simpson TL. The cell biology of sponges. New York: Springer; 1984.
[82] Okamoto K, Nakatsukasa M, Alié A, Masuda Y, Agata K, Funayama N. The active stem cell specific expression of sponge Musashi homolog EflMsiA suggests its involvement in maintaining the stem cell state. Mech Dev 2012;129:24–37.
[83] Coutinho CC, de Azevedo Maia G. Mesenchymal cells in ancestral spongiomorph urmetazoa could be the mesodermal precursor before gastrulation origin. In: Custódio MR, Lôbo-Hajdu G, Hajdu E, Muricy G, editors. Porifera research biodiversity, innovation and sustainability. Série Livros, 28. Rio de Janeiro: Museu Nacional; 2007. p. 281–95.
[84] Mah JL, Christensen-Dalsgaard KK, Leys SP. Choanoflagellate and choanocyte collar-flagellar systems and the assumption of homology. Evol Dev 2014;16:25–37.
[85] King N, Young SL, Abedin M, Carr M, Leadbeater BS. The choanoflagellates: heterotrophic nanoflagellates and sister group of the metazoa. Cold Spring Harb Protoc 2009;2009(2). pdb.emo116.
[86] Connes R. Contribution a l'Etude de la gemmulogenése chez la d'Eponge marine *Suberites domuncula* (Olivi) Nardo. Arch Zool Exp Gen 1977;118:391–407.
[87] King N. The choanoflagellate transcriptome: insights into animal origins and evolution. Dev Biol 2004;271:554–5.
[88] King N. The unicellular ancestry of animal development. Dev Cell 2004;7:313–25.
[89] Maldonado M. Choanoflagellates, choanocytes, and animal multicellularity. Invertebr Biol 2004;123:1–22.
[90] Carr M, Leadbeater BS, Hassan R, Nelson M, Baldauf SL. Molecular phylogeny of choanoflagellates, the sister group to Metazoa. Proc Natl Acad Sci USA 2008;105:16641–6.
[91] Wagner DE, Wang IE, Reddien PW. Clonogenic neoblasts are pluripotent adult stem cells that underlie planarian regeneration. Science 2011;332:811–6.
[92] Syed T, Schierwater B. *Trichoplax adhaerens*: discovered as a missing link, forgotten as a hydrozoan, re-discovered as a key to metazoan evolution. Vie Milieu 2002;52:177–87.
[93] Ender A, Schierwater B. Placozoa are not derived cnidarians: evidence from molecular morphology. Mol Biol Evol 2003;20:130–4.
[94] Schierwater B, DeSalle R. Can we ever identify the Urmetazoan? Integr Comp Biol 2007;47:670–6.
[95] DeSalle R, Schierwater B. An even "newer" animal phylogeny. Bioessays 2008;30:1043–7.
[96] Dohrmann M, Wörheide G. Novel scenarios of early animal evolution–is it time to rewrite textbooks? Integr Comp Biol 2013;53:503–11.

[97] Osigus HJ, Eitel M, Schierwater B. Chasing the urmetazoon: striking a blow for quality data? Mol Phylogenet Evol 2013;66:551–7.

[98] Buchholz K, Ruthmann A. The mesenchyme-like layer of the fiber cells of *Trichoplax adhaerens* (Placozoa), a syncytium. Z Naturforsch 1995;50c:282–5.

[99] Smith CL, Varoqueaux F, Kittelmann M, Azzam RN, Cooper B, Winters CA, et al. Novel cell types, neurosecretory cells, and body plan of the early-diverging metazoan *Trichoplax adhaerens*. Curr Biol 2014;24:1565–72.

[100] Jakob W, Sagasser S, Dellaporta S, Holland P, Kuhn K, Schierwater B. The Trox-2 Hox/ParaHox gene of Trichoplax (Placozoa) marks an epithelial boundary. Dev Genes Evol 2004;214:170–5.

[101] Guidi L, Eitel M, Cesarini E, Schierwater B, Balsamo M. Ultrastructural analyses support different morphological lineages in the phylum placozoa Grell, 1971. J Morphol March 2011;272(3):371–8.

[102] Agata K, Nakajima E, Funayama N, Shibata N, Saito Y, Umesono Y. Two different evolutionary origins of stem cell systems and their molecular basis. Semin Cell Dev Biol 2006;17:503–9.

[103] Sánchez Alvarado A, Kang H. Multicellularity, stem cells, and the neoblasts of the planarian *Schmidtea mediterranea*. Exp Cell Res June 10, 2005;306(2):299–308.

[104] Petralia RS, Mattson MP, Yao PJ. Aging and longevity in the simplest animals and the quest for immortality. Ageing Res Rev 2014;16:66–82.

[105] Niculescu VF. The stem cell biology of the protist pathogen *Entamoeba invadens* in the context of eukaryotic stem cell evolution. Stem Cell Biol Res 2015. http://dx.doi.org/10.7243/2054-717X-2-2. http://www.hoajonline.com/journals/pdf/2054-717X-2-2.pdf3.

[106] Niculescu VF. The cell system of *Giardia lamblia* in the light of the protist stem cell biology. Stem Cell Biol Res 2014;1:3. http://dx.doi.org/10.7243/2054-717X-1-3.

[107] Niculescu VF. Evidence for asymmetric cell fate and hypoxia induced differentiation in the facultative pathogen protist *Colpoda cucullus*. Microbiol Discov 2014;2:3. http://dx.doi.org/10.7243/2052-6180-2-3.

[108] Niculescu VF. The evolutionary history of eukaryotes: how the ancestral proto-lineage conserved in hypoxic eukaryotes led to protist pathogenicity. Microbiol Discov 2014;2:4. http://dx.doi.org/10.7243/2052-6180-2-4.

[109] Lund EJ. Reversibility of morphogenetic process in Bursaria. J Exp Zool 1917;24:1–33.

[110] Niculescu VF. On the origin of stemness and ancient cell lineages in single-celled eukaryotes. SOJ Microbiol Infect 2014;2:1–3.

[111] Müller M, Mentel M, van Hellemond JJ, Henze K, Woehle C, Gould SB, et al. Biochemistry and evolution of anaerobic energy metabolism in eukaryotes. Microbiol Mol Biol Rev 2012;76:444–95.

[112] Laverack MS. The physiology of earthworms: International series of monographs on pure and applied biology: zoology. Pergamon Press; 1963.

[113] Løvtrup E. Studies on planarian respiration. J Exp Zool 1953;124:427–34.

[114] Pedersen KJ. On the oxygen consumption of planaria vitta during starvation, the early phase of regeneration and asexual reproduction. J Exp Zool 1956;131:123–35.

[115] Barrett J, Butherworth PE. Carbohydrate and lipid metabolism in the planarian *Polycelis nigra*. J Comp Physiol B 1982;146:107–12.

[116] Matta JTD, Kanaan S, Matta AND. Changes in glycolytic metabolism during the regeneration of planarians dugesia tigrina girard. Comp Biochem Physiology B 1989;93:391–6.

[117] Jenkins MM. Respiration rates in planarian I. The use of the Warburg respirometer in determining oxygen consumption. Proc Okla Acad Sci 1959;40:35–40.

[118] González-Estévez C, Felix DA, Rodríguez-Esteban G, Aboobaker AA. Decreased neoblast progeny and increased cell death during starvation-induced planarian degrowth. Int J Dev Biol 2012;56:83–91.
[119] Gschwentner R, Ladurner P, Nimeth K, Rieger R. Stem cells in a basal bilaterian. S-phase and mitotic cells in *Convolutriloba longifissura* (Acoela, Platyhelminthes). Cell Tissue Res 2001;304:401–8.
[120] Lopes KA, DE Campos Velho NM, Pacheco-Soares C. Method of isolation and characterization of *Girardia tigrina* stem cells. Biomed Rep 2015;3:163–6.
[121] Lenhoff HM, Loomis WF. Environmental factors controlling respiration in hydra. J Exp Zool 1957;134:171–81.
[122] Zaheer Baquer N, McLean P, Hornbruch A, Wolpert L. Positional information and pattern regulation in hydra: enzyme profiles. J Embryol Exp Morphol 1975;33:853–67.
[123] Needham J. Biochemistry and Morphogenesis. Cambridge University Press; 1942. p. 471.
[124] Pardy RL, White BN. Metabolic relationships between green hydra and its symbiotic algae. Biol Bull 1977;153:228–36.
[125] Hand AR. Ultrastructural localization of catalase and L-alpha-hydroxy acid oxidase in microperoxisomes of hydra. J Histochem Cytochem 1976;24:915–25.
[126] Dash B, Phillips TD. Molecular characterization of a catalase from *Hydra vulgaris*. Gene 2012;501:144–52.
[127] Belousov LV, Ostroumova TV. Metabolic gradients and morphological polarization in embryonic development of hydroid polypes. J Embryol Exp Morphol 1969;22:431–47.
[128] Blackstone NW. Redox state, reactive oxygen species and adaptive growth in colonial hydroids. J Exp Biol 2001;2004(Pt 11):1845–53.
[129] Blackstone NW. Redox signaling in the growth and development of colonial hydroids. J Exp Biol 2003;206(Pt 4):651–8.
[130] Blackstone NW. Redox control in development and evolution: evidence from colonial hydroids. J Exp Biol 1999;202:3541–53.
[131] Blackstone NW, Cherry KS, Van Winkle DH. The role of polypstolon junctions in the redox signaling of colonial hydroids. Hydrobiologia 2004;530/531:291–8.
[132] Blackstone NW, Bivins MJ, Cherry KS, Fletcher RE, Geddes GC. Redox signaling in colonial hydroids: many pathways for peroxide. J Exp Biol 2005;208:383–90.
[133] Doolen JF, Geddes GC, Blackstone NW. Multicellular redox regulation in an early-evolving animal treated with glutathione. Physiol Biochem Zool 2007;80:317–25.
[134] Vogt KS, Geddes GC, Bross LS, Blackstone NW. Physiological characterization of stolon regression in a colonial hydroid. J Exp Biol 2008;211:731–40.
[135] Vogt KS, Harmata KL, Coulombe HL, Bross LS, Blackstone NW. Causes and consequences of stolon regression in a colonial hydroid. J Exp Biol 2011;214:3197–205.
[136] Harmata KL, Blackstone NW. Reactive oxygen species and the regulation of hyperproliferation in a colonial hydroid. Physiol Biochem Zool 2011;84:481–93.
[137] Morgan JB, Mahdi F, Liu Y, Coothankandaswamy V, Jekabsons MB, Johnson TA, et al. The marine sponge metabolite mycothiazole: a novel prototype mitochondrial complex I inhibitor. Bioorg Med Chem 2010;18:5988–94.
[138] Gunda VG, Janapala VR. Effects of dissolved oxygen levels on survival and growth in vitro of *Haliclona pigmentifera* (Demospongiae). Cell Tissue Res 2009;337:527–35.
[139] Tunnicliffe V. High species diversity and abundance of the epibenthic community in an oxygen-deficient basin. Nature 1981;294:354–6.

[140] Levin LA, Huggett CL, Wishner KF. Control of deep-sea benthic community structure by oxygen and organic matter gradients in the eastern Pacific Ocean. J Mar Res 1991;49:763–800.
[141] Mills DB, Ward LM, Jones C, Sweeten B, Forth M, Treusch AH, et al. Oxygen requirements of the earliest animals. Proc Natl Acad Sci USA 2014;111:4168–72.
[142] Hoffmann F, Larsen O, Thiel V, Rapp HT, Pape T, Michaelis W, et al. An anaerobic world in sponges. Geomicrobiol J 2005;22:1–10.
[143] Hoffmann F, Røy H, Bayer K, Hentschel U, Pfannkuchen M, Brümmer F, et al. Oxygen dynamics and transport in the mediterranean sponge *Aplysina aerophoba*. Mar Biol 2008;153:1257–64.
[144] Sun L, Song Y, Qu Y, Yu X, Zhang W. Purification and in vitro cultivation of archaeocytes (stem cells) of the marine sponge *Hymeniacidon perleve* (Demospongiae). Cell Tissue Res 2007;328:223–37.
[145] Loenarz C, Coleman ML, Boleininger A, Schierwater B, Holland PW, Ratcliffe PJ, et al. The hypoxia-inducible transcription factor pathway regulates oxygen sensing in the simplest animal, *Trichoplax adhaerens*. EMBO Rep 2011;12:63–70.
[146] McDonald AE, Vanlerberghe GC, Staples JF. Alternative oxidase in animals: unique characteristics and taxonomic distribution. J Exp Biol 2009;212:2627–34.
[147] Lambowitz AM, Slayman CW. Cyanide-resistant respiration in *Neurospora crassa*. J Bacteriol 1971;108:1087–96.
[148] Massabuau JC. From low arterial- to low tissue-oxygenation strategy. An evolutionary theory. Respir Physiol 2001;128:249–61.
[149] Corbari L, Carbonel P, Massabuau JC. How a low tissue O_2 strategy could be conserved in early crustaceans: the example of the podocopid ostracods. J Exp Biol 2004;207:4415–25.
[150] Massabuau JC. Primitive, and protective, our cellular oxygenation status? Mech Ageing Dev 2003;124:857–63.
[151] Corbari L, Carbonel P, Massabuau JC. The early life history of tissue oxygenation in crustaceans: the strategy of the myodocopid ostracod *Cylindroleberis mariae*. J Exp Biol 2005;208:661–70.
[152] Massabuau JC, Abele D. Principles of oxygen uptake and tissue oxygenation in water-breathing animals. In: Abele D, Vàquez-Medina JP, Zenteno-Savin T, editors. Oxydative stress in aquatic ecosystems. Blackwell Publishing Ltd; 2012. p. 141–56.
[153] Boveris A, Costa LE, Cadenas E, Poderoso JJ. Regulation of mitochondrial respiration by adenosine diphosphate, oxygen, and nitric oxide. Methods Enzymol 1999;301:188–98.
[154] Winslow RM. Oxygen: the poison is in the dose. Transfusion 2013;53:424–37.
[155] Ivanovic Z. Respect the anaerobic nature of stem cells to exploit their potential in regenerative medicine. Regen Med 2013;8:677–80.
[156] Guitart AV, Hammoud M, Dello Sbarba P, Ivanovic Z, Praloran V. Slow-cycling/quiescence balance of hematopoietic stem cells is related to physiological gradient of oxygen. Exp Hematol 2010;38:847–51.
[157] Ivanovic Z. Hypoxia or in situ normoxia: the stem cell paradigm. J Cell Physiol 2009;219:271–5.
[158] Field MC, Dacks JB. First and last ancestors: reconstructing evolution of the endomembrane system with ESCRTs, vesicle coat proteins, and nuclear pore complexes. Curr Opin Cell Biol 2009;21:4–13.
[159] Leadbeater BSC, Karpov S. Cyst formation in a freshwater strain of the choanoflagellate *Desmarella moniliformis* Kent. J Eukaryot Microbiol 2000;47:433–9.
[160] Carr M, Leadbeater BS, Baldauf SL. Conserved meiotic genes point to sex in the choanoflagellates. J Eukaryot Microbiol 2010;57:56–62.

[161] Xiao S, Yuan X, Knoll AH. Eumetazoan fossils in terminal proterozoic phosphorites? Proc Natl Acad Sci USA 2000;97:13684–9.
[162] Huldtgren T, Cunningham JA, Yin C, Stampanoni M, Marone F, Donoghue PC, et al. Fossilized nuclei and germination structures identify Ediacaran "animal embryos" as encysting protists. Science 2011;334:1696–9.
[163] Koumandou VL, Wickstead B, Ginger ML, van der Giezen M, Dacks JB, Field MC. Molecular paleontology and complexity in the last eukaryotic common ancestor. Crit Rev Biochem Mol Biol 2013;48:373–96.
[164] Forterre P. A new fusion hypothesis for the origin of Eukarya: better than previous ones, but probably also wrong. Res Microbiol 2011;162:77–91.
[165] Spang A, Saw JH, Jørgensen SL, Zaremba-Niedzwiedzka K, Martijn J, Lind AE, et al. Complex archaea that bridge the gap between prokaryotes and eukaryotes. Nature 2015;521:173–9.
[166] Haeckel E. Natürliche Schöpfungsgeschichte. Berlin: Georg Reimer; 1868. 15th lecture.
[167] Isaeva VV, Akhmadieva AV, Aleksandriova IaN, Shukaliuk AI. Morphofunctional organization of reserve stem cells providing for asexual and sexual reproduction of invertebrates. Ontogenez 2009;40:83–96.
[168] Seita J, Weissman IL. Hematopoietic stem cell: self-renewal versus differentiation. Wiley Interdiscip Rev Syst Biol Med 2010;2:640–53.
[169] Eaves CJ. Hematopoietic stem cells: concepts, definitions, and the new reality. Blood 2015;125:2605–13.
[170] Campbell KH, McWhir J, Ritchie WA, Wilmut I. Sheep cloned by nuclear transfer from a cultured cell line. Nature 1996;380:64–6.
[171] Takahashi K, Tanabe K, Ohnuki M, Narita M, Ichisaka T, Tomoda K, et al. Induction of pluripotent stem cells from adult human fibroblasts by defined factors. Cell 2007;30(131):861–72.
[172] Kapinas K, Grandy R, Ghule P, Medina R, Becker K, Pardee A, et al. The abbreviated pluripotent cell cycle. J Cell Physiol 2013;228:9–20.
[173] Yamamoto R, Morita Y, Ooehara J, Hamanaka S, Onodera M, Rudolph KL, et al. Clonal analysis unveils self-renewing lineage-restricted progenitors generated directly from hematopoietic stem cells. Cell 2013;154:1112–26.
[174] Nava MM, Raimondi MT, Pietrabissa R. Controlling self-renewal and differentiation of stem cells via mechanical cues. J Biomed Biotechnol 2012;2012:797410.
[175] Knoblich JA. Mechanisms of asymmetric stem cell division. Cell 2008;132:583–97.
[176] Miguel MP, Alcaina Y, Maza DS, Lopez-Iglesias P. Cell metabolism under microenvironmental low oxygen tension levels in stemness, proliferation and pluripotency. Curr Mol Med 2015;15:343–59.
[177] Ijima I. Untersuchunge iiber den Bau und die Entwicklungsgeschichte der Susswasser Dendrocoelen. Ztschr Wiss Zool 1884;40:359–464.
[178] Lehnhert GH. Beobachtungen an Landplanarien. Arch fur Naturgeschiehte 1891;57: 306–50.
[179] Prenant M. Recherches sur le parenchyme des platyhelminthes. Arch Morph Gén Exp 1922;5:1–174.
[180] Bartsch O. Die histiogenese der Planarienregenerate. Zool Anz 1923;56:63–7.
[181] Steinmann P. Das verhalten der Zellen und Gewebe in regenerierenden Tricladenkorper. Verch Naturf Ges Basel 1925;36:133–62.

Metabolic and Genetic Features of Ancestral Eukaryotes versus Metabolism and "Master Pluripotency Genes" of Stem Cells

Chapter Outline

11.1 Ancestral Energetic Character of the Stem Cells

In order to understand this question we will present the most common eukaryote energetic pathways with special regard to protists.

In contrast to prokaryotes, core energy metabolic reactions show very narrow biochemical diversity between major eukaryotes groups with respect to aerobic or anaerobic lifestyle or enzymes involved, in spite of the fact that they power millions of different species [1]. Whole spectrums of the eukaryotes encompass less energy metabolic diversity than can be found in the individual group of bacteria. For instance, all the eukaryotes possess an organelle of mitochondrial origin [2–4] that is crucial for both anaerobic and aerobic energy metabolism. Also, anaerobic protists have only a few different enzymes than mammals. Comparing anaerobic metabolism among major eukaryote groups, a common set of about 50 enzymes was detected. Pyruvate

Anaerobiosis and Stemness. http://dx.doi.org/10.1016/B978-0-12-800540-8.00011-9

kinase and phosphoglycerate kinase in the glycolysis associated with the glycolytic ATP production are preserved in all groups of eukaryotes!

It is accepted that this remarkable absence of diversity in eukaryotic energetic machinery indicates two things: its common origin and the presence of the same collection of enzymes in the unicellular eukaryotic common ancestor. In addition, not only are the eukaryotes energetically similar, but there is also a similarity between enzymes of eukaryotes and eubacterial energy metabolism, particularly glycolysis, proving its eubacterian origin as predicted by the hydrogen hypothesis [5].

From an energetic point of view, it is accepted that the first eukaryotic common ancestor (FECA) was able to respire in the presence of O_2 but in its absence was able to perform anaerobic respiration and fermentation (facultative anaerobe) (see Chapters 8 and 9).

Common ancestors of eukaryotes are thought to have arisen through symbiotic association of an autotrophic methanogenic archaebacterium (host) with eubacterium (symbiont) that was able to respire and generate hydrogen as a waste product of anaerobic heterotrophic metabolism (see Chapter 8). Alpha proteobacterium, recognized as host, was strictly an anaerobe that utilizes CO_2 and H_2 as the energetic fuel that produces energy ATP via methanogenesis (production of methane) and glucose by acetyl-CoA pathway. By contrast, the symbiont was a heterotrophic free-living eubacterium that, in the presence of O_2, was able to oxidase glucose via pyruvate to produce ATP, CO_2, and H_2O by the Krebs cycle and oxidative phosphorylation (OXPHOS); without O_2 it produced ATP by metabolizing glucose to pyruvate and its conversion to H_2, CO_2, and acetate.

Keeping in mind that the general notion is that appearance and diversification of the eukaryotes had taken place in an anoxic or almost anoxic Proterozoic ocean, these two organisms would have had to meet in the anaerobic environment. Here, anaerobic energetic profile of the symbiont is a more favorable variant. In this case, it is considered that the host would have become dependent on the symbiont. This, in turn, developed an irreversible association of the host and the symbiont, followed by the genetic transfer between host and symbiont, resulting in the transfer of the symbiont's importers and glycolytic metabolism to the host's membrane and in the host's cytoplasm, respectively. In this way, the host would satisfy its need for carbons and ATP and would have become heterotrophic. H_2 waste product would be produced as a part of cytoplasmic compartment [5]. This event would result in creation of the first eukaryote, the new organelle-bearing organism that is able to utilize, in the presence of O_2, the Krebs cycle, and OXPHOS in the mitochondria to produce energy in the aerobic manner, or to function in the anaerobic manner in the absence of O_2. This ability enabled anaerobic eukaryotes to widely spread in the almost anoxic environment in the Proterozoic ocean. According to this theory, origin of mitochondria accounts for the origin of the eukaryotic cell.

From a FECA point, metabolism has evolved into three directions: mitochondriate eukaryote that bases its energetic metabolism on O_2 consummation; eukaryote that utilizes mitochondria in the anaerobic fashion (anaerobic mitochondria and hydrogen-producing mitochondria); and eukaryote that lost the "canonical" mitochondria secondarily, but contains mitochondria-derived organelles (mitosomes) (see Chapter 9). All three types of metabolism exist from protists to animals [1].

In any case, glycolysis, which is an extramitochondrial process, is the backbone of eukaryotic energetic metabolism from protists to mammals [1]. Glucose and other carbohydrates are considered the major carbon source for energy production with contribution of catabolism of the certain amino acids [6]. One mole of glucose is oxidized to pyruvate yielding 2 mol of ATP and reduced NADH. Here the energy is produced by substrate-level phosphorylation, considered the most ancient energy-producing mechanism [7].

From this point, metabolism has branched out to include metabolizing pyruvate and reoxidization of the reduced cofactor (NADH), either by fermentation (anaerobic metabolism) or respiration (anaerobic or aerobic).

Eukaryotes with anaerobic organelles of mitochondrial origin base their energetic metabolism on the anaerobic fermentation. Glycolytic pyruvate is metabolized by pyruvate: ferredoxin oxidoreductase (PFO). In some anaerobic eukaryotes, cytosol PFO decarboxylates pyruvate resulting in the reduced ferredoxin and acetyl-CoA that is converted into ethanol, acetate, and CO_2, yielding up to two additional moles of ATP.

Aerobic protists Thrichomonads contain mitochondria-like organelle hydrogenosomes. Here, cytosolic glycolytic pyruvate is transported to hydrogenosome, where it is converted by PFO to CO_2, Acetyl-CoA, and reduced ferredoxin [8]. Then ferredoxin can be reoxidized by the hydrogenase, which is associated with H_2 release (which accounts for the name). ATP is synthetized during the conversion of the succinyl-CoA to succinate via substrate level phosphorylation that is catalyzed with succinyl-CoA synthetase [9] (in exchange conversion of acetyl-CoA to acetate by acetate-generating enzymes is acetate:succinate CoA transferase) [10]. Therefore, this yields two additional molecules of ATP and H_2, CO_2 and acetate.

In contrast, in eukaryotes with "canonical mitochondria," pyruvate is usually oxidized by pyruvate dehydrogenase complex (PDH) and enters the Krebs cycle in mitochondria that produces reduced cofactors. These oxidize via respiratory complexes to final acceptor O_2 that creates proton force necessary for the ATP production by the ATPase (chemiosmotic production of ATP). O_2-dependent respiration yields 34–36 molecules of ATP per molecule of glucose, CO_2 and H_2O.

In addition, the aerobic respiration mitochondria could function in anaerobic mode in various protists and animals. The common way for this is fumarate respiration or "malate dismutation" [11].

Hydrogenosomes from *Nyctotherus ovalis* anaerobic ciliate could represent a link between mitochondria and hydrogenosome, as they possess a genome and electron transport chain (features of mitochondria), but they produce hydrogen (features of hydrogenosome) [12]. These organisms produce pyruvate from glucose, and excrete acetate, succinate, lactate, and ethanol. Oxidative decarboxylation of the pyruvate to acetyl-CoA by PDH results in the NAD^+ reduction to NADH. The oxidation of the NADH is thought to happen in the mitochondrial complex I that passes the electrons from NADH through rhodoquinone to complex II, which reduces fumarate as electron acceptor to succinate. Another part of the NADH is oxidized by the iron hydrogenase that is linked to the complex I, which results in the production of the molecular H_2. This reoxygenation of the NADH by the aid of the complex I appears to generate a proton gradient [12]. But it seems that this proton force is not associated with the ATP production in ATPase since the gene for the ATP synthase is missing [13]. This process is seen in some metazoan including parasitic helminths, freshwater snails, mussels, lugworms, and other marine invertebrates. In these animals cytosolic malate enters the mitochondria and is converted to fumarate. Donor of the electrons is the reduced NADH that is oxidized in exchange for reduction of ubiquinone or rhodoquinone at complex I. Then, the reduced ubiquinone or rhodoquinone serves to donate the electron to endogenous fumarate, as a final acceptor and its reduction to succinate by fumarate reductase (mitochondrial complex II). This creates a proton gradient by the complex I in mitochondria that enables production of ATP by the ATP synthase under anoxic conditions [14].

Intriguingly, it was published that rat or bovine (ox) heart, liver, kidney, kidney proximal tubule cells, and variety of the cancer cells could ensure anaerobic ATP formation in anoxic/ischemic condition by anaerobic metabolism of the citric acid cycle intermediates and/or anaerobic mitochondrial respiration [15–21].

Also, in anaerobic fungus (*Fusarium oxysporum*), another way of the anaerobic energy production is described: They produce energy by reduction of elemental sulfur to generate H_2S as an end product of metabolism. In reaction, sulfate is used as a terminal acceptor of electrons from the most part coming from the organic acids (e.g., lactate, acetate, etc.) to generate H_2S. Interestingly, H_2S, which can be formed endogenously, enzymatically or not, in the invertebrates and vertebrates including mammals, is shown as an inorganic energetic substrate. Within physiological range it could sustain ATP production in anoxic/hypoxic conditions in human colon cells, murine smooth muscle cells [22,23]. In stress condition, it seems that in murine smooth cells H_2S could be produced in mitochondria, and the electrons from H_2S could be then transferred to ubiquinone and complex II. These electrons could be utilized for fumarate reduction to succinate [22,23] followed with ATP production by the ATP synthase.

It should be noted that H_2S could also sustain ATP production in the aerobic manner in addition to its anaerobic O_2-mediated oxidation with enzymatic mitochondrial machinery [24]. Intriguingly, H_2S was shown to protect mesenchymal or neural stem cells from hypoxia or ischemia [25,26]. It would be worthwhile to test if those effects include H_2S-mediated energetic effects.

Mitosomes are the most highly reduced forms of the mitochondria discovered in the *Entamoeba histolytica* and *Giardia*. Mitosomes appear to have no direct role in the ATP synthesis. Enzymes for the energy metabolism are localized in the cytosol out of the mitosomes [1]. The main end products are ethanol and acetate or molecular H_2 under anoxic condition [27]. The enzymes involved are PFO [28] and acetyl-CoA synthase (ADP forming ACS-ADP) [29]. Pyruvate produced via glycolysis is decarboxylated to Acetyl-CoA via PFO, and Acetyl-CoA is converted to acetate that is associated with production of the ATP.

It is interesting that many eukaryotes could live in the oxic environment without using mitochondrial OXPHOS. For instance, yeasts in the presence of the sufficient level of the fermentable substrate preferably utilize fermentation in comparison to OXPHOS: In the excess of sugar they preferably do alcoholic fermentation in the aerobic condition (Crabtree effect) [30,31].

Also, trypanosomes in the blood stream produce their ATP through substrate-level phosphorylation in glycolysis and use O_2 as a terminal acceptor in mitochondria, but without OXPHOS. This resembles in a high degree the metabolism of the circulatory stem cells in mammals, which show glycolytic metabolic orientation in the excess of O_2. Also, goldfish could survive a prolonged period of anoxia grace to fermentation that generate ethanol and CO_2 as an end product of metabolism [32].

Most of the anaerobic eukaryotes can readily survive in the low O_2 concentration (i.e., are microaerophiles), and very few are truly strict anaerobes [1]. Interestingly, they developed a mechanism to scavenge O_2 and purge their environment of O_2. For example, *Trichomonas vaginalis* and *Giardia lamblia* possess the enzyme NADH oxidase, which converts and transforms O_2 to H_2O. Homologs for the NADH oxidase reported for the protists are common among eukaryotic genomes [1,33,34].

Regarding above presented data and comparing them with the knowledge related to energetic mechanisms operating in stem cells, multiple analogies between them are evident. We will try to summarize them in the following passages.

In Chapter 6, we explained extensively from many aspects that somatic as well as pluripotent stem cells fuel their energy demands mostly by the glycolysis even when the O_2 is available and shunt pyruvate away from the oxidation in mitochondria for use in anabolic pathway [35–37]. Also, stimulation of the glycolysis in embryonic stem cells (ESCs) by low O_2 concentration [38,39] and inhibition of the mitochondrial respiration [40–43] sustain stemness, while inhibition of the glycolysis promotes cell death [44] and loss of stem cell function (SP population) [45]. Reprogramming of somatic to divert in pluripotent cells is associated with a major bioenergetic restructuring to facilitate a conversion from somatic mitochondrial oxidation to the glycolysis-dependent pluripotency state.

Even if they are mostly glycolytic, pluripotent cells possess functional mitochondria that operate both to recycle NAD^+ and keep glycolytic flux and to generate intermediaries for anabolic pathways, epigenetic control, and oxidative stress protection [46–49]. Also, the most primitive pluripotent cells that we can use in experimental studies are energetically bivalent (they can switch between OXPHOS or glycolysis on demand), followed by the next pluripotent state, which is mostly glycolytic [50]. Hence, even though the stem cells are mostly glycolytic, the mitochondria contribute to the maintenance of stemness. Also, functional mitochondria are critical for differentiation of the stem cells to somatic cells [46,51]. Differentiation to somatic cells is associated with biogenesis of mitochondria and the increase of the mitochondrial activity [52,53].

These features astonishingly resemble the hypothetical energetic profile of FECA that was basically a facultative anaerobe, with the possibility for the utilization of both the oxidative respiration and anaerobic pathways. Also, it fits well with the microaerophilic nature of the ancestral protists: they utilize a mostly anaerobic pathway that was an advantage in the anoxic environment, but they are also capable of utilizing traces of O_2 where they encounter it.

Also, a very important feature of the stem cell is huge metabolic plasticity that enables energetic reprogramming on the basis of the functional demands. One of these adaptations is revealed by the capacity to utilize mitochondria in anaerobic mode (particularly fumarate respiration) in an anoxic/ischemic condition as it is evidenced for cancer cells or mesenchymal stem cells (our work in progress). As we mentioned above, this type of metabolism that is common for the anaerobic ancestral protists or other animals indicated that stem cells evoke the same mechanism to assure their maintenance.

Finally, since they could alter their functions, metabolism of the stem cells is adapted to minimize reactive oxygen species (ROS) generation, mainly produced by the mitochondrial respiration. In order to protect themselves, the stem cells developed various mechanisms and regulators that serve for antioxidative defense and as ROS scavengers, as we reported in Chapter 7. Regarding the function of the NADH oxidase in *Trichomonas vaginalis* as an ROS scavenger enzyme, it is evident that stem cells use the same strategy that exists in the protists.

From protists to mammals these conserved anaerobic energy production pathways ensure energetic homeostasis in a low O_2 environment and protect the cell indirectly from oxidative stress. Thus, all the data presented above indicate that the anaerobic nature of the mammal stem cells (see Chapter 7) is actually an ancestral characteristic inherited from the unicellular eukaryotes. Favoring anaerobiosis favors "stemness."

Stem cell metabolism is tightly regulated by the set of the various signaling pathways, genetic, and epigenetic regulators (see Chapter 6). Astonishing preservation of energetic reactions among the major eukaryote groups implies that their genetic, epigenetic, and metabolic control is also evolutionary conserved.

In order to verify this hypothesis, we will present molecular and signaling pathways distributed among the eukaryotes, in comparison to the stem cell essential "toolkit" (set of the regulators evidenced to be essential for stem cell maintenance in mammals; see Chapter 10). This analysis would enable determination at which evolutional point the "stemness toolkit" was established.

11.2 Genetic Features of Ancestral Eukaryotes with Respect to "Master Pluripotency Genes" of Stem Cells

11.2.1 Protist

Analysis of the molecular and metabolic markers presented in the genome of unicellular eukaryote anaerobic parasite could be instructive in order to find an evolutive connection between ancestral eukaryotes and stem cells because these organisms live in an environment similar to the mammal stem cells.

These anaerobic protists share the same metabolic adaptations: absence of the tricarboxylic acid cycle (TCA) and mitochondrial electron chain, presence of the antioxidative stress enzymes, relying on glycolytic metabolism and pentose phosphate pathway, utilization of the amino acid metabolism for energy production (e.g., tryptophane and aspartate) [54]. Also, fatty acids synthesis is absent, but phospholipid and sphingolipid synthesis is presented. The latter is considered a protist's novel pathway in comparison to prokaryotes [55]. In addition they possess complex signaling transduction networks, including the enzymes existing in the higher animals: tyrosine kinase cytosolic serine/thronine phosphatase, diacylglycerol; G protein; GTPase-activating protein, inositol-1,4,5-trisphosphate system, protein kinase C; phospholipase C; PTEN, phosphatase, and tensin homolog; tyrosine phosphatase; and Ras family proteins, MAPK kinase. Furthermore, even Wnt signaling is absent, β-catenin-related gene Aardvark is identified in amoebozoans. Aardvark is implicated in signaling in *Dictyostelium*, and this occurs in a glycogen synthase kinase 3 (GSK-3)-dependent manner [56].

In comparison to prokaryotes, many protists have complex life cycles (change of the environment, form, and function) that are associated with the complex molecular regulatory network (see Chapter 10). In general, eukaryotic transcription systems appear to have a chimeric pattern—the archaea-like elements contribute to the core transcription apparatus, including the bulk of the basal or general transcription factors (TFs), and the bacteria-like elements supply some additional factors of the basal transcription apparatus [57–59]. TFs containing homeodomains and C_2H_2 zinc (Zn) fingers are dominant. In addition, novel protist TFs contain GATA, MYB, bZIP, Retinoblastoma (Rb), and p53 like FOX, bHLH, AP2, STAT, POU, and TALE domains [54]. Most of these domains have the homologs in the TFs identified in the mammal stem cells. Meiotic genes machinery appeared at the earliest branches on the eukaryotic tree and was evidenced in *Giardia lamblia* [60].

Analysis of the genome of *Entamoeba histolytica* gives more specific information about the molecular control of the life cycle associated with fluctuation in the O_2 level.

Entamoeba histolytica undergoes the reversible switch between infective cysts and invasive trophozoites. From anoxic colon, *E. histolytica* counters a relatively high-oxygen environment during invasive amebiasis in intestinum, and coping with this change is, therefore, an important virulence factor.

By analogy mammal stem cells localized in the almost anoxic niche should cope with the rise of O_2 upon their exit into the circulation.

Different stages of life cycle have different sets of the expressed genes. These stage-specific genes are also developmentally regulated. Genes enriched in cysts included cysteine proteinases and transmembrane kinase enzymes involved in DNA repair. In trophozoites, genes are involved in invasion, regulators of the differentiation, G-coupled receptors, signal transduction proteins, and TFs [61]. In *E. histolytica* in the multidrug resistance proteins Eh*PGP1* and Eh*PGP5* gene promoter consensus sequences for the binding of putative AP-1, HOX, OCT-1, 6, PIT-1, MYC, and GATA-1 TFs were found [62,63]. Also, tumor suppressor p53, C/EBP, STAT, and MYB DNA-binding proteins were identified.

In addition, epigenetic regulators as in mammal stem cells control gene expression. For instance, chromatin assembly factor-1-like protein (CAF-1) is a conserved developmental essential factor in the metazoan and protist that operates by regulating heterochromatin organization, asymmetric cell division, and specific signal transduction through epigenetic modulations of the chromatin [64]. Specific histone modifications appear during Paramecium development, which is strongly reduced in chromatin assembly factor-1-like protein (PtCAF-1) depleted cells [65]. Interestingly, very recently, totipotent embryonic cells were obtained in vitro through down-regulation of the CAF-1 [66].

Eukaryotic histones are targets of several posttranslational modifications such as acetylation/deacetylation, methylation/demethylation, phosphorylation, and ubiquitination [54,67]. S-adenosyl methionine, a main physiological methyl group donor in cellular methyltransferase reactions (e.g., mammalian DNA methyltransferase), is important for the epigenetic regulation in *E. histolytica* [54,68]. Furthermore, at least one member of the sirtuin super family deacetylases, the classical SIR2, can be traced back to the common ancestor of eukaryotes and archaea [54]. They deacetylate DNA sequences in a NAD-dependent manner just as sirtuins in mammal stem cells. Also, histone demethylation in protist is carried out by the Jumonji-related (JOR/JmjC) domain [69,70] and has a bacterial origin [69]. This is interesting due to the fact that the same family of demethylases has a crucial role in regulating epigenetic processes in animals.

11.2.2 Choanoflagellate and Last Common Ancestor of Metazoan

Phylogenetic analysis indicates that choanoflagellates are the unicellular organisms phylogenetically closest to the Metazoa [71,72].

Comparing the choanoflagellate *Monosiga brevicollis* genome to early metazoan reveals the molecular and metabolic regulators presented in the last common ancestor of choanoflagellates and metazoans. Many pathways that are the hallmarks of the mammal stem cell maintenance are missing entirely; no receptors or ligands were identified from the Wnt, or Toll signaling pathway. However, homologs of the Notch, STAT, and Hedgehog signaling pathway components are present [72]. In addition, phosphotyrosine-based signaling is found in abundance in the *M. brevicollis* genome. TGF-β signaling is restricted to the Metazoa with neither ligand nor receptor molecules being found outside of the animal kingdom [73].

The core transcriptional apparatus of *M. brevicollis* is, in many ways, typical of most eukaryotes examined to date including, for example, all 12 RNA polymerase II subunits and most of the transcription elongation factors (TFIIS, NELF, PAF, DSIF, and P-TEFb) [72].

Most of TFs are C_2H_2-type zinc fingers, FOX TFs, otherwise known only from metazoans and fungi. *M. brevicollis* contains a subset of the TF families previously thought to be specific to metazoans. Members of the p53, MYC, and SOX/TCF families were identified as well as the homolog of MYC TF whose activity is implicated in the regulation of the differentiation and mobilization of the hematopoietic stem cells (HSCs) [74]. Also, p53 regulates mammalian stem cell self-renewal; the SOX family of TFs involve the members that are part of a core transcriptional regulatory network that maintains the pluripotent state of mammal stem cells (see Chapter 7). These data implicate that the abovementioned TFs took place in the "stemness toolkit" at the choanoflagellate level.

Presence of all these factors in choanoflagellateas and the metazoan indicates that they evolved before the divergence of the choanoflagellates and metazoan and further suggests that they were present in the last common ancestor of choanoflagellates and metazoans.

In contrast, many TF families associated with metazoan patterning and development (ETS, HOX, NHR, POU, and T-box) seem to be absent [72].

Comparing the genomes of basal metazoan anthozoan cnidarian *Nematostella vectensis*, the hydrozoan cnidarian *Hydra magnipapillata*, the placozoan *Trichoplax adhaerens*, the demosponge *Amphimedon queenslandica*, and the expanding list of bilaterian genomes revealed that genome organization and content as well as TFs and components of the signaling pathways are remarkably similar [75]. Also, comparing those genomes to the genome of the choanoflagellates helps to establish a set of TFs and signaling molecules that were present in the last common ancestor of metazoan before their divergence [75]. This set is considered to have been developed at the protometazoan stage. Also, it revealed the factors that were prerequisite for the evolution of metazoan multicellularity.

The ancestral metazoan genome included TFs that are members of the bHLH, MEF2, FOX, SOX, T-BOX, ETS, nuclear receptor, Rel/NF-kB, bZIP, Smad families, and homeobox-containing classes, including ANTP, PRD-like, PAX, POU, LIM-homeodomain, SIX, and TALE. Some of these TFs have even more ancient origins (e.g., FOX, bZIP, Rel/NK-kB, TALE, typical (non-TALE) homeoboxes, and T-box) [72]. These factors could be divided into three categories. Animal TF genes that have no clear relatives outside the Metazoa are considered a type I novelty and currently include nuclear receptor families, ANTP homeobox. The POU, PAX, and SIX homeobox classes all can be classified as type II novelties, where the animal restricted POU, SIX, and PAX domains are combined with the more ancient homeodomains to produce the metazoan novelty. In contrast, type III novelties are those where ancient premetazoan domains combine in novel ways to generate metazoan specific domain architecture; an example is the animal-specific way in which the ancient LIM and homeodomain combine in the LIM-homeodomain class [76].

Also, functional Wnt, Notch, and TGF-β signaling pathways were developed on the protometazoan stage [73].

11.2.3 Porifera

The porifera are the oldest living metazoans. *Amphimedon queenslandica* (Porifera, Demospongiae, Haplosclerida, and Niphatidae) is the first poriferan representative to have its genome sequenced, assembled, and annotated [77].

Comparative analysis of the *Amphimedon* genome reveals genomic events linked to the origin and early evolution of metazoa, including the appearance, expansion, and diversification of pan-metazoan TF, signaling pathway, and structural genes [78].

Analysis of the *Amphimedon* genome allows us to assess the origin of the important vital function in the metazoan regulation of the cell cycling and growing, apoptosis, innate immunity and specialization of the cell types, developmental signaling, and gene regulation.

Considering the signaling pathways, the p53-mediated response to DNA damage is holozoan (metazoan plus choanoflagellates) specific, but MDM2 ubiquitin ligase that regulates p53 appears as a eumetazoan feature [78]. In contrast to the cell-cycle machinery, most of the apoptotic circuitry is unique to animals, increasing in complexity along metazoan, eumetazoan, and bilaterian stems. Both intrinsic and extrinsic programmed cell death pathways require caspases, a metazoan-specific family of cysteine aspartyl proteases. The intrinsic pathway-driven cell death is regulated by the BCL-2 family of pro- and antiapoptotic factors. The proapoptotic protein BAK arose in the metazoan lineage, whereas BAX and BOK seem to be eumetazoan-specific. BCL-2 and BCL-X are antiapoptotic and metazoan-specific [78]. Furthermore, spectra of TFs include PAXB, LHX genes, SOXB, MSX, MEF2, IRX and BHLH, FOS, JUN, NF-E2A, GATA, GLI, LIM, and *Amphimedon* lacks *HOX* genes [76,79,80].

When considering signaling cascades, Wnt, TGF-β, Notch, and Hedgehog pathways are functional. The ligands and receptors of all of these cascades are metazoan innovations at the cell surface apart from the eumetazoan-specific Hedgehog ligand (some signaling components of the Hedgehog pathway have ancient origin, e.g., dispatched and patched genes). The TFs specific to these pathways are also metazoan-specific (Tcf/Lef, Smads, CSL, Gli), whereas the cytosolic signal transducers generally have more ancient origins, except for the TGF-β pathway, which is metazoan novelty [78].

11.2.3.1 Molecular Analysis of Archeocytes and Choanocytes

It is well known that sponges have remarkable reconstitutive and regenerative abilities [81,82]. Sponges possess a well-developed system to regulate stem cell differentiation and probably also self-renewal [83]. Archeocytes, thought to be pluripotent stem cells in sponges, are essential for the reconstitution of dissociated sponge cells [84], and other types of cells differentiate from archeocytes, including choanocytes [85]. The presumably pluripotent archeocytes support both sexual and asexual reproduction systems. Choanocytes are considered as other stem cells in sponge. They are food-entrapping cells with morphological features similar to those of choanoflagellates (microvillus collar and a flagellum). Their known abilities to transform into archeocytes under specific circumstances and to give rise to gametes (mostly sperm) indicate that even when they are fully differentiated, choanocytes maintain pluripotent stem cell-like potential [83]. Since archeocytes and choanocytes represent the evolutionarily oldest extant metazoans, molecular analysis of these cells could help to elucidate the origin of stem cell self-renewal and differentiation [83].

This analysis revealed the existence of the genome safeguard pathway that involves PIWI proteins. These are known to function mostly in a small RNA (piRNA)-mediated manner, via several different molecular mechanisms [86]. The best-studied molecular mechanism of PIWI is to bind piRNAs, which guide PIWI to degrade and thus repress the translation of specific complementary mRNAs, including those of transposable elements. Furthermore, accumulating studies on nonconventional model animals, PIWI proteins function to maintain totipotency/multipotency in various animals. For example, PIWI homologs are expressed in multipotent stem cells of a colonial tunicate [87], pluripotent stem cells (neoblasts) of planarians [88], interstitial stem cells in hydrozoans [89], but also in human embryonic tissue and cancer [90].

Archeocytes and choanocytes express NANOS (a CCHC zinc finger RNA-binding protein) and PL10 (a Vasa-related protein), while *VASA* (a gene encoding a DEAD-box RNA helicase) is identified in archeocytes. Interestingly, these genes seem to be markers of the stem cell in other animals: *VASA* and *PL10* are expressed in planarians [91], and *VASA* and *NANOS* are expressed in interstitial cells in hydra [92,93] and in a ctenophore [94]. It is interesting that the regulatory actions of most of those proteins, including Musashi, either target or depend on RNAs (small RNAs, so far as is known), or both. Expression of Musashi homolog in demosponges is specifically restricted to archeocytes [95].

11.2.4 Placozoa

Trichoplax adhaerens is considered the simplest animal. As explained in Chapter 10, the genome analysis firmly suggested that the Placozoa represent the basal lower metazoan phylum: They would have arisen relatively soon after the evolutionary transition from unicellular to multicellular forms.

In spite of the fact that, *T. adhaerens* consists of only six cell types, its genome encodes a rich array of TFs and signaling genes, showing remarkable complexity [96].

Analysis of its genome revealed a rich repertoire of TFs commonly associated with patterning and regionalization during eumetazoan development, including homeobox-containing genes from the *ANTP*, Paired (*PRD*), *POU*, and *SIX* subfamilies, animal-specific *SOX* (Sry-related HMG-box) family, the T-box family, including brachyury (whose expression defines the blastopore in eumetazoan gastrulation), the opisthokont (animal and fungi)-specific *FOX* (forkhead/winged-helix) family, *LIM* class, *TALE* family (e.g., *MEIS*), helix loop helix class (e.g., *MYC*) zinc finger class (*GATA*, nuclear receptor, and C2H2). Considering signaling pathways, Wnt/β-catenin, TGF-β, Notch, Jak/STAT are present. In contrast, the Hedgehog ligand is still absent [73]. Meiosis-associated genes are found in the Trichoplax genome, such as the zinc-finger protein NANOS and a member of the VASA/PL10 family of DEAD-box helicases, as well as homologs of mago nashi, PAR-1, pumilio, and tudor, which are all implicated in primary germ cell development in eumetazoans.

Trichoplax has a complex life cycle [96]. The crawling presumptive adult form seen in laboratory culture reproduces asexually by fission. One or more small ciliated spheres, called "swarmers," bud off from the crawler form, and presumably revert to the adult state. The crawling form may produce sperm and eggs that undergo embryonic development under the right environmental conditions and either directly or indirectly through a larval stage, give rise to the adult form. It was suggested that the common part of the genome is inherited from the common ancestor and shared with other metazoan while there is still one part of the genome that is specific, associated with the specific life cycle, and style [96].

11.2.5 Cnidaria

11.2.5.1 Nemostella vestens

Genome analysis of *Nemostella vestens*, a sea anemone belonging to the phylum Cnidaria, served to reconstruct genome organization and gene repertoire of the ancestral eumetazoan genome. This revealed that nearly 80% of the ancestral eumetazoan genes clearly have homologs outside the animal kingdom including fungi, plants, slime molds, and ciliates [97]. These are the genes families that were established in the unicellular ancestor of metazoan (LECA/FECA) and are involved in essential eukaryotic functions [97]. The remaining 20% comprises animal novelties that were created along the evolution of the eumetazoa.

For instance, Wnt signaling has a novel component secreted frizzled related factors, which is combined with the ancient part of the pathway. The same principle is employed for the other major signaling pathways. Also, cytokine signaling has novel components, such as Inositol 1,4,5-triphosphate receptor, SOCS, arrestin, guanine nucleotide binding protein γ, regulator of G protein signaling, and so on (a detailed list of genes is reviewed in Ref [97]).

11.2.5.2 Hydra

Hydra has a multipotent stem cell system that behaves similarly to the stem cell systems in bilaterians (see Chapter 10). The availability of the Hydra genome sequence provides an opportunity to understand how this system works. Hydra contains multipotent stem cell intestitial stem cells, "I-cells," that give rise to germ cells, nerve cells, nematocytes, and secretory cells (see Chapter 10).

Looking for five genes that have been shown to induce pluripotency in differentiated somatic cells of mammals (*MYC*, *NANOG*, *KLF4*, *OCT4*, and *SOX2*) evidenced that the Hydra genome does not contain all the homologs.

MYC homolog (*MYC1* and *MYC2*) and max are identified in Hydra. *MYC1* and *MYC2* are expressed in interstitial stem cells; in contrast *MYC1* but not *MYC2* is identified in the proliferating epithelial cells and germ cell formation [98].

SOX genes are present across the cnidarian. Four paralogues representing three different *SOX* families are expressed in I-cells, but the evolutionary relationship between vertebrate *SOX2* genes and Hydra *SOX2* B genes is not clear [99]; *hySOX1/2/3* seems to represent the ancestral form of the *SOX2* B1 and *SOX2* B2 gene family. Transfection of the *hySOX1/2/3* in mESCs where *msox2* activity could be conditionally regulated evidenced that *hySOX1/2/3* could not substitute the function of *mSox* in mESCs [100].

Krüppel-like family is present, but no homolog of *kfl4* was found [99]. Mammalian OCT-4 is a member of class V of the POU family. Members of the POU family are present in Hydra, but class V is absent. Ectopic expression of the POU domain protein Polynem (PLN) in the epithelial cells induces stem cells neoplasms and the loss of the epithelial tissue (dedifferentiation) [101]. This is accompanied with the expression of stem cell marker genes *NANOS*, *VASA*, *PIWI*, and *MYC* genes [101].

However, further analysis showed that POU-5, whose member is a stem cell pluripotency marker *OCT4*, is a vertebrate novelty. This indicated that Pln is a paralogue of the vertebrate *OCT4* gene.

It is likely that vertebrate-specific core TF network (OCT-3/4, NANOG, and SOX-2) is a bilaterian invention [100]. Also, it was suggested that stem cell genetic network in Hydra probably has an evolutionary origin independent from the network used in mammalian stem cells.

Epigenetic regulation by miRNA. *HOX* and *paraHOX* gene families arose from a megacluster that included a number of other homeobox genes.

The embryonic ectoderm development (EED) homolog hyEED, a member of Polycomb repressor complex, and two PCR2 are expressed in interstitial stem cells but is lost during their differentiation of nematocytes indicating that, as in the mammalian stem cell, chromatin modification influences stem cell fate (see Chapter 7). Also, hyEED is co-expressed with EZH2 indicating that Hydra may possess PCR two complex that is similar to mammals [102].

Also, interstitial stem cells express bHLH TFs gene achaete-scute homolog *Cnash, Hyzic* homolog of Zn-finger TF zic-odd-paired, *NANOS* (*cnNANOS1* and *cnNANOS2*), vasa-related genes *cnVAS1* and *cnVAS2*. FOXO TF is also present in the interstitial stem cells. Its overexpression increases the interstitial stem cell proliferation and inhibits their differentiation [103,104].

To safeguard genome integrity, an evolutionary conserved small RNA-based silencing mechanism implicated PIWI proteins and PIWI-interacting (piRNA) suppress transposons through transcriptional and posttranscriptional silencing [105]. It was evidenced that PIWI proteins (HYWI and HYLI) are implicated in the piRNAs mediated transposon silencing in Hydra I-cells, suggesting operating piRNA pathway [106]. These genes are implicated in the stem cell regulation and differentiation in the various bilaterians including mammalian stem and progenitors cell [107–111].

Wnt, TGF-β, Notch, and Hedgehog pathways are functional in Hydra. Activated Wnt signaling is tightly linked to stemness [112]. Creation of the transgenic interstitial stem cell expressing eGFP revealed that Wnt/β-catenin pathway and Notch are activated in I-stem cells [113] where they are implicated in the stem cell fate determination [102]. High levels of frizzled transcripts (downstream component of the Wnt signaling) were detected in interstitial stem cells. Activating downstream events of the Wnt-cascade in the postmetamorphic life induced recruitment of nematocytes and nerve cells from the pool of I-cells that were preceded by the increased proliferation of the i-cells [114].

11.2.6 Platyhelminthes

Planarians, representative members of this phylum, exhibit an extraordinary ability to regenerate lost body parts, making this animal an attractive model for the study of stem cell biology (see Chapter 10). Amputated planaria regenerate missing complex body parts within 10 days, and even cutting animals into dozens of pieces, will result in dozens of regenerated healthy animals [115]. This extreme plasticity of planaria critically depends on stem cells (neoblasts) that are the only proliferating cells in asexual animals and constitute 30% of all cells [116]. Recently, transplantation studies have proven the capacity of a single neoblast to regenerate entire stem-cell-deficient animals and to transform their host into a genetic clone of the donor, indicating that neoblast are totipotent stem cells [117,118].

Molecular mechanisms for pluripotency in neoblasts revealed similarity to those in mammalian ESCs. Genes important for pluripotency in ESCs *OCT4* and *SOX2*, including their upstream regulators as well as targets of them, are well conserved and upregulated in neoblasts. Various *POU* paralogues appear to be utilized in the stem cells of planaria neoblasts, such as *POU5* and *POU6* [119].

SOX-2 belongs to the class B family of SOX proteins. Two candidates, Smed-SOXB-1 and Smed-SOXB-2, homologs of mammalian members of this class, were identified in neoblasts [117].

The apparent lack of a Nanog homolog contrasts with extensive conservation of gene expression observed for its targets and indicates that in planaria, its role might be played by a different TF.

Also, conserved expression of epigenetic regulators known to be required for maintaining pluripotency in ESCs (esBAF, MLL1, PRC2, and PAF1 complex) is critical for the maintenance of the neoblasts [117]. Neoblasts express the set of miRNAs [120], but it seems these are different from miRNAs expressed in the ESCs. The data presented here provide the evidences for global conservation of expression of pluripotency-associated genes [115].

In planaria, Wnt signaling is associated with control of the regeneration and the specification of missing body part [121]. Furthermore, planaria TOR homolog (Smed-TOR) I ubiquitously are expressed in neoblasts and in differentiated tissues. Inhibition of TOR with RNA interference impaired cellular proliferation; it is critical for tissue regeneration and is essential for the maintenance of balance between cell proliferation and cell death. It seems that TOR functions via TOR Complex one in planaria [122], similarly to mammalian stem cells (see Chapter 6). Also, homologs to mammalian tumor suppressor phosphatase and tensin homolog deleted on chromosome 10 (PTEN) are identified in planaria *Smed-PTEN-1* and *Smed-PTEN-2*; inactivation of these disrupts regeneration and leads to hyperproliferation of the neoblasts. Treatment with rapamycin inhibits this phenotype, while inhibition of the PTEN activates Akt kinase. In mammals, the PI3K/Akt kinase pathway activates the TOR pathway, which is inhibited by PTEN (see Chapter 6). Thus, the effect of the rapamycin and activation of the Akt kinase following inhibition of the PTEN indicate that in planaria PTEN homologs regulate stem cells through TOR signaling [123].

11.3 Evolution of HIF Pathway

The HIF pathway is the central regulator of the metabolic homeostasis in the low oxygen environment, not only in the mammalian stem cells (see Chapter 7).

Even the whole HIF system (HIF, PHD, and VHL) is not detected in protest homologs of the HIF pathway components existing since prokaryotes. Namely, HIF-1α protein belongs to the family of TFs containing bHLH-PAS domains (bHLH-PAS TFs) (see Chapter 7). HIFs arise from the protein sensors containing PAS domain [124] involved in the environmental sensing. These types of sensors were identified in protist and prokaryotes [124]. However, it is suggested that association of the PAS with bHLH DNA-binding domain did not appear until metazoan.

PHD acts as a true O_2-sensing HIF system component. PHD-related oxidase has existed since prokaryotes and protist [125–127], and it is hypothesized that they mediate oxygen-sensing mechanisms in these animals, and this was further established as evolutionary conserved from the eukaryotes common ancestor to mammals [125].

However, in amoeba [128] and plants [129], oxygen-sensing prolyl hydroxylases seem not to be connected to TFs [130]. Also, in protist, *Perkinsus olseni* PHD has an additional role implicated in the virulence and pathogenicity [125].

Human HIF-1 homolog is identified in primitive organisms, such as nematodes, coral (Porifera), and sea anemone (Cnidaria), and is present in all metazoan [131]. The most primitive metazoan has a single isoform similar to human HIF-1α. HIF-2 appeared after division of arthropods and fish and the third HIF-3α appeared after division of birds and mammals. HIF-β (ARNT) protein is also highly conserved and appeared in nematodes [132,133]. FIH has been identified since corals [131].

Recently, a functional HIF system was shown to exist in the simplest metazoan *T. adhaerens*. Here, as in the mammalian stem cells, the HIFs regulate expression of the glycolytic and metabolic enzymes [134].

In nonmetezoan, in fission yeast (*Schizosaccharomysec pombe*), prolyl hydroxylase family member Ofd1 regulates a TF named Sre1 (sterol regulatory element binding protein) in an oxygen-dependent manner [135]. In contrast, in budding yeast prolyl hydroxylase is not involved in the oxygen-sensing [136].

Since then, the functional HIF system appeared in the earliest animal species indicating that it is evolved in response to the rise of O_2 concentration in the pre-Cambrian period (see Chapter 9).

The data presented above evidence that the HIF pathway is the critical pan-metazoan system, enabling metabolic adaptation in the response to fluctuations in the environmental O_2 level.

11.4 Conclusion

Comparison between metabolic energetic features among major groups of eukaryotes and stem cells showed that the anaerobic nature of the mammalian stem cell is an ancestral characteristic inherited from unicellular eukaryotes. Genetic analysis revealed that stem cell metabolism is tightly regulated by the set of various signaling pathways, genetic, and epigenetic regulators that are evolutionary conserved. Also, the mechanisms that underlie stemness show astonishing narrow diversity during evolution. This is reflected through establishment of the stemness "toolkit" that involves molecular and signaling pathways enabling their maintenance (Figure 11.1). In contrast, appearance of species-specific genes is associated with particular lifestyle and cellular differentiation of protists, which is a prototype of the stem cell differentiation in metazoa. In fact, to perpetuate the protist line the return to an undifferentiated state was necessary (see Chapter 10, and Chapter 4, Figure 4.1). Hence these genes allowed in fact a coexistence of the self-renewal principle and differentiation (complex life cycle). This supports the concept of "stem cells essential genome." It would contain a "stemness toolkit" in addition to eukaryotic essential core, which includes genes required for DNA repair, transcription, RNA processing, export through the nuclear pore, RNA turnover, tRNA synthesis, translation, protein turnover, and metabolism [137]. Since the genes of the "essential core" enable the simple cell survival and division

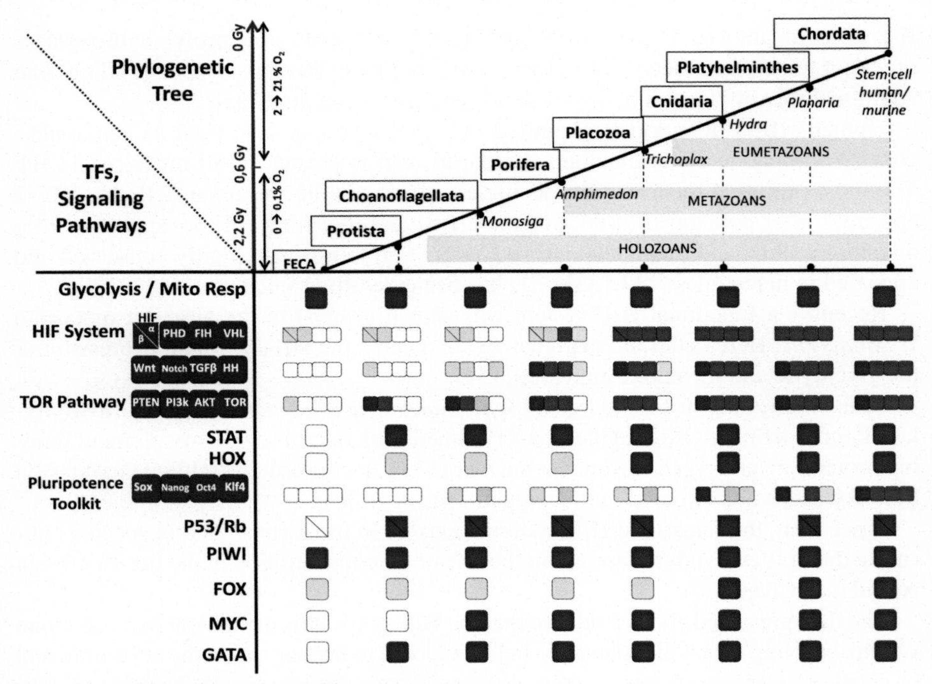

Figure 11.1 Illustration of evolutionary history of metabolic energetic features and molecular regulators implicated in the mammalian stem cell maintenance. Evolutionary analysis (from the first eukaryotic common ancestor (FECA) to mammals) of signaling pathways and transcription factors essential for the mammalian stem cell maintenance reveals their early evolutionary set up. Furthermore, it shows that bioenergetics of mammal stem cells is an ancestral characteristic, inherited either from the unicellular eukaryote (protist) or FECA stage. Dark blue squares represent the presence of the transcription factor homologs or functional signaling pathways in the corresponding eukaryotic group. Light blue squares represent the presence of either protein family whose member is the transcription factor in question or component (enzyme, receptor or ligand) of the signaling pathway analyzed. Non colored squares represent absence of the signaling pathway or absence of transcription factor or corresponding protein family. For example, forkhead box proteins have been identified since FECA stage, but the homologs of the mammalian FOX transcription factor (FOXO) have been identified since Hydra (Cnidaria). FIH, factor inhibiting HIF; FOX, forkhead homeobox; HH, Hedgehog signaling pathway; HIF, hypoxia inducible factor; Mito resp, mitochondrial respiration; PI3K, phosphotidylinositol 3-kinase; PTEN, phosphatase and tensin homolog; PHD, prolyl hydroxylase; Rb, retinoblastoma protein; STAT, Signal Transducers and Activators of Transcription; TFs, transcription factors; TOR, target of rapamycin; VHL, Von Hippel-Lindau factor.

à l'identique from protists to mammals (where it is perceived as "self-renewal"), the genes of the "stemness toolkit" aimed to prevent all pro-differentiation signals allowing this self-renewing division. The "stemness toolkit," although highly conserved, evolved in parallel with the evolution of the differentiation repertoire.

References

[1] Muller M, Mentel M, van Hellemond JJ, Henze K, Woehle C, Gould SB, et al. Biochemistry and evolution of anaerobic energy metabolism in eukaryotes. Microbiol Mol Biol Rev 2012;76(2):444–95.

[2] Embley TM, Martin W. Eukaryotic evolution, changes and challenges. Nature 2006;440(7084):623–30.

[3] Shiflett AM, Johnson PJ. Mitochondrion-related organelles in eukaryotic protists. Annu Rev Microbiol 2010;64:409–29.

[4] Simpson AG, Inagaki Y, Roger AJ. Comprehensive multigene phylogenies of excavate protists reveal the evolutionary positions of "primitive" eukaryotes. Mol Biol Evol 2006;23(3):615–25.

[5] Martin W, Muller M. The hydrogen hypothesis for the first eukaryote. Nature 1998;392(6671):37–41.

[6] Ginger ML, McFadden GI, Michels PA. Rewiring and regulation of cross-compartmentalized metabolism in protists. Philos Trans R Soc Lond B Biol Sci 2010;365(1541):831–45.

[7] Sousa FL, Thiergart T, Landan G, Nelson-Sathi S, Pereira IA, Allen JF, et al. Early bioenergetic evolution. Philos Trans R Soc Lond B Biol Sci 2013;368(1622):20130088.

[8] Steinbuchel A, Muller M. Anaerobic pyruvate metabolism of *Tritrichomonas foetus* and *Trichomonas vaginalis* hydrogenosomes. Mol Biochem Parasitol 1986;20(1):57–65.

[9] Dacks JB, Dyal PL, Embley TM, van der Giezen M. Hydrogenosomal succinyl-CoA synthetase from the rumen-dwelling fungus *Neocallimastix patriciarum*; an energy-producing enzyme of mitochondrial origin. Gene 2006;373:75–82.

[10] Lindmark DG. Energy metabolism of the anaerobic protozoon *Giardia lamblia*. Mol Biochem Parasitol 1980;1(1):1–12.

[11] Tielens AG, Rotte C, van Hellemond JJ, Martin W. Mitochondria as we don't know them. Trends Biochem Sci 2002;27(11):564–72.

[12] Boxma B, de Graaf RM, van der Staay GW, van Alen TA, Ricard G, Gabaldon T, et al. An anaerobic mitochondrion that produces hydrogen. Nature 2005;434(7029):74–9.

[13] de Graaf RM, Ricard G, van Alen TA, Duarte I, Dutilh BE, Burgtorf C, et al. The organellar genome and metabolic potential of the hydrogen-producing mitochondrion of *Nyctotherus ovalis*. Mol Biol Evol 2011;28(8):2379–91.

[14] Ginger ML, Fritz-Laylin LK, Fulton C, Cande WZ, Dawson SC. Intermediary metabolism in protists: a sequence-based view of facultative anaerobic metabolism in evolutionarily diverse eukaryotes. Protist 2010;161(5):642–71.

[15] Hunter Jr FE. Anaerobic phosphorylation due to a coupled oxidation-reduction between alpha-ketoglutaric acid and oxalacetic acid. J Biol Chem 1949;177(1):361–72.

[16] Sanadi DR, Fluharty AL. On the mechanism of oxidative phosphorylation. VII. The energy-requiring reduction of pyridine nucleotide by succinate and the energy-yielding oxidation of reduced pyridine nucleotide by fumarate. Biochemistry 1963;2:523–8.

[17] Penney DG, Cascarano J. Anaerobic rat heart. Effects of glucose and tricarboxylic acid-cycle metabolites on metabolism and physiological performance. Biochem J 1970;118(2):221–7.

[18] Gronow GH, Cohen JJ. Substrate support for renal functions during hypoxia in the perfused rat kidney. Am J Physiol 1984;247(4 Pt 2):F618–31.

[19] Pisarenko OI. Mechanisms of myocardial protection by amino acids: facts and hypotheses. Clin Exp Pharmacol Physiol 1996;23(8):627–33.

[20] Weinberg JM, Venkatachalam MA, Roeser NF, Nissim I. Mitochondrial dysfunction during hypoxia/reoxygenation and its correction by anaerobic metabolism of citric acid cycle intermediates. Proc Natl Acad Sci USA 2000;97(6):2826–31.

[21] Tomitsuka E, Kita K, Esumi H. The NADH-fumarate reductase system, a novel mitochondrial energy metabolism, is a new target for anticancer therapy in tumor microenvironments. Ann N Y Acad Sci 2010;1201:44–9.

[22] Goubern M, Andriamihaja M, Nubel T, Blachier F, Bouillaud F. Sulfide, the first inorganic substrate for human cells. FASEB J 2007;21(8):1699–706.

[23] Fu M, Zhang W, Wu L, Yang G, Li H, Wang R. Hydrogen sulfide (H2S) metabolism in mitochondria and its regulatory role in energy production. Proc Natl Acad Sci USA 2012;109(8):2943–8.

[24] Wang R. Physiological implications of hydrogen sulfide: a whiff exploration that blossomed. Physiol Rev 2012;92(2):791–896.

[25] Guo Z, Li CS, Wang CM, Xie YJ, Wang AL. CSE/H_2S system protects mesenchymal stem cells from hypoxia and serum deprivationinduced apoptosis via mitochondrial injury, endoplasmic reticulum stress and PI3K/Akt activation pathways. Mol Med Rep 2015;12(2):2128–34.

[26] Dongo E, Benko Z, Csizmazia A, Marosi G, Grottke A, Jucker M, et al. H_2S preconditioning of human adipose tissue-derived stem cells increases their efficacy in an in vitro model of cell therapy for simulated ischemia. Life Sci 2014;113(1–2):14–21.

[27] Lloyd D, Ralphs JR, Harris JC. *Giardia intestinalis*, a eukaryote without hydrogenosomes, produces hydrogen. Microbiology 2002;148(Pt 3):727–33.

[28] Townson SM, Upcroft JA, Upcroft P. Characterisation and purification of pyruvate:ferredoxin oxidoreductase from *Giardia duodenalis*. Mol Biochem Parasitol 1996;79(2):183–93.

[29] Sanchez LB, Muller M. Purification and characterization of the acetate forming enzyme, acetyl-CoA synthetase (ADP-forming) from the amitochondriate protist, *Giardia lamblia*. FEBS Lett 1996;378(3):240–4.

[30] van Dijken JP, van den Bosch E, Hermans JJ, de Miranda LR, Scheffers WA. Alcoholic fermentation by 'non-fermentative' yeasts. Yeast 1986;2(2):123–7.

[31] van Urk H, Postma E, Scheffers WA, van Dijken JP. Glucose transport in crabtree-positive and crabtree-negative yeasts. J Gen Microbiol 1989;135(9):2399–406.

[32] van den Thillart G, Modderkolk J. The effect of acclimation temperature on the activation energies of state III respiration and on the unsaturation of membrane lipids of goldfish mitochondria. Biochim Biophys Acta 1978;510(1):38–51.

[33] Rasoloson D, Tomkova E, Cammack R, Kulda J, Tachezy J. Metronidazole-resistant strains of *Trichomonas vaginalis* display increased susceptibility to oxygen. Parasitology 2001;123(Pt 1):45–56.

[34] Nixon JE, Wang A, Field J, Morrison HG, McArthur AG, Sogin ML, et al. Evidence for lateral transfer of genes encoding ferredoxins, nitroreductases, NADH oxidase, and alcohol dehydrogenase 3 from anaerobic prokaryotes to *Giardia lamblia* and *Entamoeba histolytica*. Eukaryot Cell 2002;1(2):181–90.

[35] Zhang J, Khvorostov I, Hong JS, Oktay Y, Vergnes L, Nuebel E, et al. UCP2 regulates energy metabolism and differentiation potential of human pluripotent stem cells. EMBO J 2011;30(24):4860–73.

[36] Zhang WC, Shyh-Chang N, Yang H, Rai A, Umashankar S, Ma S, et al. Glycine decarboxylase activity drives non-small cell lung cancer tumor-initiating cells and tumorigenesis. Cell 2012;148(1–2):259–72.

[37] Ito K, Suda T. Metabolic requirements for the maintenance of self-renewing stem cells. Nat Rev Mol Cell Biol 2014;15(4):243–56.
[38] Ezashi T, Das P, Roberts RM. Low O_2 tensions and the prevention of differentiation of hES cells. Proc Natl Acad Sci USA 2005;102(13):4783–8.
[39] Mohyeldin A, Garzon-Muvdi T, Quinones-Hinojosa A. Oxygen in stem cell biology: a critical component of the stem cell niche. Cell Stem Cell 2010;7(2):150–61.
[40] Varum S, Momcilovic O, Castro C, Ben-Yehudah A, Ramalho-Santos J, Navara CS. Enhancement of human embryonic stem cell pluripotency through inhibition of the mitochondrial respiratory chain. Stem Cell Res 2009;3(2–3):142–56.
[41] Maryanovich M, Zaltsman Y, Ruggiero A, Goldman A, Shachnai L, Zaidman SL, et al. An MTCH2 pathway repressing mitochondria metabolism regulates haematopoietic stem cell fate. Nat Commun 2015;6:7901.
[42] Chen T, Shen L, Yu J, Wan H, Guo A, Chen J, et al. Rapamycin and other longevity-promoting compounds enhance the generation of mouse induced pluripotent stem cells. Aging Cell 2011;10(5):908–11.
[43] Liu X, Zheng H, Yu WM, Cooper TM, Bunting KD, Qu CK. Maintenance of mouse hematopoietic stem cells ex vivo by reprogramming cellular metabolism. Blood 2015;125(10):1562–5.
[44] Kondoh H, Lleonart ME, Nakashima Y, Yokode M, Tanaka M, Bernard D, et al. A high glycolytic flux supports the proliferative potential of murine embryonic stem cells. Antioxid Redox Signal 2007;9(3):293–9.
[45] Takubo K, Nagamatsu G, Kobayashi CI, Nakamura-Ishizu A, Kobayashi H, Ikeda E, et al. Regulation of glycolysis by Pdk functions as a metabolic checkpoint for cell cycle quiescence in hematopoietic stem cells. Cell Stem Cell 2013;12(1):49–61.
[46] Chung S, Dzeja PP, Faustino RS, Perez-Terzic C, Behfar A, Terzic A. Mitochondrial oxidative metabolism is required for the cardiac differentiation of stem cells. Nat Clin Pract Cardiovasc Med 2007;4(Suppl. 1):S60–7.
[47] Shyh-Chang N, Zheng Y, Locasale JW, Cantley LC. Human pluripotent stem cells decouple respiration from energy production. EMBO J 2011;30(24):4851–2.
[48] Prigione A, Adjaye J. Modulation of mitochondrial biogenesis and bioenergetic metabolism upon in vitro and in vivo differentiation of human ES and iPS cells. Int J Dev Biol 2010;54(11–12):1729–41.
[49] Ito K, Carracedo A, Weiss D, Arai F, Ala U, Avigan DE, et al. A PML-PPAR-delta pathway for fatty acid oxidation regulates hematopoietic stem cell maintenance. Nat Med 2012;18(9):1350–8.
[50] Nichols J, Smith A. Naive and primed pluripotent states. Cell Stem Cell 2009;4(6):487–92.
[51] Folmes CD, Dzeja PP, Nelson TJ, Terzic A. Metabolic plasticity in stem cell homeostasis and differentiation. Cell Stem Cell 2012;11(5):596–606.
[52] Aad G, Abajyan T, Abbott B, Abdallah J, Abdel Khalek S, Abdelalim AA, et al. Search for dark matter candidates and large extra dimensions in events with a photon and missing transverse momentum in pp collision data at sqrt[s]=7 TeV with the ATLAS detector. Phys Rev Lett 2013;110(1):011802.
[53] Simsek T, Kocabas F, Zheng J, Deberardinis RJ, Mahmoud AI, Olson EN, et al. The distinct metabolic profile of hematopoietic stem cells reflects their location in a hypoxic niche. Cell Stem Cell 2010;7(3):380–90.
[54] Iyer LM, Anantharaman V, Wolf MY, Aravind L. Comparative genomics of transcription factors and chromatin proteins in parasitic protists and other eukaryotes. Int J Parasitol 2008;38(1):1–31.

[55] Loftus B, Anderson I, Davies R, Alsmark UC, Samuelson J, Amedeo P, et al. The genome of the protist parasite *Entamoeba histolytica*. Nature 2005;433(7028):865–8.
[56] Grimson MJ, Coates JC, Reynolds JP, Shipman M, Blanton RL, Harwood AJ. Adherens junctions and beta-catenin-mediated cell signalling in a non-metazoan organism. Nature 2000;408(6813):727–31.
[57] Dacks JB, Doolittle WF. Reconstructing/deconstructing the earliest eukaryotes: how comparative genomics can help. Cell 2001;107(4):419–25.
[58] Aravind L, Anantharaman V, Balaji S, Babu MM, Iyer LM. The many faces of the helix-turn-helix domain: transcription regulation and beyond. FEMS Microbiol Rev 2005;29(2):231–62.
[59] Aravind L, Iyer LM, Koonin EV. Comparative genomics and structural biology of the molecular innovations of eukaryotes. Curr Opin Struct Biol 2006;16(3):409–19.
[60] Ramesh MA, Malik SB, Logsdon Jr JM. A phylogenomic inventory of meiotic genes; evidence for sex in *Giardia* and an early eukaryotic origin of meiosis. Curr Biol 2005;15(2):185–91.
[61] Ehrenkaufer GM, Haque R, Hackney JA, Eichinger DJ, Singh U. Identification of developmentally regulated genes in *Entamoeba histolytica*: insights into mechanisms of stage conversion in a protozoan parasite. Cell Microbiol 2007;9(6):1426–44.
[62] Gomez C, Perez DG, Lopez-Bayghen E, Orozco E. Transcriptional analysis of the EhPgp1 promoter of *Entamoeba histolytica* multidrug-resistant mutant. J Biol Chem 1998;273(13):7277–84.
[63] Perez DG, Gomez C, Lopez-Bayghen E, Tannich E, Orozco E. Transcriptional analysis of the EhPgp5 promoter of *Entamoeba histolytica* multidrug-resistant mutant. J Biol Chem 1998;273(13):7285–92.
[64] Yu Z, Liu J, Deng WM, Jiao R. Histone chaperone CAF-1: essential roles in multi-cellular organism development. Cell Mol Life Sci 2015;72(2):327–37.
[65] Ignarski M, Singh A, Swart EC, Arambasic M, Sandoval PY, Nowacki M. *Paramecium tetraurelia* chromatin assembly factor-1-like protein PtCAF-1 is involved in RNA-mediated control of DNA elimination. Nucleic Acids Res 2014;42(19):11952–64.
[66] Ishiuchi T, Enriquez-Gasca R, Mizutani E, Boskovic A, Ziegler-Birling C, Rodriguez-Terrones D, et al. Early embryonic-like cells are induced by downregulating replication-dependent chromatin assembly. Nat Struct Mol Biol 2015;22(9):662–71.
[67] Martens JA, Winston F. Recent advances in understanding chromatin remodeling by Swi/Snf complexes. Curr Opin Genet Dev 2003;13(2):136–42.
[68] Paraskevopoulou C, Fairhurst SA, Lowe DJ, Brick P, Onesti S. The Elongator subunit Elp3 contains a Fe_4S_4 cluster and binds S-adenosylmethionine. Mol Microbiol 2006;59(3):795–806.
[69] Aravind L, Koonin EV. Prokaryotic homologs of the eukaryotic DNA-end-binding protein Ku, novel domains in the Ku protein and prediction of a prokaryotic double-strand break repair system. Genome Res 2001;11(8):1365–74.
[70] Anantharaman V, Koonin EV, Aravind L. Regulatory potential, phyletic distribution and evolution of ancient, intracellular small-molecule-binding domains. J Mol Biol 2001;307(5):1271–92.
[71] King N. The unicellular ancestry of animal development. Dev Cell 2004;7(3):313–25.
[72] King N, Westbrook MJ, Young SL, Kuo A, Abedin M, Chapman J, et al. The genome of the choanoflagellate *Monosiga brevicollis* and the origin of metazoans. Nature 2008;451(7180):783–8.
[73] Richards GS, Degnan BM. The dawn of developmental signaling in the metazoa. Cold Spring Harb Symp Quant Biol 2009;74:81–90.

[74] Wilson A, Murphy MJ, Oskarsson T, Kaloulis K, Bettess MD, Oser GM, et al. c-Myc controls the balance between hematopoietic stem cell self-renewal and differentiation. Genes Dev 2004;18(22):2747–63.

[75] Degnan BM, Vervoort M, Larroux C, Richards GS. Early evolution of metazoan transcription factors. Curr Opin Genet Dev 2009;19(6):591–9.

[76] Larroux C, Luke GN, Koopman P, Rokhsar DS, Shimeld SM, Degnan BM. Genesis and expansion of metazoan transcription factor gene classes. Mol Biol Evol 2008;25(5):980–96.

[77] Degnan BM, Adamska M, Craigie A, Degnan SM, Fahey B, Gauthier M, et al. The demosponge *Amphimedon queenslandica*: reconstructing the ancestral metazoan genome and deciphering the origin of animal multicellularity. CSH Protoc 2008;2008. pdb emo108.

[78] Srivastava M, Simakov O, Chapman J, Fahey B, Gauthier ME, Mitros T, et al. The *Amphimedon queenslandica* genome and the evolution of animal complexity. Nature 2010;466(7307):720–6.

[79] Simionato E, Ledent V, Richards G, Thomas-Chollier M, Kerner P, Coornaert D, et al. Origin and diversification of the basic helix-loop-helix gene family in metazoans: insights from comparative genomics. BMC Evol Biol 2007;7:33.

[80] Larroux C, Fahey B, Degnan SM, Adamski M, Rokhsar DS, Degnan BM. The NK homeobox gene cluster predates the origin of Hox genes. Curr Biol 2007;17(8):706–10.

[81] Wilson HV. A new method by which sponges may be artificially reared. Science 1907;25(649):912–5.

[82] Fortunato SA, Adamski M, Adamska M. Comparative analyses of developmental transcription factor repertoires in sponges reveal unexpected complexity of the earliest animals. Mar Genomics 2015;S1874-7787(15)30014-3.

[83] Funayama N. The stem cell system in demosponges: insights into the origin of somatic stem cells. Dev Growth Differ 2010;52(1):1–14.

[84] Coutinho CC, Fonseca RN, Mansure JJ, Borojevic R. Early steps in the evolution of multicellularity: deep structural and functional homologies among homeobox genes in sponges and higher metazoans. Mech Dev 2003;120(4):429–40.

[85] Muller WE. The stem cell concept in sponges (Porifera): metazoan traits. Semin Cell Dev Biol 2006;17(4):481–91.

[86] Grimson A, Srivastava M, Fahey B, Woodcroft BJ, Chiang HR, King N, et al. Early origins and evolution of microRNAs and Piwi-interacting RNAs in animals. Nature 2008;455(7217):1193–7.

[87] Sunanaga T, Inubushi H, Kawamura K. Piwi-expressing hemoblasts serve as germline stem cells during postembryonic germ cell specification in colonial ascidian, *Botryllus primigenus*. Dev Growth Differ 2010;52(7):603–14.

[88] Shibata N, Rouhana L, Agata K. Cellular and molecular dissection of pluripotent adult somatic stem cells in planarians. Dev Growth Differ 2010;52(1):27–41.

[89] Seipel K, Yanze N, Schmid V. The germ line and somatic stem cell gene Cniwi in the jellyfish Podocoryne carnea. Int J Dev Biol 2004;48(1):1–7.

[90] Navarro A, Tejero R, Vinolas N, Cordeiro A, Marrades RM, Fuster D, et al. The significance of PIWI family expression in human lung embryogenesis and non-small cell lung cancer. Oncotarget 2015;6(31):31544–56.

[91] Shibata N, Umesono Y, Orii H, Sakurai T, Watanabe K, Agata K. Expression of vasa(vas)-related genes in germline cells and totipotent somatic stem cells of planarians. Dev Biol 1999;206(1):73–87.

[92] Mochizuki K, Sano H, Kobayashi S, Nishimiya-Fujisawa C, Fujisawa T. Expression and evolutionary conservation of nanos-related genes in Hydra. Dev Genes Evol 2000;210(12):591–602.

[93] Mochizuki K, Nishimiya-Fujisawa C, Fujisawa T. Universal occurrence of the vasa-related genes among metazoans and their germline expression in Hydra. Dev Genes Evol 2001;211(6):299–308.

[94] Alie A, Leclere L, Jager M, Dayraud C, Chang P, Le Guyader H, et al. Somatic stem cells express *Piwi* and *Vasa* genes in an adult ctenophore: ancient association of "germline genes" with stemness. Dev Biol 2011;350(1):183–97.

[95] Okamoto K, Nakatsukasa M, Alie A, Masuda Y, Agata K, Funayama N. The active stem cell specific expression of sponge Musashi homolog EflMsiA suggests its involvement in maintaining the stem cell state. Mech Dev 2012;129(1–4):24–37.

[96] Srivastava M, Begovic E, Chapman J, Putnam NH, Hellsten U, Kawashima T, et al. The *Trichoplax* genome and the nature of placozoans. Nature 2008;454(7207):955–60.

[97] Putnam NH, Srivastava M, Hellsten U, Dirks B, Chapman J, Salamov A, et al. Sea anemone genome reveals ancestral eumetazoan gene repertoire and genomic organization. Science 2007;317(5834):86–94.

[98] Hartl M, Glasauer S, Valovka T, Breuker K, Hobmayer B, Bister K. Hydra myc2, a unique pre-bilaterian member of the myc gene family, is activated in cell proliferation and gametogenesis. Biol Open 2014;3(5):397–407.

[99] Chapman JA, Kirkness EF, Simakov O, Hampson SE, Mitros T, Weinmaier T, et al. The dynamic genome of Hydra. Nature 2010;464(7288):592–6.

[100] Hemmrich G, Khalturin K, Boehm AM, Puchert M, Anton-Erxleben F, Wittlieb J, et al. Molecular signatures of the three stem cell lineages in hydra and the emergence of stem cell function at the base of multicellularity. Mol Biol Evol 2012;29(11):3267–80.

[101] Millane RC, Kanska J, Duffy DJ, Seoighe C, Cunningham S, Plickert G, et al. Induced stem cell neoplasia in a cnidarian by ectopic expression of a POU domain transcription factor. Development 2011;138(12):2429–39.

[102] Khalturin K, Anton-Erxleben F, Milde S, Plotz C, Wittlieb J, Hemmrich G, et al. Transgenic stem cells in Hydra reveal an early evolutionary origin for key elements controlling self-renewal and differentiation. Dev Biol 2007;309(1):32–44.

[103] Boehm AM, Bosch TC. Migration of multipotent interstitial stem cells in Hydra. Zoology 2012;115(5):275–82.

[104] Boehm AM, Khalturin K, Anton-Erxleben F, Hemmrich G, Klostermeier UC, Lopez-Quintero JA, et al. FoxO is a critical regulator of stem cell maintenance in immortal Hydra. Proc Natl Acad Sci USA 2012;109(48):19697–702.

[105] Aravin AA, Hannon GJ, Brennecke J. The Piwi-piRNA pathway provides an adaptive defense in the transposon arms race. Science 2007;318(5851):761–4.

[106] Lim RS, Anand A, Nishimiya-Fujisawa C, Kobayashi S, Kai T. Analysis of Hydra PIWI proteins and piRNAs uncover early evolutionary origins of the piRNA pathway. Dev Biol 2014;386(1):237–51.

[107] Sharma AK, Nelson MC, Brandt JE, Wessman M, Mahmud N, Weller KP, et al. Human CD34(+) stem cells express the hiwi gene, a human homologue of the Drosophila gene piwi. Blood 2001;97(2):426–34.

[108] Nolde MJ, Cheng EC, Guo S, Lin H. Piwi genes are dispensable for normal hematopoiesis in mice. PloS One 2013;8(8):e71950.

[109] Rinkevich Y, Rosner A, Rabinowitz C, Lapidot Z, Moiseeva E, Rinkevich B. Piwi positive cells that line the vasculature epithelium, underlie whole body regeneration in a basal chordate. Dev Biol 2010;345(1):94–104.

[110] Cox DN, Chao A, Lin H. piwi encodes a nucleoplasmic factor whose activity modulates the number and division rate of germline stem cells. Development 2000;127(3):503–14.

[111] De Mulder K, Pfister D, Kuales G, Egger B, Salvenmoser W, Willems M, et al. Stem cells are differentially regulated during development, regeneration and homeostasis in flatworms. Dev Biol 2009;334(1):198–212.

[112] Wend P, Holland JD, Ziebold U, Birchmeier W. Wnt signaling in stem and cancer stem cells. Semin Cell Dev Biol 2010;21(8):855–63.

[113] Muller W, Frank U, Teo R, Mokady O, Guette C, Plickert G. Wnt signaling in hydroid development: ectopic heads and giant buds induced by GSK-3beta inhibitors. Int J Dev Biol 2007;51(3):211–20.

[114] Teo R, Mohrlen F, Plickert G, Muller WA, Frank U. An evolutionary conserved role of Wnt signaling in stem cell fate decision. Dev Biol 2006;289(1):91–9.

[115] Onal P, Grun D, Adamidi C, Rybak A, Solana J, Mastrobuoni G, et al. Gene expression of pluripotency determinants is conserved between mammalian and planarian stem cells. EMBO J 2012;31(12):2755–69.

[116] Baguna J, Salo E, Romero R. Effects of activators and antagonists of the neuropeptides substance P and substance K on cell proliferation in planarians. Int J Dev Biol 1989;33(2):261–6.

[117] Wagner DE, Ho JJ, Reddien PW. Genetic regulators of a pluripotent adult stem cell system in planarians identified by RNAi and clonal analysis. Cell Stem Cell 2012;10(3):299–311.

[118] Wagner DE, Wang IE, Reddien PW. Clonogenic neoblasts are pluripotent adult stem cells that underlie planarian regeneration. Science 2011;332(6031):811–6.

[119] Resch AM, Palakodeti D, Lu YC, Horowitz M, Graveley BR. Transcriptome analysis reveals strain-specific and conserved stemness genes in *Schmidtea mediterranea*. PloS One 2012;7(4):e34447.

[120] Sasidharan V, Lu YC, Bansal D, Dasari P, Poduval D, Seshasayee A, et al. Identification of neoblast- and regeneration-specific miRNAs in the planarian *Schmidtea mediterranea*. RNA 2013;19(10):1394–404.

[121] Petersen CP, Reddien PW. A wound-induced Wnt expression program controls planarian regeneration polarity. Proc Natl Acad Sci USA 2009;106(40):17061–6.

[122] Peiris TH, Weckerle F, Ozamoto E, Ramirez D, Davidian D, Garcia-Ojeda ME, et al. TOR signaling regulates planarian stem cells and controls localized and organismal growth. J Cell Sci 2012;125(Pt 7):1657–65.

[123] Oviedo NJ, Pearson BJ, Levin M, Sanchez Alvarado A. Planarian PTEN homologs regulate stem cells and regeneration through TOR signaling. Dis Model Mech 2008; 1(2–3):131–43. discussion 141.

[124] Taylor BL, Zhulin IB. PAS domains: internal sensors of oxygen, redox potential, and light. Microbiol Mol Biol Rev 1999;63(2):479–506.

[125] Leite RB, Brito AB, Cancela ML. An oxygen molecular sensor, the HIF prolyl 4-hydroxylase, in the marine protist *Perkinsus olseni*. Protist 2008;159(3):355–68.

[126] McDonough MA, Li V, Flashman E, Chowdhury R, Mohr C, Lienard BM, et al. Cellular oxygen sensing: crystal structure of hypoxia-inducible factor prolyl hydroxylase (PHD2). Proc Natl Acad Sci USA 2006;103(26):9814–9.

[127] van der Wel H, Ercan A, West CM. The Skp1 prolyl hydroxylase from Dictyostelium is related to the hypoxia-inducible factor-alpha class of animal prolyl 4-hydroxylases. J Biol Chem 2005;280(15):14645–55.

[128] West CM, van der Wel H, Wang ZA. Prolyl 4-hydroxylase-1 mediates O_2 signaling during development of *Dictyostelium*. Development 2007;134(18):3349–58.

[129] Mustroph A, Lee SC, Oosumi T, Zanetti ME, Yang H, Ma K, et al. Cross-kingdom comparison of transcriptomic adjustments to low-oxygen stress highlights conserved and plant-specific responses. Plant Physiol 2010;152(3):1484–500.

[130] Rytkonen KT, Williams TA, Renshaw GM, Primmer CR, Nikinmaa M. Molecular evolution of the metazoan PHD-HIF oxygen-sensing system. Mol Biol Evol 2011;28(6): 1913–26.

[131] Hampton-Smith RJ, Peet DJ. From polyps to people: a highly familiar response to hypoxia. Ann N Y Acad Sci 2009;1177:19–29.

[132] Powell-Coffman JA, Bradfield CA, Wood WB. *Caenorhabditis elegans* orthologs of the aryl hydrocarbon receptor and its heterodimerization partner the aryl hydrocarbon receptor nuclear translocator. Proc Natl Acad Sci USA 1998;95(6):2844–9.

[133] Taylor CT, McElwain JC. Ancient atmospheres and the evolution of oxygen sensing via the hypoxia-inducible factor in metazoans. Physiology 2010;25(5):272–9.

[134] Loenarz C, Coleman ML, Boleininger A, Schierwater B, Holland PW, Ratcliffe PJ, et al. The hypoxia-inducible transcription factor pathway regulates oxygen sensing in the simplest animal, *Trichoplax adhaerens*. EMBO Rep 2011;12(1):63–70.

[135] Hughes BT, Espenshade PJ. Oxygen-regulated degradation of fission yeast SREBP by Ofd1, a prolyl hydroxylase family member. EMBO J 2008;27(10):1491–501.

[136] Grahl N, Cramer Jr RA. Regulation of hypoxia adaptation: an overlooked virulence attribute of pathogenic fungi? Med Mycol 2010;48(1):1–15.

[137] Strobel GL, Arnold J. Essential eukaryotic core. Evol Int J Org Evol 2004;58(2):441–6.

Other Features Concerning the Analogy "Stem Cells: Primitive Eukaryotes": ABC Transporters' Anaerobiosis/Stemness Link

12

Chapter Outline

12.1 ATP-Binding Cassette Transporters

The discovery of the first ATP-binding cassette (ABC) superfamily member [1] occurred three years after the first evidence that the active export from the cells can result in drug resistance [2]. Using Chinese hamster ovary cells resistant to colchicine and displaying pleiotropic cross-resistance to a wide range of amphiphilic drugs, the authors observed the overexpression of a 170 kDa glycoprotein that was related to the drug resistance and "altered drug permeability." It was termed "P glucoprotein—P-gp" [1]. Later, P-gp gene was cloned and identified and termed *P-gp/multidrug resistance (MDR)* or ABCB1 (reviewed in Ref. [3]). Seven subfamilies of ABC transporters were established (ABCA to ABCG) [4] with a total number of 18 members, and two others were later added to the family "C" and "G," respectively: multidrug resistance protein 1 (MRP1/ABCC1) (in the cells of a human lung cancer cell line) [5] and breast cancer resistance protein (BCRP/ABCG2) [6–8]. The first one (P-gp/ABCB1) and the last

Anaerobiosis and Stemness. http://dx.doi.org/10.1016/B978-0-12-800540-8.00012-0

two (MRP1/ABCC1 and BCRP/ABCG2) turned out to be the most important drug transporters in cancer biology and therapy.

The term "MDR" is obviously related to the cancer cells; their resistance to drugs and the ABC transporters were intuitively associated with a cancer-related phenomenon and even considered one of the hallmarks of cancer. However, further research has shown that these transporters are not only expressed, but are also functional, in the cells exhibiting mainly an absorptive/secretory function. For example, ABCG2 is strongly expressed by enterocytes (apical surface) of colon and syncytiothrophoblast cells in placenta; its expression was also evidenced in enterocytes of ileum, cells of bile canaliculi in the liver, vein endothelium and capillaries and, to a lesser extent, in cells of lobules and lactiferous ducts (mammary gland) [9]. In fact, the cells presenting the "barriers" between the blood and some tissues (blood–brain, blood–testis, blood–retinal, and blood–placental barriers) are expressing ABCG2 and other transporters. For example, in normal testis, ABCB1 and ABCG2 were strongly expressed by myoid cells and luminal capillary endothelial wall and ABCB1 also by Leydig cells. MRP1 was observed at the basal side of Sertoli cells and on Leydig cells. MRP2 was only weakly expressed by myoid cells [10]. Also, ABCG2 is expressed on the luminal membrane of retinal capillary endothelial cells [11], and so forth. However, further research has shown that the ABC transporters are not only highly expressed but also highly functionally active in primitive cell populations (stem cells/primitive progenitors) both in normal and cancer tissues.

In fact, the ABC transporters are highly evolutionary-conserved structures and are expressed in prokaryotes and protists, especially in some species that we identified as the prototype models of stemness (see Chapter 10).

12.1.1 ABC Transporters in Hematopoietic Cells, Side Population, Hematopoietic Stem Cells, and ABCG2

During 1992 to 1993, several papers evidenced expression of a "multidrug resistance gene" (MDR1; i.e., ABCB1 or P-gp) in normal leukocytes (lymphocytes in particular) and bone marrow cells [12]. A possible role of MDR1 in maintenance of the primitive cells was suggested on the basis of data showing that its expression is down-regulated along with the myeloid commitment and differentiation [13]. However, it has turned out that MDR1 (ABCB1) and MRP family transporters, although detected in variety of hematopoietic cells, were not highly expressed in primitive hematopoietic stem cells (HSCs). These cells are characterized by a high expression of ABCG2 [14]. In fact, in early experiments of cell sorting, it was found that the murine bone marrow cell population with the capacity to efflux the fluorescent dye Hoechst 33342 (as well as Rhodamine 123) is highly enriched in HSCs exhibiting the in vivo repopulating potential [15]. A thoughtful approach to the acquisition and analysis of flow-cytometry data obtained with Hoechst 33342 (using simultaneously red and blue fluorescent wavelengths) enabled visualization, on cytometry dot plots, a cell population with the high dye efflux, which was termed "side population" (SP) [16]. This cell population contains the vast majority of HSC activity and is enriched at least 1000-fold for HSCs exhibiting the in vivo reconstitution ability [16]. The same features were shown for rhesus and miniature swine bone marrow cells [17], placental (cord) blood [18], and even in CD34$^+$ cells from steady-state human blood [19]. In the first studies it was shown that SP cells do not initially possess hematopoietic colony-forming capability (absence of committed progenitors in SP), but can produce committed progenitors after ex vivo of in vivo differentiation assays as LTC-IC and SRC, which is the basic principle to detect HSCs (see Chapter 1).

In order to determine which of the ABC transporters is responsible for the SP phenomenon, the sorted SP cells were analyzed by Northern blot and PCR approaches. Using the steady-state cells, several groups found a predominant expression of ABCG2 in the distal region of SP, representative for the highest enrichment in HSC activity [14,20,21]. ABCG2 (BCRP1) mRNA is sharply down-regulated with hematopoietic differentiation; furthermore, enforced expression of the ABCG2 cDNA directly conferred the SP phenotype to bone marrow cells and caused a reduction in maturing progeny HSC assays [14]. It should be mentioned that some results that are not in line with previous ones were published: Alt et al. [22] claim that ABCG2 mRNA was not enriched in the SP and was specifically diminished ninefold in CD133 cells, which were eightfold enriched for MDR1 mRNA. However, enforced expression of P-gp (MDR1/(ABCB1)) in murine bone marrow cells results in the expanded SP, which is associated with enhancing the in vivo repopulating ability [23]. Also ex vivo amplification of murine Lin$^-$Sca-1$^+$ cells and their LTC-IC potential is correlated with the MDR1 (ABCB1) gene expression and proportion of cells that effluxed Hoechst 33342 dye. In addition to the preclinical models, in order to enhance the resistance of the normal HSC to hemiotherapy [24] the unexpanded MDR1-transduced cells were tested in clinical trials (phase I) [25–27]. However, association of MDR1 with an extensive ex vivo amplification may result in cells able to induce, after in vivo injection, a myeloproliferative syndrome in transplanted mice that may be related to the rapid self-renewal divisions and consequent increase in probability of replication errors [28]. In line with the data and points presented in this book (see Chapters 2, 4 and 13), it will be interesting to perform similar experiments in physiologically relevant oxygen concentrations.

The fact that SP enriches the functional HSC was exploited in combination with the surface membrane markers (SP CD34$^-$ KSL) to achieve a high-resolution enrichment of HSC ([29]; see also Chapter 1). However, the ABC transporter overexpression and SP phenomenon in stem/progenitor cells is specific neither for bone marrow nor for stem cells with hematopoietic differentiation potential.

12.1.2 ABC Transporters in Embryonic/Fetal and Prenatal Cell Populations

Murine yolk sac and embryonic tissue cells contain a discrete SP [30] as does the human trophoblast [31]. SP cells, present in the second-trimester human fetal liver, are enriched in HSC exhibiting in vivo reconstituting capacity confined in CD34$^+$ CD38$^-$ SP$^+$ fraction, which includes all the transplantable HSC activity detectable in NOD/SCID mice and also certain other, more differentiated hematopoietic cell types [32]. While a tenfold enrichment of transplantable HSC is evidenced in this population it is not coenriched in hematopoietic progenitors detectable by either short- or long-term in vitro assays [32]. SP cells are present early during human liver development and contribute to hematopoietic and epithelial lineage generation: First trimester SP cells are more enriched for HSC/HPC and endothelial progenitors than those from the second trimester. Oval/progenitor cells, responsible for liver regeneration after massive or chronic hepatic injury, are probably also present in SP of developing human liver [33]. Spermatogonial stem cells displaying the SP phenotype have been found in mouse prenatal testes as well as in the immature and adult ones; in both cases SP fraction is composed of both somatic and germ cells. Both cell types expressed the ABC transporters ABCG2, ABCB1a, ABSB1b, and ABCC1 [34]. SP cells were found in human placental tissue and a subpopulation of these cells ($CD45^+$ c-Kit^+ SP) exhibits hematopoietic differentiation potential [35]. On the contrary, unlike the murine ESC, which present an $ABCG2^+$ population as well as an SP phenomenon [36], human ESC do not express ABCG2 and cannot efflux Hoechst and, consequently, do not exhibit SP phenomenon [37]. We believe that this feature may be a result of an adaptation to the nonphysiological conditions (especially with respect to oxygenation and antioxidative defense) experienced by the embryonic lines during their numerous passages ex vivo.

In 2000, Hulspas and Quesenberry [38] showed that the cells from embryonic mouse neurosphere contain a subpopulation that is characterized by a low emission of Hoechst fluorescence and gives rise to a phenotype similar to freshly isolated, uncultured neural cells. This way, a discrete SP (3.6% of all neurosphere-derived cells), which is a highly enriched population of neural stem cells, can be isolated, while a non-SP (Hoechst+) cell population contains mostly committed progenitors [39]. The low fluorescent SP comprises only 3.6% of all live cells sorted and contains >99% of all the neural stem cells as assayed by the formation of neurospheres in culture.

In human amnion mesenchymal layer SP (AMC-SP) cells were found in low frequency (0.1–0.2%). Inside this population MSC capable of differentiating, upon appropriate differentiation protocols, to neuroectodermal, osteogenic, chondrogenic, and adipogenic cells were detected. AMC-SP fraction is also characterized by expression of Oct-4, Sox-2, and Rex-1 genes considered “stemness markers” [40].

12.1.3 SP Derived from Skeletal Muscle

Concerning the skeletal muscle, in addition to hematopoietic potential documented for the cells of SP phenotype, they can undergo myogenic commitment and differentiation after coculture with myoblasts and even give rise to both myocytes and satellite cells after intramuscular transplantation. These data provided the evidence that satellite cells and muscle-derived stem cells represent distinct populations [41,42]. Recently, it was shown that the growth of SP cells in suspension (combined with feeder cells cultured on spherical microcarriers) is possible in a specially designed spinner-flask culture system [43]. The determinant of the adult skeletal muscle SP cell phenotype is ABCG2 [44]. SP, enriched in repopulating activity, contains the cells expressing Foxk1; furthermore, in Foxk1 mutant skeletal muscle SP cells are decreased in number. Examining the molecular signature of muscle SP cells from normal, injured, and Foxk1 mutant skeletal muscle allowed the authors to propose that muscle SP cells represent a “progenitor” cell population that participates in repair and regeneration of adult skeletal muscle [44]. As already stated, in the normal nonregenerating muscle SP is heterogeneous from the phenotypical and functional viewpoint [45]. A fraction of SP cells expresses CD31 and exhibits endothelial characteristics (thus representing the mature cells with a specific function); the other subsets are able to differentiate in hematopoietic cells, but also more primitive subsets as one characterized by ABCG2, Syndecan-4, and Pax7 expression highly enriched in self-renewing muscle stem cell capable of generating both satellite cells and their myonuclear progeny in vivo [46]. Integrity of SP and ABCG2 is important in muscle regeneration since in ABCG2-null mice (lost SP fraction) muscle regeneration is impaired resulting in fewer centrally nucleated myofibers, reduced myofiber size, and fewer satellite cells [47]. It was reported that, in skeletal muscle SP cells, a micro-RNA (miR-128a) contributes to the maintenance of the quiescent state and regulates cellular differentiation by repressing individual genes in SP cells [48]. Also, bone morphogenetic protein 4 (BMP4) is highly expressed in SP cells of human fetal skeletal muscle but not in the main population (MP) mononuclear muscle-derived cells; in contrast, the BMP4 antagonist Gremlin is specifically up-regulated in MP cells, suggesting that both BMP4 and Gremlin are regulators of myogenic progenitor proliferation [49].

12.1.4 Cardiac Muscle SP

Since the first reports suggesting that the responsive stem cell pool that may influence adaptation of the postnatal heart resides in the adult myocardium, “myocardial stem cells” were related to SP [42]. SP from cardiac muscle was a subject of many publications until it became a promising tool of the future cell therapy of heart ischemic diseases [50]. This SP, in the developing and adult mouse heart [51], as well as in the human heart [52], also results from a persistent expression of ABCG2. On the basis of their molecular signature, it was proposed that the cardiac SP cells have function of “progenitors” in context of the development, maintenance, and repair of the heart [51]. Pfister et al. [53] demonstrated that among cardiac SP cells, the greatest potential for cardiomyogenic differentiation is restricted to cells positive for stem cell antigen 1 (Sca-1) expression and negative for CD31 expression ($Sca1^{+}/CD31^{-}$). So, $Sca1^{+}/CD31^{-}$ cardiac SP cell population enrich a distinct cardiac progenitor cell population, capable of cardiomyogenic differentiation into mature cardiomyocytes [53,54]. Interestingly, $SCA1^{+}/CD31^{+}$ CSP isolated from the mouse heart are highly enriched in endothelial progenitors, as concluded by Liang et al. [55] on the basis of gene expression profile, their localization, ability to proliferate, differentiate, migrate, and vascularize in vitro and in vivo.

Rat cardiac SP cells have a potential to migrate and differentiate into cardiomyocytes in vitro and in vivo (shown via cardiac SP transplantation in cryoinjured heart model) [56]. Using postnatal and adult mice, it was demonstrated that the cardiac SP cell phenotype in the neonatal heart is dependent on ABCG2 while in adult heart the main contributor to the SP phenotype is MDR1 (ABCB1) [57]. Interestingly, it appears that the aging of murine cardiac $Sca1^{+}$ $CD31^{-}$ SP cells is distinct from the aging of mature cells and perhaps progenitors from other tissues: the aging seems to not affect telomerase activity and expression of cell cycle genes. Surprisingly, the aging significantly reduces expression of differentiation-associated genes. Furthermore, these age effects were not altered by antiaging caloric-restriction diet [58]. Genetic ablation of ABCG2 in adult cardiac SP showed blunted proliferation capacity and increased cell death, while overexpression of ABCG2 enhanced cell proliferation [57]. It was shown that Urotensin II, a vasoactive ligand, inhibits proliferation of cardiac SP cells possibly through increase of JNK phosphorylation without affecting their commitment and differentiation [59] as does Wnt signaling: injected after myocardial infarction, r-Wnt3a limits cell renewal, blocks endogenous cardiac regeneration, and impairs cardiac performance [60].

12.1.5 Renal Tissue SP

As in mice kidney tissue [41,61,62], SP is present in the renal tissue in humans [63]. SP isolated from the renal tissue is heterogeneous. In the SP, the cells exhibiting a high proliferative potential, long-term culture, and spheroids assays are concentrated and were shown as capable of maintenance upon tissue culture passage. In term of cycling, SP is enriched for quiescent cells with a high proliferative capacity and stem cell-like properties [64]. Certain authors concluded that SP cells in the kidney are not stem cells for renal tissues, especially mesangial and tubular ones, but may have potential for hematopoietic and nonhematopoietic lineages [65]. The others suggest, in addition to a hematopoietic-oriented stem-cell population, the existence of nonhematopoietic ones, including the renal one. Indeed, the adult kidney SP cells demonstrated multilineage differentiation in vitro, whereas microinjection into mouse metanephroi showed that SP cells had a 3.5- to 13-fold greater potential to contribute to developing kidney than non-SP main population cells. However, although reintroduction of SP cells into an Adriamycin-nephropathy model reduced albuminuria/creatinine ratios, this was without significant tubular integration, suggesting a humoral role for SP cells in renal repair [66].

12.1.6 Lung Tissue SP

The results were published, suggesting that the nonhematopoietic SP isolated from lung are enriched so-called "airway" stem cells [67]. In fact, the adult mouse lung SP has epithelial and mesenchymal potential that resides within a $CD45^-$ mesenchymal subpopulation, as well as limited hematopoietic ability, which resides in the bone marrow-derived $CD45^+$ subpopulation [68]. This mesenchymal potential corresponds to MSCs, which are contained in lung $CD45^-$ SP. These cells, characterized by a normal chromosomal structure and expression of a high levels of telomerase, after being cultured ex vivo in a differentiation medium exhibit adipogenic, osteogenic, and chrondrogenic potential [69]. Irwin et al. [70] demonstrated that the mouse neonatal lung SP contains the cells exhibiting endothelial potential. They also showed that the hyperoxia-induced changes in lung SP cells may limit their ability to effectively contribute to tissue morphogenesis during room air recovery [70].

12.1.7 Intestinal Tissue

The $CD45^-$ SP fraction isolated from mouse jejunum seems to be significantly enriched in small intestinal epithelial stem cells [71]. It was confirmed later that in this SP fraction putative "progenitor" cells are enriched while immune, mesenchymal, and differentiated epithelial cells are depleted as established by gene ontology and pathway mapping analyses [72].

12.1.8 Mesenchymal Stem Cells in SP

The "traditional" sources of MStroC are bone marrow (BM) and fat (adipose) tissue (FT). Comparing phenotypical and structural criteria SP from these two tissues, Fujimura et al. [73] concluded that the SP-FT cells exhibiting nonhematopoietic potential are more primitive than their SP-BM analogs. The SP from human dental pulp, studied after ischemic culture, contains the cells exhibiting the mesenchymal stem cell characteristics (colony-forming efficiency, self-renewal, multilineage differentiation capabilities, ability to differentiate into odontoblast/osteoblast-like cells, adipocytes, neural-like cells, and endothelial cells) [74]. Ishizaka et al. [75] demonstrated that dental pulp stem cells (DP) have higher angiogenic, neurogenic, and regenerative potential compared to the regenerative potential of porcine DP, BM, and FT-derived $CD31^-$ SP cells. The authors used a mouse hind limb ischemia model and showed that pulp $CD31^-$ SP cells transplanted into ischemic zone produced higher blood flow and capillary density than BM and FT $CD31^-$ SP cells and enabled better motor function recovery and infarct size reduction comparing to the SP of other two sources [75]. The SP isolated from other tissues contains the mesenchymal stem cells [76].

12.1.9 SP as a Marker of Stemness

This list of tissues in which the SP is demonstrated and related to the primitive stem/progenitor cell populations is far from complete. For example, SP is evidenced in the murine endometrium [77,78], pituitary gland [79], murine [80,81], dog [82] and human [83] liver (dog and human liver SP are able to generate hepatocyte-like cells in vitro), mouse hair follicle [84], epidermal [85–87] as well as in human epidermal [88] keratinocytes, and so on. A 2015 publication [89] reveals that the ABCG2 is expressed in 2–3% keratinocytes in human interfollicular epidermis and that most of these ABCG2 positive keratinocytes are capable of forming "holoclones" (clones originating from a cell exhibiting a higher proliferative potential) and of generating a stratified differentiating epidermis in organotypic culture models [89].

It is obvious that the discrete SP isolated from a variety of adult tissues contain cells exhibiting hematopoietic differentiation potential [90,91,41] but this does not mean that SP cells from the same sources do not contain the stem/progenitor cells of other cell type differentiation potentials. In fact, the SP from different tissues contains the stem/progenitor cells of various differentiation potential and/or multipotent stem cells: SP cells from skeletal muscle exhibit the hematopoietic differentiation potential [92]; bone marrow SP contains the "progenitors" capable of skeletal muscle [93] and myocardial [94] differentiation; murine bone marrow SP cells ($CD34^-$/low, c-Kit^+, Sca-1^+; fraction highly enriched in HSC), transplanted into lethally irradiated mice (subsequently rendered ischemic by coronary artery occlusion followed by reperfusion), migrated into ischemic cardiac muscle and blood vessels, differentiated to cardiomyocytes and endothelial cells, and contributed to the formation of functional tissue [95]; the cardiac SP contains mesenchymal stem/progenitor cells of various differentiation potential including cardiomyogenic [76]; bone marrow SP cells (or rather "some bone marrow SP cells") are capable of functional cardiomyogenic differentiation [96]; adult murine liver SP 45+ cells do not express c-Kit but contains the HSC exhibiting a potent hematopoietic–reconstitution activity [97]; and so on.

These facts rather come from the multipotency and/or functional heterogeneity of the cells inside SP than from the "plasticity" and "transdifferentiation" [98].

The above reviewed data show that the side population from all tissues is composed of some mature cells (endothelial par example) and a heterogeneous stem/progenitor cell population that is either multipotent and/or exhibits more restricted differentiation potential. Some authors consider "SP phenotype" cells simply "stem cells." In our opinion, this is not correct, since not only mature cells of SP phenotype, but also the committed progenitors, should be removed from SP to get highly enriched stem cell populations inside SP. This is quite possible applying, in parallel with SP labeling (Hoechts or some of new alternative stains), immunofluorescent antibodies-based detection of the membrane differentiation and/or so-called "stem cells markers," followed by cell sorting (applying the "positive" or "negative" principle, or both). Using this approach, practically pure HSC population was isolated from murine bone marrow [29]. Others, on the basis of the fact that the hematopoietic, small intestine, testicular germ, skeletal, among other stem cells reside inside ABCG2 positive population propose the use of ABCG2 as an exclusive stem cell marker, providing a new model for studying stem cell activity that does not require transplant-based assays [99]. In our opinion ABCG2 can be used alone or in combination with other markers only to enrich the stem cells in phenotypically defined populations, but ABCG2 positive-cell count, although in some situations may be correlated with the stem cell activity, cannot always "match" it.

12.2 The Physiological Role of ABC Transporters

The physiological role of ABC transporters in mature cells is usually related to their function and/or to a specific protection against toxins. For example, hepatic cell ABCA1 is involved in high density lypoproteins' metabolism [100]. Hepatocytes and enterocytes express on the apical membrane ABCG5 and ABCG8 dimer that limits intestinal absorption and facilities biliary secretion of cholesterol and phytosterols [101]. In corneal endothelial cells ABCG2 mediates the efflux of phototoxins [11] and it has a similar role in corneal epithelial cells [102]. In cells representing blood–brain "barrier" the ABC transporters play a key role in increasing the clearance of potentially harmful organic ions from brain and, in the opposite direction, in preventing brain entry of blood-borne molecules (drugs, environmental toxins, endogenous metabolites, etc.) [103].

The physiological role of ABC proteins in stem cells is less evident. As usually pointed out, a stem cell does not have a specific function, it has a potential. Can overexpression of ABC in stem cells be related to protection of stem cell potential?

Indirectly, on the basis of the evidence that expression of MDR1 (ABCB1) is indispensable for an efficient long-term regeneration of tissues [104,105] it can be deduced that this transporter is indispensable for stem cell maintenance. Indeed, the expression of ABC transporters seems to be of vital importance for stem cells [57].

As presented in Chapters 6 and 7, the stem cells behave as anaerobic/facultative aerobic "organisms" if we consider a cell as a unicellular organism; these features seem to be closely related to their "primitive" status; they are reminiscent of the first eukaryotes, whose existence was promoted by an anoxic/nanooxic environment (Chapters 8–10). Following this logic, the possibility that ABC transporters are primarily related to the anaerobic/nano/microaerophilic metabolic properties should be considered.

12.3 ABC Transporters and Their Connection with Anaerobiosis and Stemness

Applying an in vivo mice model of renal ischemia/reperfusion, Liu et al. [106] showed that ABCG2 not only protects kidney SP cells from hypoxia/reoxygenation-induced cytotoxicity, but also confers to these cells a potential to participate in renal regeneration in vivo. In fact, hypoxia activates ERK phosphorylation in kidney SP and the MEK/ERK pathway is involved in ABCG2 expression in these cells, [106]. The same group showed a crucial role of ABCG2 expression for vital functions (survival, proliferation, and paracrine actions) of kidney SP cells in "ischemia", in an experimental design providing ex vivo ischemic conditions (medium without glucose; 0% O_2 during 1–6 h). In view of the short time of incubation some dissolved oxygen may still be present in the medium, so this condition represents probably a severe hypoxia rather than pure anoxia, followed by "reoxygenation"; that is, exposure of culture to atmosphere O_2 concentration (21%) in a glucose supplemented medium [107]. The proportion of SP cells in dental pulp cells (DPC) cultured for 24 and 48 h in a low O_2 concentration (2%) and low serum concentration (2%) increases. Furthermore, ABCG2 and OCT4 messenger RNA levels increased more than threefold in the 48 h "ischemic" group compared with the control group [108].

The SP cells isolated after such a low O_2 and low serum concentrations culture are highly enriched in cells that exhibit the features of mesenchimal stem cell phenotype (colony-forming efficiency, self-renewal, and multilineage differentiation capabilities and are able to differentiate into odontoblast/osteoblast-like cells, adipocytes, neural-like cells, and endothelial cells). Further culture of these SP in standard conditions (21% O_2) leads to a decrease in ABCG2 expression along with cell passage, which would be consistent with an enhanced differentiation occurring in this situation [74].

It is interesting that IL-6, which synergizes in stem cells maintenance either with low O_2 concentration or with hypoxia-response mimicking conditions ([109,110]; see Chapter 13), increases expression level of ABCG2 [111]. The same effect is exhibited by the endoplasmic reticulum (ER) stress inducers, while knock-down of ABCG2 by siRNA or ABCG2 inhibitor reduced plasma cell viability under ER stress [111].

The number of SP cells increases in heart muscle after injury (mice transmural cryo-injury model, resulting into hypoxic/ischemic zones) and these cells overexpress ABCG2, which enhances their capacities of survival following exposure to hydrogen peroxide [112]. In fact, the mechanism of this overexpression in adult murine heart SP goes via HIF-2α, which binds an evolutionary-conserved HIF-2α response element in the murine ABCG2 promoter [112]. Thus ABCG2 seems to be a direct downstream target of HIF-2α.

It should be mentioned that HIF-2α is required for the maintenance of primitive HSC endowed by the long-term repopulating ability [113]. By using chemical inhibitors of ABCG2 and by studying ABCG2 null mice cells, it was concluded that ABCG2 has a physiological role in the survival of HSC in hypoxia (0.1% O_2). In the absence of ABCG2 the HSC become extremely sensitive to hypoxia, but their ability to survive in hypoxia is entirely recovered by inhibition of heme synthesis. Since the porphirins (including heme) are effluxed through ABCG2, the sensibility to hypoxia of cells in which ABCG2 is not functional results from intracellular accumulation of heme. It was shown that ABCG2 is activated via HIF-1α signaling pathway [114]. Thus, the up-regulation of the ABCG2 promoter by HIF1α and consequent enhancing of ABCG2 expression reveals a mechanism allowing regulation of the accumulation of detrimental porphirins (including heme) in cells avoiding their participation in ROS generation and consequent mitochondria damages [114]. Maintenance of porphyrin homeostasis by ABCG2 was also shown in murine embryonic stem cells (ESCs) [115].

Of interest for this discussion is transcription factor NF-E2-related factor 2 (Nrf2). Among Nrf2-inducible genes are those implied in antioxidant (NQO1, NQO2, HO-1, GCLC), antiapoptotic (Bcl-2), metabolic (G6PD, TKT, PPARγ), and drug efflux transporter (ABCG2, MRP3, and MRP4) function [116]. It was shown in lung and prostate cancer cells that Nrf2 interacts with antioxidant-response element (5′ promoter flanking region of ABCG2). This interaction seems to be critical for ABCG2 expression: disruption of Nrf2 in these cells attenuated expression of both ABCG2 transcript and protein and dramatically reduced SP fraction [117]. Using HepG2 cells (a human hepatocellular carcinoma line) as a model, the Nrf2-dependent induction of ABC transporters ABCG2 and ABCC2 under oxidative stress was evidenced [118]. Similar features were shown for the normal functional mature cells: activating this "oxidative/electrophilic stress sensor" Nrf2 at the blood–brain and blood–spinal cord barrier increases protein expression and transport activity of three ABC transporters (ABCB1, MRP2, and ABCG2). Although the data are not yet available for the role of Nrf2 in normal stem cells enriched in SP, it might be similar to cancer and mature cells mentioned. In fact, the ROS profile of normal SP—a low level of ROS [87]—is reminiscent of one found in the primitive stem cell populations enriched on the basis of the membrane markers [119].

As elaborated in Chapters 3–7, the stemness is closely related to anaerobic/microaerophilic metabolic set-up. The data reviewed in this paragraph provide enough material to associate the ABC transporters (most data concerns ABCG2) both with phenomenon of stemness and anaerobic/microaerophilic metabolic set-up. We identified (Chapters 9 and 10) the association of the stemness and anaerobic/aerophilic metabolic set-up as the most primitive single-cell eukaryote property, enabling only a simple cell division without a parallel commitment and/or differentiation. This is proposed to be an ancestral feature dating from the first eukaryotic common ancestor (FECA) and a prototype of self-renewal (Chapter 10). Is it possible to reveal the same association anaerobiosis/microaerophilia versus ABC transporters in protists considered to be relevant models for most primitive single-cell eukaryotes?

12.4 ABC Transporters, Protist Anaerobiosis/Microaerophilia, and Life Cycle

As explained in Chapter 10, during the life cycle of protist, the primitive nondifferentiated forms capable of full/complete self-renewal appeared only in "strong hypoxic"/nanoaerobic environmental conditions conserving the features of ancient "hidden stemness." The example of the Entamoeba family (most studied *Entamoeba invadens* and *Entemoeba histolytica*) was elaborated in detail: strong hypoxic forms of *E. invadens* undergo full/complete self-renewal by slow cycling and symmetric division, while rather anoxic forms switch into a nonproliferating state of stemness [120,121]. Environmental dependent switching from symmetric to asymmetric division and from slow-proliferating to nonproliferating state in protist is proposed by Vladimir Niculescu as evolutionary prototype of self-renewal and stemness. As explained in Chapter 10, we consider the symmetrical self-renewing division in a nanooxic environment in protist as an evolutionary prototype of the self-renewal phenomenon and most ancestral feature of stemness.

In fact, a phenomenon similar to the "multidrug resistance" one, already described for tumor cells [122] and some other anaerobic and/or microaerophilic ("hypoxia tolerant") protozoa as *Plasmodium falciparum* [123], *Leishmania tarentolae* [124], *Trichomonas vaginalis* [125], and so on, was evidenced also in *E. histolytica* on the basis of the phenomenon of drug clearance by a calcium-dependent mechanism (reviewed in Ref. [126]). This "MDR phenotype" in *E. histolytica* involves the membrane overexpression of P-glycoprotein (170 kDa), the ABC transporter exhibiting a high sequence homology with bacterial homologs, as one found in facultative anaerobic bacterium *E. coli* [127]. Four complete *mdr*-like genes were initially cloned in *E. histolytica* (EhPgp1, 2, 5, and 6; EhPgp 3 and 4 have been found to be pseudogenes) and one later (Eh *abc1* gene [128]) representing, in fact, one of the largest *mdr* gene families found in one organism. Some of these genes (as EhPgp1 gene) are expressed constitutively and for others (as EhPgh5 gene), expression is induced in the presence of certain compounds (reviewed in Ref. [126]). Transmission microscopy revealed that the membrane of "multidrug-resistant trophozoites" presented four- to fivefold more Eh Pgp_s compared to drug-sensitive trophozoites [129].

Thus, the ABC transporters are revealed to be one of the most primitive functional molecular structures in evolution, conserved from prokaryotes to the primates. In view of this fact, it is almost certain that the similar mechanisms were expressed by the first eukaryotic common ancestor (FECA) and last eukaryotic common ancestor (LECA).

Both these stages in eukaryote evolution appeared and evolved in anoxic/nanooxic conditions (Chapters 8 (Figure 8.1), 9 and 10), which suggest that the anaerobic metabolism provided enough energy to build ATP whose energy was harnessed for these "pumps." A similar situation is in anaerobic protozoa, in which the inhibition of ABC transporters induces programmed cell death as shown after inhibition of EhPgp in *E. histolytica* trophozoites [130]. Similarly, the overexpression of ABC transporters seems to allow the survival of vertebrate stem cells in anoxia/ischemia/hypoxia [107], and as explained above, the overexpression is induced by "hypoxic cell response" pathways directing toward anaerobic metabolism. Thus, it appears that the anaerobic metabolic set-up is closely related to the "MDR phenotype" in protozoa and stem cells (normal and cancer ones), which, in turn, provides an advantage for survival and maintenance of the lineage in an ecological or microenvironmental niche. The "anaerobic home for stem cell genome" seems to be evident in view of highly expressed anaerobic glycolysis enzymes and a high lactate production in stem cells as well as a loss of self renewal if they cannot overcome oxidative damage [131]. Furthermore, the complete (symmetric) self-renewal in adult stem cells conditioned by an anaerobic metabolic state [132] seems to represent an ancestral feature conserved from the FECA (Chapter 10). Similar to the way the microenvironment supports the maintenance of stemness, the anaerobic/nanooxic/microaerophilic ecological niche supports the proliferative capacity of protozoa (and, in the case of parasites their pathogenicity and virulence) [133], which is paralleled with their "mdr" capacities [134].

We propose that the ABC transporter system was from the beginning (FECA stage) associated with symmetric self-renewal and anaerobiosis representing the prototype feature of stemness and that this feature is still conserved in modern stem cells. In addition to the arguments showing the interplay between the anaerobic regulation in cell machinery and ABC transporters discussed above, we should raise the question of the ABC transporters' role in the ancestral primitive eukaryotes and modern primitive self-renewing stem cells. To give a response to this question, it is necessary to evoke the fact that the "energy of life's origin"—ATP synthesis—comes from anaerobic processes [135]. Since ABC transporters, as the term suggests, are ATP-dependent, the anaerobic ATP synthesis is obviously qualitatively and quantitatively sufficient for their function. This feature is also conserved in stem cells. In the context of Proterozoic life of the first eukaryotes, the oxygen-independent generation of ATP in mitochondria (i.e., anaerobic mitochondrial respiration) is considered to be of paramount importance for their development [136]. Several forms of the anaerobic mitochondrial respiration still exist (see Chapter 9) in single-cell eukaryotes and multicellular eukaryotes.

Concerning single-cell anaerobic eukaryotes, a one-phrase summary of Mentel and Martin's review [136], paraphrasing the title of the famous Dobzhansky's [137] assay, should be cited: "Nothing in evolution of eukaryotic anaerobes makes sense except in light of Proterozoic ocean chemistry" (see Chapters 8–10). In this phrase, the authors summarize the article that provides a line of evidence about mitochondria in eukaryote cells being involved in anaerobic ATP synthesis from the very beginning (first endosymbiotic event). Concerning the modern protist organisms, in which ABC transporters have vital importance, some examples can be mentioned: *Entamoeba* (posess mitosomes) during anaerobic ATP synthesis produces lactic and succinic acid [138] as well as acetate and ethanol [139,140]; *Trichomonas* (*T. vaginalis* and *T. fetus* that possess the hydrogenosomes) produces molecular hydrogen, CO_2, and acetate (1 mol of each), along with one mol of ATP per mol of pyruvate that enters the hydrogenosome, but, among its metabolic products are also glycerol and lactate or succinate [141,142]; *P. falciparum* possesses mitochondria in anaerobic conditions, in glycolytic pathway, produces, in addition to lactate, several molecules including glycerol and pyruvate [143], but also by anaerobic mitochondria respiration (complex II "respiration" or "fumarate respiration"), the succinate [144,145]. *Leishmania* that has mitochondria is particularly sensitive to the complex II-specific inhibitors [146], possesses fumarate reductase (capable of reducing fumarate to succinate), and is capable of "fumarate respiration" [147], yielding succinate as a metabolic endproduct. In addition, at the state of promastigote, it exhibits a complex metabolic network with apparent lack of feedback regulation of glycolytic fluxes. In glucose-rich conditions, the capacity of mitochondrial metabolism to completely oxidize internalized glucose to CO_2 is insufficient, leading to excessive secretion of partially catabolized intermediates such as succinate, acetate and alanine [148], among others.

The anaerobic mitochondrial respiration was found to be essential for cancer cells (especially cancer-initiating cells), which can extensively use "fumarate respiration" and produce succinate [149–151]. However, some normal human cells can also produce succinate upon ischemic insult [152]; the fumarate respiration, in fact, is not specific for cancer cells since some normal somatic cells (as renal tubules) are able, upon ischemic/hypoxic stress, to activate the "fumarate respiration" [153,154]. The experimental data showing that the MSC effective proliferation in anoxia and ischemia is stopped upon blockade of fumarate-reductase (our unpublished data) suggest that the mammalian mitochondria (at least in the stem cells) may revert to a primitive functional state and respire in an anaerobic way where fumarate (instead of oxygen) as the ultimate electron acceptor being reduced in succinate. It should be stressed that the prototype of fumarate respiration exists in anaerobic bacteria *E. coli* [155] expressing, as mentioned above, a P-gp prototype.

If the primary role of ABC transporters in ancestral prokaryotes was the export of peptides and sugars and import of nutrients and ions [128], is it possible that, in first eukaryotes, the main role for these transporters becomes elimination of the products of anaerobic metabolism and, in particular, of anaerobic mitochondrial respiration? A role that is conserved until now not only in protist resuming the functional modalities of the first eukaryotes, but also in normal and cancer stem cells, exhibiting anaerobic features and maintaining in their stemness phenomenon the ancestral features of the first eukaryotes? This hypothesis remains to be tested.

12.5 Conclusion

In normal and cancer stem biology, the SP phenotype resulting from an overexpression of ABC transporters (mainly ABCG2 and ABCB1) is considered as a marker of stemness. These ABC transporters are related to the anaerobic metabolic properties of stem cells. Since the simple life cycle (low-energy proliferation *à l'identique*) of FECA, proposed to be a prototype of self-renewal and the first hallmark of stemness, is also conditioned by anaerobiosis, we considered ABC overexpression and function in some anaerobic protist exhibiting the most primitive modalities of eukaryote life, including the self-replication in the form of self-renewal in anaerobic conditions. The vital importance of ABC transporter overexpression for these organisms is in favor of the association of these mechanisms with anaerobiosis and stemness. The common feature spanning from the most primitive anaerobic protist to the human stem cells is the anaerobic metabolic set-up and, probably, capacity of anaerobic mitochondrial respiration resulting in specific metabolic products as succinate. In that respect, the primary role of ABC transporters in these cells might be to enable efflux of these products from the cell. This hypothesis remains to be tested.

References

[1] Juliano RL, Ling V. A surface glycoprotein modulating drug permeability in Chinese hamster ovary cell mutants. Biochim Biophys Acta 1976;455:152–62.
[2] Dano K. Active outward transport of daunomycin in resistant ehrlich ascites tumor cells. Biochim Biophys Acta 1973;323:466–83.
[3] Dean M, Allikmets R. Complete characterization of the human ABC gene family. J Bioenerg Biomembr 2001;33:475–9.
[4] Allikmets R, Gerrard B, Hutchinson A, Dean M. Characterization of the human ABC superfamily: isolation and mapping of 21 new genes using the expressed sequence tags database. Hum Mol Genet 1996;5:1649–55.
[5] Cole SP, Bhardwaj G, Gerlach JH, Mackie JE, Grant CE, Almquist KC, et al. Overexpression of a transporter gene in a multidrug-resistant human lung cancer cell line. Science 1992;258:1650–4.
[6] Doyle LA, Yang W, Abruzzo LV, Krogmann T, Gao Y, Rishi AK, et al. A multidrug resistance transporter from human MCF-7 breast cancer cells. Proc Natl Acad Sci USA 1998;95:15665–70.
[7] Allikmets R, Schriml LM, Hutchinson A, Romano-Spica V, Dean M. A human placenta-specific ATP-binding cassette gene (ABCP) on chromosome 4q22 that is involved in multidrug resistance. Cancer Res 1998;58:5337–9.
[8] Miyake K, Mickley L, Litman T, Zhan Z, Robey R, Cristensen B, et al. Molecular cloning of cDNAs which are highly overexpressed in mitoxantrone-resistant cells: demonstration of homology to ABC transport genes. Cancer Res 1999;59:8–13.
[9] Maliepaard M, Scheffer GL, Faneyte IF, van Gastelen MA, Pijnenborg AC, Schinkel AH, et al. Subcellular localization and distribution of the breast cancer resistance protein transporter in normal human tissues. Cancer Res 2001;61:3458–64.
[10] Bart J, Hollema H, Groen HJ, de Vries EG, Hendrikse NH, Sleijfer DT, et al. The distribution of drug-efflux pumps, P-gp, BCRP, MRP1 and MRP2, in the normal blood-testis barrier and in primary testicular tumours. Eur J Cancer 2004;40:2064–70.
[11] Asashima T, Hori S, Ohtsuki S, Tachikawa M, Watanabe M, Mukai C, et al. ATP-binding cassette transporter G2 mediates the efflux of phototoxins on the luminal membrane of retinal capillary endothelial cells. Pharm Res 2006;23:1235–42.
[12] Baccarani M, Damiani D, Michelutti A, Michieli M. Expression of multidrug resistance gene (MDR1) in normal hematopoietic cells. Blood 1993;81:3480–1.

[13] Chaudhary PM, Roninson IB. Expression and activity of P-glycoprotein, a multidrug efflux pump, in human hematopoietic stem cells. Cell 1991;66:85–94.
[14] Zhou S, Schuetz JD, Bunting KD, Colapietro AM, Sampath J, Morris JJ, et al. The ABC transporter Bcrp1/ABCG2 is expressed in a wide variety of stem cells and is a molecular determinant of the side-population phenotype. Nat Med 2001;7:1028–34.
[15] Wolf NS, Koné A, Priestley GV, Bartelmez SH. In vivo and in vitro characterization of long-term repopulating primitive hematopoietic cells isolated by sequential Hoechst 33342-rhodamine 123 FACS selection. Exp Hematol 1993;21:614–22.
[16] Goodell MA, Brose K, Paradis G, Conner AS, Mulligan RC. Isolation and functional properties of murine hematopoietic stem cells that are replicating in vivo. J Exp Med 1996;183:1797–806.
[17] Goodell MA, Rosenzweig M, Kim H, Marks DF, DeMaria M, Paradis G, et al. Dye efflux studies suggest that hematopoietic stem cells expressing low or undetectable levels of CD34 antigen exist in multiple species. Nat Med 1997;3:1337–45.
[18] Storms RW, Goodell MA, Fisher A, Mulligan RC, Smith C. Hoechst dye efflux reveals a novel CD7(+)CD34(-) lymphoid progenitor in human umbilical cord blood. Blood 2000;96:2125–33.
[19] Brunet de la Grange P, Vlaski M, Duchez P, Chevaleyre J, Lapostolle V, Boiron JM, et al. Long-term repopulating hematopoietic stem cells and "side population" in human steady state peripheral blood. Stem Cell Res 2013;11:625–33.
[20] Scharenberg CW, Harkey MA, Torok-Storb B. The ABCG2 transporter is an efficient Hoechst 33342 efflux pump and is preferentially expressed by immature human hematopoietic progenitors. Blood 2002;99:507–12.
[21] Kim M, Turnquist H, Jackson J, Sgagias M, Yan Y, Gong M, et al. The multidrug resistance transporter ABCG2 (breast cancer resistance protein 1) effluxes Hoechst 33342 and is overexpressed in hematopoietic stem cells. Clin Cancer Res 2002;8:22–8.
[22] Alt R, Wilhelm F, Pelz-Ackermann O, Egger D, Niederwieser D, Cross M. ABCG2 expression is correlated neither to side population nor to hematopoietic progenitor function in human umbilical cord blood. Exp Hematol 2009;37:294–301.
[23] Bunting KD, Zhou S, Lu T, Sorrentino BP. Enforced P-glycoprotein pump function in murine bone marrow cells results in expansion of side population stem cells in vitro and repopulating cells in vivo. Blood 2000;96:902–9.
[24] Fruehauf S, Wermann K, Buss EC, Hundsdoerfer P, Veldwijk MR, Haas R, et al. Protection of hematopoietic stem cells from chemotherapy-induced toxicity by multidrug-resistance 1 gene transfer. Recent Results Cancer Res 1998;144:93–115.
[25] Hesdorffer C, Ayello J, Ward M, Kaubisch A, Vahdat L, Balmaceda C, et al. Phase I trial of retroviral-mediated transfer of the human MDR1 gene as marrow chemoprotection in patients undergoing high-dose chemotherapy and autologous stem-cell transplantation. J Clin Oncol 1998;16:165–72.
[26] Hanania EG, Giles RE, Kavanagh J, Fu SQ, Ellerson D, Zu Z, et al. Results of MDR-1 vector modification trial indicate that granulocyte/macrophage colony-forming unit cells do not contribute to posttransplant hematopoietic recovery following intensive systemic therapy. Proc Natl Acad Sci USA 1996;93:15346–51.
[27] Moscow JA, Huang H, Carter C, Hines K, Zujewski J, Cusack G, et al. Engraftment of MDR1 and NeoR gene-transduced hematopoietic cells after breast cancer chemotherapy. Blood 1999;94:52–61.
[28] Bunting KD, Galipeau J, Topham D, Benaim E, Sorrentino BP. Transduction of murine bone marrow cells with an MDR1 vector enables ex vivo stem cell expansion, but these expanded grafts cause a myeloproliferative syndrome in transplanted mice. Blood 1998;92:2269–79.

[29] Matsuzaki Y, Kinjo K, Mulligan RC, Okano H. Unexpectedly efficient homing capacity of purified murine hematopoietic stem cells. Immunity 2004;20:87–93.
[30] Nadin BM, Goodell MA, Hirschi KK. Phenotype and hematopoietic potential of side population cells throughout embryonic development. Blood 2003;102:2436–43.
[31] Takao T, Asanoma K, Kato K, Fukushima K, Tsunematsu R, Hirakawa T, et al. Isolation and characterization of human trophoblast side-population (SP) cells in primary villous cytotrophoblasts and HTR-8/SVneo cell line. PLoS One 2011;6(7):e21990.
[32] Uchida N, Fujisaki T, Eaves AC, Eaves CJ. Transplantable hematopoietic stem cells in human fetal liver have a CD34(+) side population (SP)phenotype. J Clin Invest 2001;108:1071–7.
[33] Terrace JD, Hay DC, Samuel K, Payne C, Anderson RA, Currie IS, et al. Side population cells in developing human liver are primarily haematopoietic progenitor cells. Exp Cell Res 2009;315:2141–53.
[34] Scaldaferri ML, Fera S, Grisanti L, Sanchez M, Stefanini M, De Felici M, et al. Identification of side population cells in mouse primordial germ cells and prenatal testis. Int J Dev Biol 2011;55:209–14.
[35] Ramadan A, Nobuhisa I, Yamasaki S, Nakagata N, Taga T. Cells with hematopoietic activity in the mouse placenta reside in side population. Genes Cells 2010;15:983–94.
[36] Vieyra DS, Rosen A, Goodell MA. Identification and characterization of side population cells in embryonic stem cell cultures. Stem Cells Dev 2009;18:1155–66.
[37] Zeng H, Park JW, Guo M, Lin G, Crandall L, Compton T, et al. Lack of ABCG2 expression and side population properties in human pluripotent stem cells. Stem Cells 2009;27:2435–45.
[38] Hulspas R, Quesenberry PJ. Characterization of neurosphere cell phenotypes by flow cytometry. Cytometry 2000;40:245–50.
[39] Kim M, Morshead CM. Distinct populations of forebrain neural stem and progenitor cells can be isolated using side-population analysis. J Neurosci 2003;23:10703–9.
[40] Kobayashi M, Yakuwa T, Sasaki K, Sato K, Kikuchi A, Kamo I, et al. Multilineage potential of side population cells from human amnion mesenchymal layer. Cell Transplant 2008;17:291–301.
[41] Asakura A, Rudnicki MA. Side population cells from diverse adult tissues are capable of in vitro hematopoietic differentiation. Exp Hematol 2002;30:1339–45.
[42] Hierlihy AM, Seale P, Lobe CG, Rudnicki MA, Megeney LA. The post-natal heart contains a myocardial stem cell population. FEBS Lett 2002;530:239–43.
[43] Pacak CA, Cowan DB. Growth of bone marrow and skeletal muscle side population stem cells in suspension culture. Methods Mol Biol 2014;1210:51–61.
[44] Meeson AP, Hawke TJ, Graham S, Jiang N, Elterman J, Hutcheson K, et al. Cellular and molecular regulation of skeletal muscle side population cells. Stem Cells 2004;22:1305–20.
[45] Uezumi A, Ojima K, Fukada S, Ikemoto M, Masuda S, Miyagoe-Suzuki Y, et al. Functional heterogeneity of side population cells in skeletal muscle. Biochem Biophys Res Commun 2006;341:864–73.
[46] Tanaka KK, Hall JK, Troy AA, Cornelison DD, Majka SM, Olwin BB. Syndecan-4-expressing muscle progenitor cells in the SP engraft as satellite cells during muscle regeneration. Cell Stem Cell 2009;4:217–25.
[47] Doyle MJ, Zhou S, Tanaka KK, Pisconti A, Farina NH, Sorrentino BP, et al. Abcg2 labels multiple cell types in skeletal muscle and participates in muscle regeneration. J Cell Biol 2011;195:147–63.

[48] Motohashi N, Alexander MS, Casar JC, Kunkel LM. Identification of a novel microRNA that regulates the proliferation and differentiation in muscle side population cells. Stem Cells Dev 2012;21:3031–43.

[49] Frank NY, Kho AT, Schatton T, Murphy GF, Molloy MJ, Zhan Q, et al. Regulation of myogenic progenitor proliferation in human fetal skeletal muscle by BMP4 and its antagonist Gremlin. J Cell Biol 2006;175:99–110.

[50] Unno K, Jain M, Liao R. Cardiac side population cells: moving toward the center stage in cardiac regeneration. Circ Res 2012;110:1355–63.

[51] Martin CM, Meeson AP, Robertson SM, Hawke TJ, Richardson JA, Bates S, et al. Persistent expression of the ATP-binding cassette transporter, Abcg2, identifies cardiac SP cells in the developing and adult heart. Dev Biol 2004;265:262–75.

[52] Meissner K, Heydrich B, Jedlitschky G, Meyer Zu Schwabedissen H, Mosyagin I, Dazert P, et al. The ATP-binding cassette transporter ABCG2 (BCRP), a marker for side population stem cells, is expressed in human heart. J Histochem Cytochem 2006;54:215–21.

[53] Pfister O, Mouquet F, Jain M, Summer R, Helmes M, Fine A, et al. $CD31^-$ but not $CD31^+$ cardiac side population cells exhibit functional cardiomyogenic differentiation. Circ Res 2005;97:52–61.

[54] Pfister O, Oikonomopoulos A, Sereti KI, Liao R. Isolation of resident cardiac progenitor cells by Hoechst 33342 staining. Methods Mol Biol 2010;660:53–63.

[55] Liang SX, Khachigian LM, Ahmadi Z, Yang M, Liu S, Chong BH. In vitro and in vivo proliferation, differentiation and migration of cardiac endothelial progenitor cells ($SCA1^+/CD31^+$ side-population cells). J Thromb Haemost 2011;98:1628–37.

[56] Oyama T, Nagai T, Wada H, Naito AT, Matsuura K, Iwanaga K, et al. Cardiac side population cells have a potential to migrate and differentiate into cardiomyocytes in vitro and in vivo. J Cell Biol 2007;176:329–41.

[57] Pfister O, Oikonomopoulos A, Sereti KI, Sohn RL, Cullen D, Fine GC, et al. Role of the ATP-binding cassette transporter Abcg2 in the phenotype and function of cardiac side population cells. Circ Res 2008;103:825–35.

[58] Mulligan JD, Schmuck EG, Ertel RL, Brellenthin AG, Bauwens JD, Saupe KW. Caloric restriction does not alter effects of aging in cardiac side population cells. Age (Dordr) 2011;33:351–61.

[59] Gong H, Ma H, Liu M, Zhou B, Zhang G, Chen Z, et al. Urotensin II inhibits the proliferation but not the differentiation of cardiac side population cells. Peptides 2011;32:1035–41.

[60] Oikonomopoulos A, Sereti KI, Conyers F, Bauer M, Liao A, Guan J, et al. Wnt signaling exerts an antiproliferative effect on adult cardiac progenitor cells through IGFBP3. Circ Res 2011;109:1363–74.

[61] Hishikawa K, Marumo T, Miura S, Nakanishi A, Matsuzaki Y, Shibata K, et al. Musculin/MyoR is expressed in kidney side population cells and can regulate their function. J Cell Biol 2005;169:921–8.

[62] Imai N, Hishikawa K, Marumo T, Hirahashi J, Inowa T, Matsuzaki Y, et al. Inhibition of histone deacetylase activates side population cells in kidney and partially reverses chronic renal injury. Stem Cells 2007;25:2469–75.

[63] Inowa T, Hishikawa K, Takeuchi T, Kitamura T, Fujita T. Isolation and potential existence of side population cells in adult human kidney. Int J Urol 2008;15:272–4.

[64] Addla SK, Brown MD, Hart CA, Ramani VA, Clarke NW. Characterization of the Hoechst 33342 side population from normal and malignant human renal epithelial cells. Am J Physiol Renal Physiol 2008;295:F680–7.

[65] Iwatani H, Ito T, Imai E, Matsuzaki Y, Suzuki A, Yamato M, et al. Hematopoietic and nonhematopoietic potentials of Hoechst(low)/side population cells isolated from adult rat kidney. Kidney Int 2004;65:1604–14.
[66] Challen GA, Bertoncello I, Deane JA, Ricardo SD, Little MH. Kidney side population reveals multilineage potential and renal functional capacity but also cellular heterogeneity. J Am Soc Nephrol 2006;17:1896–912.
[67] Giangreco A, Shen H, Reynolds SD, Stripp BR. Molecular phenotype of airway side population cells. Am J Physiol Lung Cell Mol Physiol 2004;286:L624–30.
[68] Majka SM, Beutz MA, Hagen M, Izzo AA, Voelkel N, Helm KM. Identification of novel resident pulmonary stem cells: form and function of the lung side population. Stem Cells 2005;23:1073–81.
[69] Martin J, Helm K, Ruegg P, Varella-Garcia M, Burnham E, Majka S. Adult lung side population cells have mesenchymal stem cell potential. Cytotherapy 2008;10:140–51.
[70] Irwin D, Helm K, Campbell N, Imamura M, Fagan K, Harral J, et al. Neonatal lung side population cells demonstrate endothelial potential and are altered in response to hyperoxia-induced lung simplification. Am J Physiol Lung Cell Mol Physiol 2007;293:L941–51.
[71] Dekaney CM, Rodriguez JM, Graul MC, Henning SJ. Isolation and characterization of a putative intestinal stem cell fraction from mouse jejunum. Gastroenterology 2005;129:1567–80.
[72] Gulati AS, Ochsner SA, Henning SJ. Molecular properties of side population-sorted cells from mouse small intestine. Am J Physiol Gastrointest Liver Physiol 2008;294:G286–94.
[73] Fujimura J, Sugihara H, Fukunaga Y, Suzuki H, Ogawa R. Adipose tissue is a better source of immature non-hematopoietic cells than bone marrow. Int J Stem Cells 2009;2:135–40.
[74] Wang J, Wei X, Ling J, Huang Y, Gong Q, Huo Y. Identification and characterization of side population cells from adult human dental pulp after ischemic culture. J Endod 2012;38:1489–97.
[75] Ishizaka R, Hayashi Y, Iohara K, Sugiyama M, Murakami M, Yamamoto T, et al. Stimulation of angiogenesis, neurogenesis and regeneration by side population cells from dental pulp. Biomaterials 2013;34:1888–97.
[76] Yamahara K, Fukushima S, Coppen SR, Felkin LE, Varela-Carver A, Barton PJ, et al. Heterogeneic nature of adult cardiac side population cells. Biochem Biophys Res Commun 2008;371:615–20.
[77] Hu FF, Xu J, Cui YG, Qian XQ, Mao YD, Liao LM, et al. Isolation and characterization of side population cells in the postpartum murine endometrium. Reprod Sci 2010;17:629–42.
[78] Hyodo S, Matsubara K, Kameda K, Matsubara Y. Endometrial injury increases side population cells in the uterine endometrium: a decisive role of estrogen. Tohoku J Exp Med 2011;224:47–55.
[79] Chen J, Gremeaux L, Fu Q, Liekens D, Van Laere S, Vankelecom H. Pituitary progenitor cells tracked down by side population dissection. Stem Cells 2009;27:1182–95.
[80] Wulf GG, Luo KL, Jackson KA, Brenner MK, Goodell MA. Cells of the hepatic side population contribute to liver regeneration and can be replenished with bone marrow stem cells. Haematologica 2003;88:368–78.
[81] Shimano K, Satake M, Okaya A, Kitanaka J, Kitanaka N, Takemura M, et al. Hepatic oval cells have the side population phenotype defined by expression of ATP-binding cassette transporter ABCG2/BCRP1. Am J Pathol 2003;163:3–9.
[82] Arends B, Vankelecom H, Vander Borght S, Roskams T, Penning LC, Rothuizen J, et al. The dog liver contains a "side population" of cells with hepatic progenitor-like characteristics. Stem Cells Dev 2009;18:343–50.

[83] Hussain SZ, Strom SC, Kirby MR, Burns S, Langemeijer S, Ueda T, et al. Side population cells derived from adult human liver generate hepatocyte-like cells in vitro. Dig Dis Sci 2005;50:1755–63.

[84] de Marval PL, Kim SH, Rodriguez-Puebla ML. Isolation and characterization of a stem cell side-population from mouse hair follicles. Methods Mol Biol 2014;1195:259–68.

[85] Redvers RP, Li A, Kaur P. Side population in adult murine epidermis exhibits phenotypic and functional characteristics of keratinocyte stem cells. Proc Natl Acad Sci USA 2006;103:13168–73.

[86] Oberley CC, Gourronc F, Hakimi S, Riordan M, Bronner S, Jiao C, et al. Murine epidermal side population possesses unique angiogenic properties. Exp Cell Res 2008;314:720–8.

[87] Carr WJ, Oberley-Deegan RE, Zhang Y, Oberley CC, Oberley LW, Dunnwald M. Antioxidant proteins and reactive oxygen species are decreased in a murine epidermal side population with stem cell-like characteristics. Histochem Cell Biol 2011;135:293–304.

[88] Larderet G, Fortunel NO, Vaigot P, Cegalerba M, Maltère P, Zobiri O, et al. Human side population keratinocytes exhibit long-term proliferative potential and a specific gene expression profile and can form a pluristratified epidermis. Stem Cells 2006;24:965–74.

[89] Ma D, Chua AW, Yang E, Teo P, Ting Y, Song C, et al. Breast cancer resistance protein identifies clonogenic keratinocytes in human interfollicular epidermis. Stem Cell Res Ther 2015;6:43.

[90] Jackson KA, Mi T, Goodell MA. Hematopoietic potential of stem cells isolated from murine skeletal muscle. Proc Natl Acad Sci USA 1999;96:14482–6.

[91] McKinney-Freeman SL, Jackson KA, Camargo FD, Ferrari G, Mavilio F, Goodell MA. Muscle-derived hematopoietic stem cells are hematopoietic in origin. Proc Natl Acad Sci USA 2002;99:1341–6.

[92] Asakura A, Seale P, Girgis-Gabardo A, Rudnicki MA. Myogenic specification of side population cells in skeletal muscle. J Cell Biol 2002;159:123–34.

[93] Luth ES, Jun SJ, Wessen MK, Liadaki K, Gussoni E, Kunkel LM. Bone marrow side population cells are enriched for progenitors capable of myogenic differentiation. J Cell Sci 2008;121(Pt 9):1426–34.

[94] Sadek HA, Martin CM, Latif SS, Garry MG, Garry DJ. Bone-marrow-derived side population cells for myocardial regeneration. J Cardiovasc Transl Res 2009;2:173–81.

[95] Jackson KA, Majka SM, Wang H, Pocius J, Hartley CJ, Majesky MW, et al. Regeneration of ischemic cardiac muscle and vascular endothelium by adult stem cells. J Clin Invest 2001;107:1395–402.

[96] Yoon J, Choi SC, Park CY, Choi JH, Kim YI, Shim WJ, et al. Bone marrow-derived side population cells are capable of functional cardiomyogenic differentiation. Mol Cells 2008;25:216–23.

[97] Kotton DN, Fabian AJ, Mulligan RC. A novel stem-cell population in adult liver with potent hematopoietic-reconstitution activity. Blood 2005;106:1574–80.

[98] Cavazzana-Calvo M, Lagresle C, André-Schmutz I, Hacein-Bey-Abina S. The bone marrow: a reserve of stem cells able to repair various tissues? Ann Biol Clin (Paris) 2004;62:131–8.

[99] Fatima S, Zhou S, Sorrentino BP. Abcg2 expression marks tissue-specific stem cells in multiple organs in a mouse progeny tracking model. Stem Cells 2012;30:210–21.

[100] Pedrelli M, Davoodpour P, Degirolamo C, Gomaraschi M, Graham M, Ossoli A, et al. Hepatic ACAT2 knock down increases ABCA1 and modifies HDL metabolism in mice. PLoS One 2014;9:e93552.

[101] Yu XH, Qian K, Jiang N, Zheng XL, Cayabyab FS, Tang CK. ABCG5/ABCG8 in cholesterol excretion and atherosclerosis. Clin Chim Acta 2014;428:82–8.

[102] Kubota M, Shimmura S, Miyashita H, Kawashima M, Kawakita T, Tsubota K. The anti-oxidative role of ABCG2 in corneal epithelial cells. Invest Ophthalmol Vis Sci 2010; 51:5617–22.

[103] Strazielle N, Ghersi-Egea JF. Efflux transporters in blood–brain interfaces of the developing brain. Front Neurosci 2015;9:21.

[104] Israeli D, Ziaei S, Gjata B, Benchaouir R, Rameau P, Marais T, et al. Expression of mdr1 is required for efficient long term regeneration of dystrophic muscle. Exp Cell Res 2007;313:2438–50.

[105] Schumacher T, Krohn M, Hofrichter J, Lange C, Stenzel J, Steffen J, et al. ABC transporters B1, C1 and G2 differentially regulate neuroregeneration in mice. PLoS One 2012;7(4):e35613.

[106] Liu WH, Liu HB, Gao DK, Ge GQ, Zhang P, Sun SR, et al. ABCG2 protects kidney side population cells from hypoxia/reoxygenation injury through activation of the MEK/ERK pathway. Cell Transplant 2013;22:1859–68.

[107] Liu HB, Meng QH, Du DW, Sun JF, Wang JB, Han H. The effects of ABCG2 on the viability, proliferation and paracrine actions of kidney side population cells under oxygen-glucose deprivation. Int J Med Sci 2014;11:1001–8.

[108] Wang J, Wei X, Ling J, Huang Y, Gong Q. Side population increase after simulated transient ischemia in human dental pulp cell. J Endod 2010;36:453–8.

[109] Kovacevic-Filipovic M, Petakov M, Hermitte F, Debeissat C, Krstic A, Jovcic G, et al. Interleukin-6 (IL-6) and low O(2) concentration (1%) synergize to improve the maintenance of hematopoietic stem cells (pre-CFC). J Cell Physiol 2007;212:68–75.

[110] Duchez P, Rodriguez L, Chevaleyre J, Lapostolle V, Vlaski M, Brunet de la Grange P, et al. Interleukin-6 enhances the activity of in vivo long-term reconstituting hematopoietic stem cells in "hypoxic-like" expansion cultures ex vivo. Transfusion May 27, 2015. http://dx.doi.org/10.1111/trf.13175. Epub ahead of print.

[111] Nakamichi N, Morii E, Ikeda J, Qiu Y, Mamato S, Tian T, et al. Synergistic effect of interleukin-6 and endoplasmic reticulum stress inducers on the high level of ABCG2 expression in plasma cells. Lab Invest 2009;89:327–36.

[112] Martin CM, Ferdous A, Gallardo T, Humphries C, Sadek H, Caprioli A, et al. Hypoxia-inducible factor-2alpha transactivates Abcg2 and promotes cytoprotection in cardiac side population cells. Circ Res 2008;102:1075–81.

[113] Rouault-Pierre K, Lopez-Onieva L, Foster K, Anjos-Afonso F, Lamrissi-Garcia I, Serrano-Sanchez M, et al. HIF-2α protects human hematopoietic stem/progenitors and acute myeloid leukemic cells from apoptosis induced by endoplasmic reticulum stress. Cell Stem Cell 2013;13:549–63.

[114] Krishnamurthy P, Ross DD, Nakanishi T, Bailey-Dell K, Zhou S, Mercer KE, et al. The stem cell marker Bcrp/ABCG2 enhances hypoxic cell survival through interactions with heme. J Biol Chem 2004;279:24218–25.

[115] Susanto J, Lin YH, Chen YN, Shen CR, Yan YT, Tsai ST, et al. Porphyrin homeostasis maintained by ABCG2 regulates self-renewal of embryonic stem cells. PLoS One 2008;3(12):e4023.

[116] Shelton P, Jaiswal AK. The transcription factor NF-E2-related factor 2 (Nrf2): a protooncogene? FASEB J 2013;27:414–23.

[117] Singh A, Wu H, Zhang P, Happel C, Ma J, Biswal S. Expression of ABCG2 (BCRP) is regulated by Nrf2 in cancer cells that confers side population and chemoresistance phenotype. Mol Cancer Ther 2010;9:2365–76.

[118] Adachi T, Nakagawa H, Chung I, Hagiya Y, Hoshijima K, Noguchi N, et al. Nrf2-dependent and -independent induction of ABC transporters ABCC1, ABCC2, and ABCG2 in HepG2 cells under oxidative stress. J Exp Ther Oncol 2007;6:335–48.

[119] Jang YY, Sharkis SJ. A low level of reactive oxygen species selects for primitive hematopoietic stem cells that may reside in the low-oxygenic niche. Blood 2007;110:3056–63.
[120] Niculescu VF. The evolutionary history of eukaryotes: how the ancestral proto-lineage conserved in hypoxic eukaryotes led to protist pathogenicity. Microbiol Discov 2014;2(4). http://dx.doi.org/10.7243/2052-6180-2-4.
[121] Niculescu FM. The stem cell biology of the protist pathogen *Entamoeba invadens* in the context of eukaryotic stem cell evolution. Stem Cell Biol Res 2015. http://dx.doi.org/10.7243/2054-717X-2-2. http://www.hoajonline.com/journals/pdf/2054-717X-2-2.pdf3.
[122] Zinzi L, Contino M, Cantore M, Capparelli E, Leopoldo M, Colabufo NA. ABC transporters in CSCs membranes as a novel target for treating tumor relapse. Front Pharmacol 2014;5:163.
[123] Foote SJ, Thompson JK, Cowman AF, Kemp DJ. Amplification of the multidrug resistance gene in some chloroquine-resistant isolates of *P. falciparum*. Cell 1989;57:921–30.
[124] Ouellette M, Fase-Fowler F, Borst P. The amplified H circle of methotrexate-resistant *Leishmania tarentolae* contains a novel P-glycoprotein gene. EMBO J 1990;9:1027–33.
[125] Johnson PJ, Schuck BL, Delgadillo MG. Analysis of a single-domain P-glycoprotein-like gene in the early-diverging protist *Trichomonas vaginalis*. Mol Biochem Parasitol 1994;66:127–37.
[126] Orozco E, López C, Gómez C, Pérez DG, Marchat L, Bañuelos C, et al. Multidrug resistance in the protozoan parasite *Entamoeba histolytica*. Parasitol Int 2002;51:353–9.
[127] Chang G, Roth CB. Structure of MsbA from *E. coli*: a homolog of the multidrug resistance ATP binding cassette (ABC) transporters. Science 2001;293:1793–800.
[128] Zhang WW, Samuelson J. Molecular cloning of the gene for a novel ABC superfamily transporter of *Entamoeba histolytica*. Mol Biochem Parasitol 1993;62:131–4.
[129] Bañuelos C, Orozco E, Gómez C, González A, Medel O, Mendoza L, et al. Cellular location and function of the P-glycoproteins (EhPgps) in *Entamoeba histolytica* multidrug-resistant trophozoites. Microb Drug Resist 2002;8:291–300.
[130] Medel Flores O, Gómez García C, Sánchez Monroy V, Villalba Magadaleno JD, Nader García E, Pérez Ishiwara DG. *Entamoeba histolytica* P-glycoprotein (EhPgp) inhibition, induce trophozoite acidification and enhance programmed cell death. Exp Parasitol 2013;135:532–40.
[131] Moore K. An anaerobic home for stem cell proteome. Blood 1996;107:4578.
[132] Ito K, Suda T. Metabolic requirements for the maintenance of self-renewing stem cells. Nat Rev Mol Cell Biol 2014;15:243–56.
[133] Santos F, Nequiz M, Hernández-Cuevas NA, Hernández K, Pineda E, Encalada R, et al. Maintenance of intracellular hypoxia and adequate heat shock response are essential requirements for pathogenicity and virulence of *Entamoeba histolytica*. Cell Microbiol 2015;17:1037–51.
[134] Messaritakis I, Christodoulou V, Mazeris A, Koutala E, Vlahou A, Papadogiorgaki S, et al. Drug resistance in natural isolates of *Leishmania donovani* s.l. promastigotes is dependent of Pgp170 expression. PLoS One 2013;8:e65467.
[135] Martin WF, Sousa FL, Lane N. Energy of life's origin. Science 2014;344:1092–3.
[136] Mentel M, Martin W. Energy metabolism among eukaryotic anaerobes in light of Proterozoic ocean chemistry. Philos Trans R Soc Lond B Biol Sci 2008;363:2717–29.
[137] Dobzhansky T. Nothing in biology makes sense except in the light of evolution. Am Biol Teach 1973;35:125–9.
[138] Entner N, Anderson HH. Lactic and succinic acid formation by *Entamoeba histolytica* in vitro. Exp Parasitol 1954;3:234–9.
[139] Montalvo FE, Reeves RE, Warren LG. Aerobic and anaerobic metabolism in *Entamoeba histolytica*. Exp Parasitol 1971;30:249–56.

[140] Müller M. Energy metabolism of protozoa without mitochondria. Annu Rev Microbiol 1988;42:465–88.

[141] Steinbüchel A, Müller M. Anaerobic pyruvate metabolism of *Tritrichomonas foetus* and *Trichomonas vaginalis* hydrogenosomes. Mol Biochem Parasitol 1986;20:57–65.

[142] Steinbüchel A, Müller M. Glycerol, a metabolic end product of *Trichomonas vaginalis* and *Tritrichomonas foetus*. Mol Biochem Parasitol 1986;20:45–55.

[143] Lian LY, Al-Helal M, Roslaini AM, Fisher N, Bray PG, Ward SA, et al. Glycerol: an unexpected major metabolite of energy metabolism by the human malaria parasite. Malar J 2009;8:38.

[144] Fry M, Beesley JE. Mitochondria of mammalian *Plasmodium* spp. Parasitology 1991; 102:17–26.

[145] Kita K, Hirawake H, Miyadera H, Amino H, Takeo S. Role of complex II in anaerobic respiration of the parasite mitochondria from *Ascaris suum* and *Plasmodium falciparum*. Biochim Biophys Acta 2002;1553:123–39.

[146] Luque-Ortega JR, Reuther P, Rivas L, Dardonville C. New benzophenone-derived bisphosphonium salts as leishmanicidal leads targeting mitochondria through inhibition of respiratory complex II. J Med Chem 2010;53:1788–98.

[147] Chen M, Zhai L, Christensen SB, Theander TG, Kharazmi A. Inhibition of fumarate reductase in *Leishmania major* and *L. donovani* by chalcones. Antimicrob Agents Chemother 2001;45:2023–9.

[148] Saunders EC, DE Souza DP, Naderer T, Sernee MF, Ralton JE, Doyle MA, et al. Central carbon metabolism of *Leishmania* parasites. Parasitology 2010;137:1303–13.

[149] Tomitsuka E, Kita K, Esumi H. Regulation of succinate-ubiquinone reductase and fumarate reductase activities in human complex II by phosphorylation of its flavoprotein subunit. Proc Jpn Acad Ser B Phys Biol Sci 2009;85:258–65.

[150] Tomitsuka E, Kita K, Esumi H. The NADH-fumarate reductase system, a novel mitochondrial energy metabolism, is a new target for anticancer therapy in tumor microenvironments. Ann N Y Acad Sci 2010;1201:44–9.

[151] Esumi H. "S'ils n'ont pas de pain, qu'ils mangent de la brioche." Focus on "anaerobic respiration sustains mitochondrial membrane potential in a prolyl hydroxylase pathway-activated cancer cell line in a hypoxic microenvironment". Am J Physiol Cell Physiol 2014;306:C320–1.

[152] Chouchani ET, Pell VR, Gaude E, Aksentijević D, Sundier SY, Robb EL, et al. Ischaemic accumulation of succinate controls reperfusion injury through mitochondrial ROS. Nature 2014;515:431–5.

[153] Weinberg JM, Venkatachalam MA, Roeser NF, Nissim I. Mitochondrial dysfunction during hypoxia/reoxygenation and its correction by anaerobic metabolism of citric acid cycle intermediates. Proc Natl Acad Sci USA 2000;97:2826–31.

[154] Weinberg JM, Venkatachalam MA, Roeser NF, Saikumar P, Dong Z, Senter RA, et al. Anaerobic and aerobic pathways for salvage of proximal tubules from hypoxia-induced mitochondrial injury. Am J Physiol Renal Physiol 2000;279:F927–43.

[155] Iverson TM, Luna-Chavez C, Cecchini G, Rees DC. Structure of the *Escherichia coli* fumarate reductase respiratory complex. Science 1999;284:1961–6.

Harnessing Anaerobic Nature of Stem Cells for Use in Regenerative Medicine

Chapter Outline

13.1 Ex Vivo Approximation of Physiological Oxygenation

As discussed earlier (see Chapters 2 and 4), the exposure of a cell culture to the atmospheric O_2 concentration results in a hyperoxygenation. It can be estimated that, at 37°C, the dissolved O_2 in culture stands ~180–210 μM (for sake of simplicity we will consider the value of 200 μM); thus, a value 7- to 14-fold higher than the values encountered by the cells in tissues (provided that the gradients from capillaries are in the range from 10 to 20 mmHg (12.5–25 μM); reviewed in Ref. [1]). In fact, to reproduce these physiological O_2 tissue levels, for the different cell types the cultures should be exposed to a gas-phase O_2 concentration ~1.5% to 3%. However, for some organs and tissues, the O_2 concentration can be only 5 μM and in the particular niches even much lower, close to the anoxia [2]. Hence to mimic this dissolved O_2 concentration an ex vivo culture should be incubated in ~0.6% or lower atmosphere O_2 concentration. Depending on the cell type, the aerobic metabolism shifts to the anaerobic one when the pericellular O_2 concentration drops below 2–6 μM [3]. That means that, in most cases, the cultures exposed to 1–5% O_2 atmosphere

Anaerobiosis and Stemness. http://dx.doi.org/10.1016/B978-0-12-800540-8.00013-2

provide, in fact, the aerobic environment, which approximates the physiological one and not "hypoxia" as usually considered. In other words, these cultures provide an ex vivo environment, avoiding an excessive O_2 level when cultures are exposed to atmospheric O_2 concentration (reviewed in Ref. [1]), resulting in a "burning" oxidative phosphorylation and an increased ROS generation. These conditions, although still aerobic, allow a better balance of self-renewal and commitment of stem cells, contributing to their long-term maintenance in culture and definitive exhaustion. In fact, according to the concept presented in the previous chapters, the fact that the stem cells undergo commitment and consequent differentiation in a hyperoxygenated environment, producing the cell types well adapted to high O_2 concentration parallels the appearance of differentiation and the complex life cycle of ancestral eukaryotes in response to the oxygenation of environment (see Chapter 10).

13.2 Cultures Exposed to Physiologically Relevant Oxygenation or to Hypoxia in Cell Therapy

13.2.1 Hematology

13.2.1.1 Ex Vivo Expansion of Stem and Progenitor Cells

Amplifying ex vivo committed progenitors and stem cells may be an option either to overcome a limited number of these cells (cord blood, apheresis product of "bad mobilizers," a limited number of bone marrow punctions) or to shorten the period of posttransplant agranulocytosis related to immaturity of progenitor and stem cell subpopulations. Typically, an ex vivo expansion culture results in the production of precursors and mature cells (from an accelerated differentiation in culture of committed progenitors (CP) in more mature cell populations) and, in most cases, in the simultaneous amplification of committed progenitors that results from two simultaneous events: the amplification of committed progenitors by their own division, and by their production from more primitive cells differentiating rapidly in culture and, hence, exhausting themselves. As explained in the previous chapters, this "typical" scenario, due, at least in part, to excessive oxygenation of cultures exposed to the atmospheric O_2 concentration (21% O_2) much higher compared to those found in vivo [4], can be avoided if the cultures were exposed to a lower O_2 concentration [5–11].

It is interesting that only 3 years after the first study with murine hematopoietic cells showing a relative "resistance" of hematopoietic stem cell (HSC) to "hypoxia" (1% O_2) [5], the cultures exposed to an O_2 concentration lower that atmospheric one were used to attempt ex vivo expansion of HSCs and hematopoietic progenitor cells (HPCs) for the autologous hematopoietic transplantation [12]. The $CD34^+$ cells mobilized to peripheral blood were selected and ex vivo expanded for 12 days in cultures based on "X-vivo" medium stimulated by PIXI 321 (100 ng) (fusion protein; GM-CSF/IL-3). The cultures of cells from five patients were exposed to 5% O_2 while those from other the four experienced atmospheric O_2 concentration (21% O_2). Today's knowledge allows us to deduce that the GM-CSF and IL-3 make a combination far from an optimal one needed to ensure a simultaneous amplification of HPC and HSC, whichever

O_2 concentration is applied. Indeed, even though this allowed for 26-fold total cell expansion, the amplification of $CD34^+$ cells was negative (0.65-fold). We conclude that the absence of a significant shortening of posttransplantation neutropenia was not surprising while the impact of this expansion on long-term reconstituting cells cannot be evaluated (the autologous transplantation; the in vivo tests on immunodeficient mice were not performed).

As reviewed in Chapter 4, the understanding of the maintenance of HSC at the low O_2 concentration progressed a few years later. The first cues implying that the low O_2 concentration favors the self-renewing HSC divisions that appeared both in experiments with murine and human cells [7,8] were formally confirmed [9,11]. Furthermore, it was demonstrated in a preclinical level culture set-up performed with the clinical-scale cytokines and a serum-free medium, that association of an appropriate cytokine combination and a low-O_2 concentration (3% in this case) allows a simultaneous amplification of HSC and HPC [10]. This study suggested that the appropriate metabolic set-up of HSC in ex vivo cultures may resolve the problem of HSC exhaustion usually occurring in parallel with a massive HPC production. It enlightened the future direction for advancements of the ex vivo HSC and HPC expansion protocols since it provided a conceptual ground to reanalyze the results of prior attempts of ex vivo expansion in view of the "hypoxic response" (in most cases, in fact, the physiological "in situ normoxia response" [4]) induced by some cytokines as well as the role of antioxidants in the medium. This a posterior analysis corroborated the concept associating the anaerobiosis/facultative aerobiosis with the maintenance of the HSC primitiveness. Furthermore, it strengthened the pertinence of the "hypoxia-response mimicking culture" (HMRC) approach, allowing a further improvement of the ex vivo expansion procedures enabling a massive expansion of committed progenitors in parallel with prevention of HSC exhaustion. This approach was necessary from the practical point of view since the low-oxygen cultures were rather complicated to perform in the clinical-scale GMP conditions (N_2 supply, accuracy of O_2 probes, traceability of O_2 concentrations, etc.).

13.2.1.1.1 "Hypoxia-Response Mimicking Cultures" as the Solution for Expansion of Committed Hematopoietic Progenitors without Exhaustion of Hematopoietic Stem Cells

Ex vivo culture of hematopoietic cells represents a complex system (medium composed of several compounds, cytokines/growth factors in various concentrations, physical/chemical factors to which a culture is exposed including the O_2 concentration) in which variation of a single factor can significantly change the cellular output [13]. In fact, an ex vivo expansion culture, aimed to produce as many as possible of the committed progenitors while simultaneously maintaining HSC potential, can rather be compared with the hematopoiesis in regeneration than to steady state hematopoiesis. Thus, the ex vivo conditions have to provide at least the minimal factors ensuring these two opposite processes: generation of committed progenitors and maintenance of HSC. While the first process was easily realized ex vivo, the second one failed until appearance of the Thrombopoietin (Tpo) (reviewed in Ref. [14]) that enabled a relative maintenance of the HSC activity in the course of ex vivo expansion. This effect was enhanced if Tpo was associated with the Flt-3L [15] resulting in a numerical amplification of the HSC (only the cells with reconstitutive capacity are considered by this term; see Chapter 1), which was even more pronounced if the stem cell factor (SCF) was added to two cytokines [16]. In this way it appeared clear that failing to maintain the HSC ex vivo was not due to a natural limit related to the stem cell nature [17] but rather to a lack of knowledge concerning the culture conditions allowing this goal to be achieved. In fact, all mentioned factors represent the integral part of stem cell microenvironmental niche (see Chapter 3) that should be, at least roughly, reproduced in an ex vivo expansion culture.

Thrombopoietin. Tpo as part of this microenvironmental niche is much more than a lineage-specific megakaryocyte growth factor (reviewed in Ref. [18]). In addition to the fact that its action concerns erythropoietic and granulocyte-monocyte progenitors [19], Tpo enables the HSC maintenance as well (reviewed by Huang and Cantor [20]). Since the study of Piacibello et al. [15], in all successful preclinical and clinical studies concerning ex vivo expansion of HSC and HPC, Tpo has been part of the cytokine cocktail [21–29], and so on. Furthermore, up to now, any significant acceleration of hematopoietic reconstitution in clinical trials was achieved only with ex vivo expansion protocols containing Tpo (reviewed in Ref. [14]). It should be stressed that Tpo induces HIF-1α in cultures, even in atmospheric (21%) O_2 concentration. This effect is operated through the regulation of the mitochondrial reactive oxygen species (ROS) and is tightly coupled with glucose metabolism [30,31]. Thus, addition of Tpo mimics a classic "hypoxic" cell response.

Stem Cell Factor. Similar to Tpo, the SCF is part of a basic cytokine cocktail in every successful ex vivo expansion procedure [32] and is related, in addition to its enhancing action on erythroid progenitors [33], to the maintenance in ex vivo culture of HSC exhibiting the hematopoietic reconstituting capacity [34]. SCF stimulation of the hematopoietic cells induces HIF-1α and HIF-2α protein accumulation in cultures exposed to atmospheric O_2 concentration. Many of the genes induced by SCF were found to be related to a hypoxic response [35]. SCF seems to link the cell cycle with nutrient availability and regulates the glycolytic-gluconeogenic switch [36]. In fact, activation of SCF receptor, c-kit (tyrosin kinase), upon binding of SCF, elicits two major signaling pathways: PI-3-kinase/Akt and Ras/MEK/Erk. Their activation results in HIF-1α protein expression. Stabilized HIF-1α as a part of the HIF-1 transcription factor, in turn, stimulates expression of HIF-1 target genes (reviewed in Ref. [14]).

Interleukin-6. In a standard ex vivo expansion culture of murine bone-marrow cells exposed to a physiologically low O_2 concentration (1%), Interleukin-6 (IL-6) improves maintenance of murine HSC subpopulations [37]. Also, IL-6 is secreted in cocultures of human $CD34^+$ cells with mesenchymal stromal cells (MStroC) in physiologically relevant O_2 concentrations (1.5% and 5%) [38], which enhances the HSC maintenance [38–40]. IL-6 secretion via MStroC in coculture with the $CD34^+$ cells is induced by IL-17 in a physiologically relevant O_2 concentration (3%) [41]. Since the O_2 concentrations applied in the experiments cited approximately the pericellular O_2 concentrations in situ in the hematopoietic tissues (~15–30 μM dissolved O_2 in cultures), these data strongly suggest the physiological role for IL-6 in the HSC maintenance. A similar effect of IL-6 was obtained on long-term repopulating HSC (detected on the basis of serial NSG mice transplantation) in the course of ex vivo expansion of CB $CD34^+$ cells in cultures mimicking "hypoxic response" by a combination of antioxidant-supplied, serum-free, xeno-free medium (see further text) with an appropriate cytokine cocktail including, among other cytokines, Tpo and SFC [42]. It is interesting also that interplay between IL-6 and HIF-1α was evidenced in the cross-talk between multiple myeloma plasma cells and tumor microenvironment [43]. Also, in some cancer cell models it was shown that the IL-6 induces a Notch-3-dependent up-regulation of the carbonic anhydrase IX gene and promotes a hypoxia-resistant/invasive phenotype typical for cancer stem cells [44]. However, the similar features were not yet reported for the normal HSC.

Notch Ligands. Notch is a signaling molecule in the context of "hypoxic response," which is related to the maintenance of stemness [45,46]. Notch-receptor-ligand engagement is related to HSC niche retention and cell cycle regulation and obviously represents an important aspect of the HSC niche [47]. The Notch signaling pathway is highly evolutionary conserved and related to promotion of undifferentiated cell state by physiologically relevant low O_2 concentrations, usually termed "hypoxia," as shown with the neuronal and myogenic primitive cell populations [45]. In these cell populations [45], as well as in medulloblastoma "precursors" [48] and *Drosophila* hematopoietic cells [49], it was demonstrated that HIF-1α interacts with the Notch intracellular domain helping to transcriptionally up-regulate Notch-target genes. Furthermore, HIF-1α-induced activation of Notch pathway is essential for hypoxia-mediated maintenance of glioblastoma stem cells. Thus, Notch synergizes with the factors stabilizing HIF-1α (among these factors, for our discussions, "hypoxia," Tpo, and SCF should be pointed out).

Using Antioxidants to Attenuate the Hyperoxic Environment in the Cultures Exposed to Atmospheric O_2 Concentrations (20–21%). A portion of the hyperoxic effects in a culture exposed to nonphysiologically elevated O_2 concentration of atmosphere air comes from excessive production of ROS. In general, the hypoxic response-mimicking cultures exhibiting an antioxidant capacity could be of interest not only because they could enhance the cell viability but also to reduce the risk of chromosomal aberrations that may appear during the ex vivo expansion of hematopoietic cells in standard cultures under atmospheric O_2 concentration (20%) [50–52]. Furthermore, decreasing overproduction of ROS at 20% O_2 is especially important for the primitive cell populations since they are relatively more sensitive to harmful ROS action [53–55]. Indeed, the standard cultures exposed to physiologically relevant O_2 concentration (3%), as well as an appropriate antioxidant dosage, prevent appearance of these chromosomal anomalies [51].

Improving the antioxidative system in the ex vivo expansion media seems to be a good idea to compensate for these hyperoxic effects in standard cultures [56]. Based on this idea, the new generation of ex vivo expansion media appeared, which exhibited a powerful antioxidant system. This way, both the amplification of committed hematopoietic progenitors and maintenance of HSC were significantly improved in cultures in hyperoxia (20–21% O_2) [23,57].

One of these media is serum-free and xenogen-free HP01 medium (Macopharma, France), whose antioxidant battery is not published (an industrial secret) but whose efficiency has been proven both in preclinical and clinical studies [23,24,58–60].

Another one, supplemented with the copper chelator tetraethylenepentamine (TEPA), by decreasing the chelatable cooper, is able to enhance the maintenance of primitive HSC in culture [61,62]. Using blood-derived $CD34^+CD38^-$ cells, Prus and Fibach [63] showed that TEPA decreases the ROS generation previously increased by chelatable copper. On the basis of these data, the authors consider that, in fact, TEPA affects expansion of HSC and CP by lowering oxidative stress. It should be mentioned, however, that decreasing the cellular copper content with TEPA in murine hematopoietic cells results in preferential expansion or maintenance of HPC that are biased for erythroid differentiation in vivo, but does not enhance the maintenance of HSC activity in culture [64]. Nevertheless, using the cord blood-derived $CD133^+$ cells, Pelled et al. [65] developed a large-scale preclinical protocol for ex vivo expansion allowing, after 3 weeks of culture, the median 89-fold increase in $CD34^+$ cell number and 172-fold increase in committed progenitors (CFC), which was paralleled with an enhanced HSC (SRC–NOD/Scid mice) activity over the input value. This protocol was used for a clinical trial [66].

Nicotinamide (NAM; a form of vitamin B3) has been shown to inhibit protein oxidation and lipid peroxidation in rat brain cells [67] and to suppress ROS production in primary human fibroblasts, extending their lifespan when added to the medium during long-term cultivation [68]. Using the MCF-7 cells and human fibroblasts, Kwak et al. [69] showed that NAM has potent antioxidative as well as antisenescent effects.

NAM seems to favor the self-renewal of cord blood HSC ex vivo, resulting in enhanced engraftment capacity in xenogenic models [70]. The protocol based on cytokine-supplemented cultures with NAM has been translated to the clinical trial level [28].

13.2.1.1.2 Clinical Trials Utilizing Transplantation of the Hematopoietic Cells Expanded Ex Vivo in Hypoxia-Response Mimicking Cultures

The first ex vivo expansion protocol enabling an abrogation of posttransplantation agranulocytosis in a clinical trial (phase I/II) was performed with the selected $CD34^+$ cells mobilized to peripheral blood by a standard G-CSF protocol [22]. On the other hand, the ability of ex vivo expanded cells to give a long-term hematopoietic reconstitution in this trial was not possible to evaluate due to two major limitations: (1) autologous context (absence of markers) and (2) coinjection of expanded cells with nonexpanded cells. The culture conditions were based on an Irwin Scientific serum-free medium and three cytokines in a high dose (100 ng/ml): SCF, MGDF (pegylated form of Tpo), and G-CSF. Thus, an association of SCF and Tpo has to be noted. This culture enabled maintenance of HSC (SRC–NOD/Scid mice) after 10 days ex vivo expansion [24]. When the Irwin Scientific medium was replaced by a medium exhibiting a high antioxidant power (HP01; Macopharma, France) both the amplification of CP and HSC (SRC–NOD/Scid mice) enhanced [23]. These cultures, associating an antioxidative medium with two cytokines stabilizing the HIF-1α, which improved ex vivo expansion efficiency both of CF and HSC, were used to produce cells transplanted to the patients without nonexpanded cell fraction. This approach gave similar results as one considering cotransplantation of expanded and nonexpanded cells [24].

As a matter of fact, in 1999 the first results concerning the transplantation of ex vivo expanded cord blood cells were published as well. This case report [71] describes the clinical evolution and outcome of one child with acute lymphoid leukemia that received an injection of the ex vivo expanded cells. The cytokine cocktail used to amplify $CD34^+$ cells selected from 1/8 CBU cell content (FLT3-L, Tpo, and G-CSF (100 ng/ml each)) contained the Tpo, which is, as explained above, one of the main levers of the "hypoxia-mimicking response" since stabilizing HIF1α. This 10-day culture provided interesting results concerning amplification of HPC and HSC: CFC were expanded 21-fold and LTC-IC (a class of HSC) 3.9-fold (it should be noted that the total nucleated cell expansion was 600-fold). In spite of the fact that only the low proportion of total umbilical cord blood (UCB) content was expanded (only 12.5%), which can partially or completely abrogate the results of this successful expansion procedure, the authors estimated that the expanded part of graft contributed to the acceleration of the hematopoietic reconstitution [71].

Another trial on five patients evaluated the "double" UCB transplantation, but in two patients the second UCB was expanded ex vivo [72]. This expansion was performed with a cytokine cocktail (SCF, FLT-3L, and IL-3) that exhibits some elements of "hypoxia-mimicking culture" (presence of SCF). Apart the expansion-fold of TNC of 11.4- and 24.3-fold, the data concerning HPC and HSC were not provided, so we cannot evaluate the ex vivo efficiency of these cultures. It can be assumed, however, with respect to the presence of serum in culture and to the extremely high dose of IL-3 (100 ng/ml) (reviewed in Ref. [73]) that these cultures were not able to prevent the exhaustion of the HSC exhibiting the repopulation potential. Thus, in addition to competition between two cord blood units for repopulation of recipient hematopoiesis, the lack of engraftment of cultured cord blood cells (chimerism ascertained by analysis of DNA polymorphisms) in this study [72] is not surprising from a graft hematopoietic capacity viewpoint.

A serious cohort of patients (37:27 adults and 12 children) transplanted by ex vivo expanded CB cells was published by Shpall et al. [74], but without showing an advantage of expanded cells. For this assay, 40–60% of $CD34^+$ cell content of a USB was expanded for 10 days in serum-free cultures supplemented with SCF, G-CSF, and MGDF, resulting in a relatively low expansion-fold (only four times for $CD34^+$ cells). In fact, it is possible that the HSC with reconstitution capacity were maintained and maybe amplified, but the weak stimulation of HPC amplification did not allow acceleration in hematopoietic reconstitution. Since neither the data concerning the content in HSC with repopulating capacity in expanded grafts (immunodeficient mice engraftment) nor those concerning the donor chimerism in transplanted patients were provided, we cannot conclude on the HSC activity in these grafts.

The efficient "hypoxia-mimicking cultures" were employed to expand the CB cells for the clinical I/II phase clinical trial of de Lima et al. [66]. The $CD133^+$ cells selected from thawed UCB (20%, 40%, and 60% of a UCB) were expanded for 3 weeks in Minimal Essential Medium α (MEMα) supplemented with 10% FCS and TEPA and containing Tpo, IL-6, Flt3L, and SCF. As explained above, this is a strong "hypoxia-mimicking" combination, which is expected to prevent the loss of primitive HSC in the course of expansion. This point, however, cannot be confirmed in spite of the fact that 8/9 patients had the complete donor chimerism at day +42 (one at day +46), as well as those who survived over 6 months due to the fact that from 80% to 50% of UCB units injected were nonmanipulated cells. This culture, rather efficient since it resulted in ~220- and ~38-fold expansion of TNC and CFC, respectively lacked acceleration of the neutrophil and platelet recovery with respect to nonmanipulated UCB transplantation literature data [66]. Even if only a fraction of UCB was expanded, the number of injected progenitors might have suggested a positive effect on the neutrophil and platelet recovery, but this did not occur. We believe that the long duration of culture (3 weeks) might have had a negative impact on the proliferative capacity of the committed progenitors, preventing the positive effect on the hematopoietic recovery.

The typical example of "hypoxia-mimicking culture" was used to produce the cells for a clinical trial comprising eight patients, both pediatric and adult [27]. The 16-day-long cultures were performed in a serum-free medium supplemented with SCF (300 ng/ml), Tpo (100 ng/ml), I-6 (100 ng/ml), Flt-L (300 ng/ml), and Il-3 (10 ng/ml). The originality of this culture is in addition to an immobilized engineered Notch ligand Delta $1^{ext\text{-}IgG}$ (extracellular domain of Delta 1 fused to the Fc domain of IgG1). This approach allowed not only maintaining the excellent amplification of $CD34^+$ cells, committed progenitors, and total cells (which are obtained with the cytokine cocktail used even without Notch ligand Delta $1^{ext\text{-}IgG}$), but also to realize the best amplification of the short- and long-term repopulating HSC ever obtained in clinical-scale cultures [27]. The design of this clinical trial comprised the infusion of two CBU: one nonmanipulated and one ex vivo expanded. As expected, the neutrophil recovery (coming from the expanded CBU) was accelerated after injection of expanded cells (a decrease in the median time to neutrophil recovery by more than 1 week was evidenced comparing to the infusion of two unmanipulated CBU). However, only a transitory short-term engraftment was obtained from the expanded CBU while the long-term chimerism was established by a nonexpanded one. In view of excellent hematopoietic potential of the expanded CBU, it is difficult to suppose that expansion abrogated the long-term repopulating capacity. It is more pertinent to propose that the disappearing of expanded CBU is due to a relatively higher allogeneic potential of nonexpanded UCB (containing the immunocompetent cells), resulting in "extermination" of the expanded one.

Another clinical trial based on infusion of one nonmanipulated and one ex vivo expanded CBU was realized with the CB cells expanded in cocultures with MStroC and injected to 31 adult patients [75]. From a technical viewpoint this was a rather laborious ex vivo expansion protocol: after 7 days of coculture of thawed CB total nucleated cells with the bone marrow MStroC, the nonadherent cells were transferred to the culture bags and cultured for additional 7 days. Both for cocultures and cultures a serum-free medium supplemented by SCF (100 ng/ml), Tpo (100 ng/ml), Flt-3L (100 ng/ml), and G-CSF (100 ng/ml) was used. This protocol, in addition to the presence of SCF and Tpo, as essential elements of a "hypoxia-mimicking culture," comprised a coculture with CSM, characterized by a relative decrease of pericellular oxygen concentration in culture due to the O_2 consumption (see Chapter 4). In cultures used for the clinical trial [75], the expansion-fold of 12.2-, 30.1-, and 17.5-fold was obtained for total nucleated cells, $CD34^+$ cells, and CFC, while no data were presented for the short- and long-term repopulating HSC. The posttransplantation neutrophil recovery was accelerated (median: 15 days). In this trial, 28 from 31 patients presented a donor chimerism. Three to 4 weeks after transplantation, 15 patients (54% of engrafted patients) had the chimerism originating from nonexpanded CBU and 13 from both expanded and nonexpanded CBU (only in four patients predominated expanded CBU). However, in all patients, the long-term engraftment (>1 year) was produced by nonmanipulated CBU [75]. Again, the same problem related to the design of the trial—transplantation of two CBUs—does not allow evaluating the long-term reconstituting capacities of expanded CBU.

Manipulation of the so-called "stem cell self-renewal gene" HOXB4 turned out to be interesting for ex vivo HSC expansion: Murine bone marrow cells transduced with HOXB4 undergo a dramatic multiplication in culture [76]. Using a nonhuman primate model, the same authors demonstrated that transduction of HOXB4 to $CD34^+$ bone marrow cells dramatically enhances the early hematopoietic engraftment; that is, it exhibits a more substantial effect on short-term repopulating HSC than long-term repopulating ones [77]. These results were confirmed later in the same nonhuman primate model with the CB $CD34^+$ cells [78]. Interestingly, when combined with the Notch ligand Delta 1, HOXB4 enabled a more enhanced expansion of HSC (NSG repopulating cells) inside CB $CD34^+$ cell population [79]. A rapid (7-day) neutrophil reconstitution and stable hematopoietic engraftment were obtained in a nonhuman-primate model using the combination of Delta-1 and HOXB4 for ex vivo expansion of CB cells [80], but so far, this approach was not translated to a clinical trial in human medicine. It should be stressed that the HOX gene family belongs to the most primitive genes related to the maintenance of stemness since it existed in the most primitive animals (a proto-Hox gene is expressed in the stem cell candidates in *Trichoplax adhaerens* [81] as well as in the first protist exhibiting a complex life cycle) and most probably existed in the last common ancestor of all living animal lineages [82]. A part of Tpo effect on HSC is related to HOXB4 since it regulates the abundance or subcellular localization of homeodomain proteins HOXB4 among other transcription factors [83].

The most recent clinical trial (11 adult patients) with ex vivo expanded CB cells is also based on infusion of two CBU: expanded and nonexpanded [28]. The expansion cultures used in this study fully correspond to "hypoxia-mimicking cultures." The "ordinary" MEMα supplemented with 10% FCS was used in the presence of SCF, Tpo, Flt-3L, and IL-6 (50 ng/ml each), and nicotinamide (2.5 mM) (in a lower concentration nicotinamide exists in many ex vivo expansion culture media). This way, the antioxidative environment was combined with the factors stabilizing HIF-1α (SCF, Tpo) and those acting synergistically in maintenance of HSC in such a situation (IL-6). The cultures were initiated with $CD133^+$ cells, and their duration was 21 ± 2 days. The median expansion fold was 486 and 72 for total nucleated cells and $CD34^+$ cells while the data for CFC (functional colony-forming assay) and short- and long-term reconstituting HSC (SRC, immunodeficient mice primary and secondary recipients) were not provided. In this trial, however, the CD133 negative fraction (containing the immunocompetent cells) was injected in addition to the expanded cells, which obviously improved the chances of expanded CBU to prevail in long-term reconstitution with respect to the nonexpanded one. Indeed, complete or partial neutrophil or T cell chimerism issued from ex vivo expanded CBU was evidenced in eight patients and remained stable for at least 21 months. These results clearly demonstrate that the HSC were at least maintained during the 3-week ex vivo expansion. The expanded CBU also accelerated the neutrophil recovery (median 13 days), which is significantly lower comparing to “double” nonexpanded CBUs (25 days) [28].

The real challenge, an appropriate ex vivo expansion of CB hematopoietic progenitor cells without decline in HSC potential, was approached also by our group using the HRMC. The cytokine cocktail composed of SCF, Flt3-L, MGDF, and G-CSF was shown to be able to maintain, in parallel with amplification of committed progenitors, the lymphohematopoietic stem cells [84]. After introducing some modifications to enable an efficient $CD34^+$ cell selection and expansion from frozen CB units [85], this culture, already presenting the main elements of “hypoxia mimicking culture” (SCF, MGDF (pegylated Tpo)) was further improved by using the antioxidative medium HP01 [59]. This allowed achieving a massive expansion of committed progenitors simultaneously with a modest amplification of short-term repopulating HSC (SRC; as evaluated with the primary NOD/Scid mice recipients) and full maintenance of long-term repopulating HSC (as evaluated in the secondary NOD/Scid mice recipients) [59]. The final set-up of this procedure concerned the replacement of MGDF (a clinical-grade molecule was not further available on the market) by Tpo adapting its dose (20 ng/ml instead 100 used for MGDF and 100 ng/ml for other cytokines) and shortening duration (12 instead of 14 days) of the two-step culture [86]. This approach was used for a phase I/II clinical trial in which 14 adult patients with lymphohematopoietic malignancies were transplanted with a single cord blood unit cells ex vivo expanded from previously selected $CD34^+$ cells together with the CD34 negative fraction from the same CBU. The results showed a rapid (median 8.5 days for neutrophil recovery) and long-term reconstitution. The rate of donor cell engraftment at day 42 was 86% and 11 patients had sustained full donor hematopoiesis up to 50 months after transplantation (preliminary data [59], definitive rapport submitted). This is the first trial showing a rapid and sustained donor-type hematopoietic reconstitution of adult patients with hematologic malignancies obtained by injection of a single allogeneic ex vivo expanded CBU. These results, at the same time show a real possibility to perform an efficient ex vivo expansion of HSC and CP with only cytokines present in culture that was objected before [76].

The recent experimental studies on murine bone marrow cells showing a possibility to conserve/enhance the HSC capacities in the course of ex vivo cultures either by a pharmacologic stabilization of HIF-1α (compounds derived from isoquinoline mimicking 2-oxoglutarate resulting in an efficient block of prolyl hydroxylase domain (PHD) enzymatic activity) [87] or by activation of anaerobic glycolysis/attenuation of oxidative phosphorylation [88]. In this respect, particularly interesting are new PHD inhibitors developed by several companies (reviewed in Ref. [89]). Since several clinical trials aimed to correct anemia caused by kidney diseases targeting the hypoxia-sensing pathways and consequent induction of Epo production (reviewed in Ref. [89]) are in course, we can expect that a similar approach will be rapidly translated at the clinical trial level for ex vivo HSC and HPC expansion too. The most recent experimental work showing that the inhibition of oxidative phosphorylation with alexidine dihydrochloride (inhibitor of the mitochondrial phosphatase Ptpmt1), and consequent shift to anaerobic glycolysis, results in remarkable preservation of long-term repopulating HSC [90]. These experiments, performed on murine LSK cells, were not yet reproduced with human hematopoietic cells.

13.2.2 Cell Therapy and Cardiovascular Regeneration

Extensive experimentation in the field of cellular therapy of acute myocardial infarction (AMI), ischemic, and nonischemic heart disease comprised the different cell sources (bone marrow nucleated cells (BMNC)), $CD34^+$ cells from bone marrow, mobilized in peripheral blood and of cord blood, MStroC, cardiac myoblasts, skeletal myoblasts, and so on, as well as a different way of cell application. For cell therapy of skeletal muscle ischemia, apart from the cells issued from cardiac muscle, all of these cell categories were assayed in experimental work. In general, the cells were used either with a minimal manipulation (without ex vivo amplification) or with ex vivo amplification. These experiments resulted in a large number of phase I and II clinical trials, whose design and results are very heterogeneous. For example, concerning the BMNC ischemic heart disease therapy, trialed in 48 eligible RCTs (enrolling 2602 patients), it can be concluded that it improves left ventricle ejection fraction, reduces infarct size, and ameliorates remodeling in patients with ischemic heart disease [91]. The best defined and most numerous are clinical trials concerning ex vivo produced MStroC treatment of these diseases, which reached level III in some cases (reviewed in Ref. [92]).

In most clinical trials the cells were shortly manipulated or expanded for several days in cultures exposed to atmosphere (20–21%) O_2 concentration. As discussed in the previous chapters, this approach is not optimal to prepare the cells for implantation even in normal heart tissue (in steady state the O_2 concentrations of 20–30 mmHg (2.6–3.9%) were measured in the epicardial microvasculature and 4–6 mmHg (0.5–0.8%) within the cardiac myocytes [93]) and certainly not in hypoxic and ischemic tissue. Indeed, even a brief exposure to ambient oxygen decreases recovery of long-term repopulating HSCs and increases progenitor cells, a phenomenon termed by Boroxmeyer's group [94] as "extraphysiologic oxygen shock/stress (EPHOSS)." It is pertinent to suppose that the similar phenomenon affects the stem cells with other differentiation capacities than hematopoietic, and that the exposure of the cells intended to be transplanted in both normal and ischemic heart muscle will at least preserve or enhance their effect. In order to evaluate this hypothesis, we present a brief review of the experimental data concerning different cell types studied in the context of cell transplantation in myocard and skeletal muscle.

13.2.2.1 Skeletal Myoblast and Other Cells Derived from Skeletal Muscle

Skeletal myoblasts were the first cells assayed in clinical trials for myocard regeneration and some of these studies have shown improved functional outcomes after transplantation of these cells in postinfarcted scar (reviewed in Ref. [95]). The important issue related to the skeletal muscle-issued cells in the context of myocardial regeneration is risk of arrhythmia. In fact, after injection into the heart, the skeletal myoblasts differentiate into multinucleated myotubes and not in cardiomyocytes. These myotubes form islands perturbing and sometimes blocking the conduction of the electric signal [96]. Whatever perspective of the use of skeletal myoblast in myocardium regeneration, the preconditioning of these cells by stimulating their "hypoxic response" enhanced the beneficial effects in experimental conditions. For example, pharmacologically preconditioned (30 min with 200 μM diazoxide) skeletal myoblasts are resistant to oxidative stress and promote angiomyogenesis via release of paracrine factors in the infarcted heart [97]. Another experimental work showed that the administration of adenovirus-encoded HIF-1-α in skeletal myoblasts ensured an enhanced beneficial effect of these cells injected in infarcted myocard of rats: Ejection fraction increased dramatically (by 27%) with respect to the control groups [98].

However, the skeletal muscle tissue contains very primitive stem cells, which are different from "satellite cells," whose differentiation potential includes skeletal muscle, bone, tendon, nerve, endothelial, hematopoietic, and smooth muscle cells (reviewed in Ref. [99]). These cells, which repair infarcted heart with much more efficiency than skeletal myoblasts [100], are very resistant to O_2 shortage (reviewed in Ref. [95]). If cultured in a three-dimensional engineered tissue construct (the conditions approximating the in vivo situation, in particular with respect to O_2 concentration that is significantly lower at the pericellular level in these cultures compared to two-dimensional ones) these stem cells differentiate into an immature functioning cardiomyocyte phenotype cells, enabling the tissue engineering protocols, which provide a better functional recovery of infarcted myocardium without arrhythmic events (reviewed in Ref. [95]).

Concerning the regeneration of ischemic skeletal muscle, skeletal myoblasts, as cells issued from homologous tissue, seem to be an interesting option. Culturing skeletal muscle myoblasts for 48 h in 1% O_2 seems to affect myoblast fate decision by promoting self-renewal and inhibiting differentiation. This effect of 1% O_2 culture is related to activation of Notch signaling pathway and consequent repressing of miR1 and miR206 [52], and is "translated" in improved grafting efficiency in regenerating muscle. Another study with skeletal muscle-derived stem cells showed that cobalt protoporphyrin efficiently protect these cells for the hypoxia/reoxygenation stress (exposition to $<0.5\%$ O_2 followed by 24 h of reoxygenation) [101]. Using a mouse model, compelling evidence was provided that the HIF-1α is essential for skeletal muscle regeneration [102].

13.2.2.2 Cardiac Stem/Progenitor Cells

Unlike the dog and some other animals [103] the regenerative capacity of heart in humans is very limited [104]. Nevertheless, a heterogeneous population of stem cells was evidenced in the heart tissue. It can be roughly classified in more primitive cardiac stem cells able to form the clusters of cardiac progenitors and the cardiac progenitors themselves [105]. These cell populations, heterogeneous with respect to the proliferative capacity and differentiation potential, are enriched in some phenotypically defined cell populations (most used are c-kit and Scay-1 markers) and highly concentrated in "side population" (cells expressing a calcium-dependent pump—an active mechanism able to reject some molecules; see Chapter 12) [105]. Thus the relationship between "phenotype" and "function" is reminiscent of a situation in other stem/progenitor cell systems (as hematopoietic and mesenchymal ones; see Chapter 1). In spite of this, due to the fact that the most cardiac progenitors belongs to the c-kit positive cell population, in most studies, c-kit$^+$ cells are simply considered as cardiac stem/progenitor cells (as CD34$^+$CD38$^-$ cells or CD133$^+$ cells are often considered HSC and HPC).

Cardiac stem/progenitor cells either isolated directly or derived from cardiospheres are assayed for myocardial repair following myocardial infarction (reviewed in Ref. [106]) and in patients with chronic ventricular dysfunction. In a later case, autologous transplantation of c-kit-positive cardiac stem cells resulted in an outstanding outcome with long-lasting effects without increasing major adverse events [107]. In the context of ischemic-zone myocardial implantation, several studies pointed out the positive impact of their ex vivo preconditioning either with physiologically relevant or really low O_2 concentrations or with the induction of "hypoxic response" in these cells.

In one of these studies the cardiosphere-derived Lin-ckit$^+$ cells (containing heart progenitor cells) isolated from mouse heart were preconditioned during 6h in 5% CO_2 atmosphere containing 0.1% O_2 [108]. Although short, exposition to the really low O_2 concentration in gas phase (0.1%) in these cultures (dissolved O_2 concentration can be estimated to ~1 μM), which, combined with the consumption, should induce a real hypoxic response in cells. When these hypoxia-preconditioned cells were transplanted in the infarcted zone (myocardial infarction in mice was induced by ligation of the middle LAD coronary artery), the results evaluated 4 weeks later were compelling: Infarct size and heart function were significantly better than in mice treated with cells cultured under "normoxic" conditions. The authors consider that this positive impact of hypoxic preconditioning results from stromal cell-derived factor 1 (SDF-1)/CXCR4axis since it was largely abolished by the addition of a CXCR4 inhibitor [108]. These results were confirmed later by a similar approach (6h 0.1% O_2) but using c-kit$^+$ cells to initiate the cardiospheres providing the new details: Prosurvival genes significantly increased in cells cultured in hypoxic conditions as well as their antiapoptosis, migration, and cardiac repair potential, leading to a better survival and cardiac function upon transplantation into acute myocardial infarction (MI) mice in vivo. These beneficial effects could be blocked by a selective antagonist of CXCR4, AMD3100 [109]. Using human c-kit$^+$ cell population and applying a similar preconditioning of 6h in 0.1% O_2 obtained from adult human heart tissue that was removed during cardiosurgery (the valve replacement surgery for instance), the protective effect of hypoxia preconditioning was confirmed [110]. Furthermore, it was evidenced that the most beneficial effects (promotion of cells migration, decrease in apoptosis, enhanced expression of proteins related to antiapoptosis action, prevention of the Cytochrome-C release from mitochondria, elevation of mitochondrial potential, and attenuation of mitochondria damages) are related to Pim-1 kinase [110], which is a crucial factor of cardioprotection downstream of Akt [111].

If human cardiac progenitors were derived from cardiospheres, (issued from endomyocardial heart tissue biopsies) culturing under permanent physiologically relevant O_2 concentration (5%) doubled cell production, markedly diminished the frequency of aneuploidy, showed lower intracellular level of ROS, less cell senescence, and higher resistance to oxidative stress compared to those derived in 21% O_2 cultures [112]. Furthermore, the injection of cells produced ex vivo in 5% O_2 cultures into infarcted hearts of mice provided improved engraftment and functional recovery compared to injection of cells issued from 20% O_2 cultures [112]. Even under 2% O_2 which is rather more of a physiologically relevant concentration than "hypoxia," human heart c-kit$^+$ cells (obtained from infants who had a congenital heart defect after surgery) are maintained better than under 21% O_2, as estimated after 5- to 7-day cultures [113]. The positive impact of "hypoxia" provided efficiently maintained expression of the c-kit, enhanced "tubular-forming capacity" (vessel-like structures formation on Matrigel), and myogenic differentiation potential [113]. However, the most interesting point of this study is that U0126, a specific inhibitor of oxidative stress-induced ERK1/2 activation [114], exhibits similar effects as hypoxia and, in combination with the hypoxic preconditioning, significantly enhanced these functional effects. The most recent publication also optimizes the derivation of cardiac progenitors from cardiospheres applying a physiologically relevant 2% O_2 concentration, resulting in better amplification rate of these cells with a higher expression of "cardiac stem cell markers" and gene markers of stemness [115]. This precondition by a physiologically relevant O_2 concentration was mimicked by treatment with two types of HIF PHD inhibitors: dimethyloxaloylglycine (DMOG) and 2-(1-chloro-4-hydroxyisoquinoline-3-carboxamido) acetic acid (BIC). Both PHD inhibitors activated HIF, Epo, and CXCR-4 and significantly increased c-Kit expression, as well cell proliferation (after 24h of treatment). They decreased oxygen consumption by cardiac progenitors derived from cardiospheres and increased glycolytic metabolism [115].

13.2.2.3 Mesenchymal Stromal Cells

Experimental data on animal models showed the interest of a "preconditioning" of MStroC by exposure to low O_2 concentration before transplantation in ischemic heart tissue. Even a relatively short exposition of murine bone marrow MStroC to a reduced O_2 concentration (0.5% O_2, 24h) significantly enhances the capacity of mesenchymal stem cells to repair infarcted myocardium (murine model), which is attributable to reduced cell death and apoptosis of implanted cells, their increased angiogenesis/vascularization capacities, and paracrine effects [116]. Similar preconditioning of the MStroC derived from umbilical cord (1% O_2, 24h) improved significantly the effect on cardiac function and remodeling, compared to 21% O_2, upon transplantation in infarcted murine heart [117]. Another approach, consisting of preconditioning of rat bone marrow MStroC with two cycles of 30-min ischemia/reoxygenation, supported their survival under subsequent longer exposure to anoxia and following engraftment in the infarcted rat heart. This study demonstrated that the HIF-1α—its downstream micro RNA miR-210 (regulating caspase-8-associated protein-2)—is a critical regulator of MStroC survival in ischemia [118]. Our preclinical study performed on the pig model of chronic myocardial ischemia showed much better effects of human bone marrow MStroC produced during the last passage (7 days) in 1.5% O_2 cultures on the improvement of global systolic and diastolic function, regional cardiac deformation, reduction of left-ventricle pressure, and increase in left ventricle contractility compared to MStroC produced in 20% O_2 cultures [119]. The MStroC produced at 1.5% O_2 exhibited better survival rate and homing in ischemic heart tissue.

Using a murine in vivo model of hind limb ischemia, it demonstrated a spectacular effect of murine bone marrow MStroC preconditioning by low O_2 concentration (36 h at 1% O_2) on skeletal muscle regeneration related to an improved blood flow and vascular formation compared to injected nonpreconditioned MStroC. This effect is elaborated via a paracrine Wnt-dependent mechanism [120]. A recent publication showed that the murine bone marrow MStroC exposed for 24 h to 0.5% O_2 atmosphere prevented cardiac fibroblast activation and collagen production, thus, potentially preventing replacing of myocytes with fibrotic scar tissue [121]. It should be stressed that "hypoxic response" of MStroC includes Notch signaling, which is required for regulation of glycolysis in physiologically relevant O_2 concentrations. For example, the proliferative and differentiation capacities of human-adipose tissue-derived MStroC are better maintained in 5% O_2 cultures, which are associated with an enhanced glycolysis rate [122]. These effects are HIF-independent and are related to the activation of Notch1 and expression of its downstream gene HES1 that is involved in induction of GLUT3, TPI, and PGK1 in addition to reduction of TIGAR and SCO2 expression [122].

In addition to exposing the MStroC to a low (physiologically relevant or really hypoxic) O_2 concentration to precondition them for in vivo transplantation in ischemic or nonischemic heart or skeletal muscle tissue, it is possible to get a similar effect directly by agents acting on the mechanism of "hypoxic" cell response, as explained above for HSC. The positive effect of hydrogen sulfide (H_2S) on in vitro hypoxia-induced cell apoptosis of rat bone marrow is associated with an improvement of the survival rate of the transplanted MStroC in infarcted myocardium 4 days after myocardial infarct (the ligation of the left anterior descending of coronary artery), compared with the untreated MStroC [123]. Also, the transplantation of the H_2S-pretreated MStroC reduced the infarct size and increased left ventricular function [123]. H_2S interferes with several mechanisms, but we believe that effects obtained in the Xie et al. [123] study at least in part come from a partial or total inhibition of mitochondrial respiration [124], which may induce the shift to anaerobiosis and, hence, prepare the MStroC for hypoxic or even anoxic conditions. Another approach, the inhibition of hexakisphosphate kinases (IP6Ks) (which are physiological inhibitors of Akt) in bone marrow-derived MStroC during 2 h before exposing them to low O_2 concentrations and serum deprivation significantly improved the viability and enhanced the paracrine effect of MStroC and their survival upon transplantation into infarcted heart muscle, which promoted the antiapoptotic and proangiogenic efficacy of MStroC in vivo [125]. This preconditioning of MStroC also significantly decreased fibrosis and preserved heart function. As discussed above for HSC, the inhibition of prolyl hydroxylase and consequent HIF1α stabilization efficiently mimics the "hypoxic response" and can enhance the stem cell capacities. This strategy was tested in rat model by preconditioning bone marrow-derived MStroC with the PHD inhibitor dimethyloxalylglycine (DMOG). This allowed enhancement of the MStroC viability and angiogenesis-capacity while transplantation of DMOG-preconditioned MStroC in the peri-infarct region reduced heart infarct size and improved cardiac function [126]. Also, an appropriate manipulation of the storage of glucose in glycogen and its liberation in combination with "hypoxic" preconditioning (1% O_2, 24 h) confers a clear advantage in terms of survival and consequent beneficial effects to these cells once injected in ischemic zone (mouse limb ischemia model) [127], providing to these cells more glucose from their own reserves.

All these approaches are still in experimental phase and were not yet translated to the clinical trial level.

It should be stressed that MStroC are a very heterogeneous population with respect to proliferation capacity and differentiation potential (see Chapters 1 and 4; [128,129]). In that respect the angiogenic commitment and differentiation at low O_2 concentration, ascribed to MStroC in general (e.g., Ref. [130]), can be the result of a specific sub-population responding to the low O_2 concentration or to the triggering of one or more mechanisms of "hypoxic response." In the same manner, the enhancing of stemness in the same conditions can consider only the most primitive—the real stem cell population of MStroC (see Chapter 4, Figures 4.1 and 4.2). It is necessary to stress that a subpopulation of MStroC is able not only to survive but also to proliferate in total anoxia (0% O_2) ([131], our unpublished results). In fact, a subpopulation of MStroC can be maintained in a multipotent state with very low ATP contents ensured by glycolysis only [132], and in the course of anaerobic proliferation, these cells, in absence of glucose, in addition to mobilizing the glucose reserve from glycogen [127], may provide the energy (ATP) from anaerobic mitochondrial respiration ("fumarate respiration") (our unpublished data) traditionally being considered to be specific for cancer cells. These preliminary data may be of paramount importance to design the new approaches to precondition the MStroC for transplantation in an anoxic and ischemic environment.

13.2.2.4 Endothelial Cells

Rat bone marrow derived, ex vivo amplified cells capable of incorporating DiI-acLDL and bind UEA-1, expressing CD34 and CD133 (which authors consider "endothelial progenitors"; no data are presented concerning the frequency of the endothelial-colony forming cells; i.e., real functional endothelial progenitors in this cell population) were "preconditioned" in 1% O_2 atmosphere during 24 h. This caused significant enhancements in the formation of tube-like structure and motility as well as mRNA expressions of CXCR4, PI3K, AKT, and NF-κB. In vivo transplantation experiments demonstrated the beneficial effect of hypoxic EPCs on left ventricular functions after acute myocardial ischemia (provoked by ligation of the left anterior descending coronary artery) [133].

13.2.3 Cell Therapy and Neuronal Restoration

Several decades ago the idea to use the cell replacement for the central nervous system tissue repair was considered science fiction. However, the experimental work, started in the late 1970s when it was demonstrated that the intrastriatal grafts of dopamine-neurons-rich mesencephalic tissue may induce the functional recovery of damaged nigrostriatal dopaminergic systems, through the extensive research using the rodent and nonhuman primate models, resulted in the first clinical trials with neural transplantation in Parkinson's disease in 1987 (cited in Ref. [134]). Parkinson's disease, however, proved to be a very complex issue for cell therapy for several reasons, including the facts that disease process destroys grafted neurons as well and that the nonmotor symptoms will not be influenced by intrastriatal dopaminergic grafts [135]. The fetal neuronal cells were also assayed in patients with Huntington's disease in the late 1990s and in other disorders (stroke, epilepsy, and spinal cord injury), but the success of these trials was not compelling. The basic and preclinical research performed over the last decade provided the proofs that replacement and reconstitution (at least partial) of neural network is feasible for motoric and sensorial restoration in the context of posttraumatic lesions and neurodegenerative disorders of neural tissue (reviewed in Refs. [136–139]), but also cognitive impairment like learning and memory deficits [140,141]. As for other cell-based regenerative therapies, the different cell populations (bone marrow nucleated cells, mesenchymal stem cells, neuronal stem/progenitor cells, either isolated from adult tissue or produced from ES or iPS) are assayed in the context of neuronal tissue regeneration. Again, the physiologically low O_2 concentrations in the nervous system tissue, of the real hypoxia in the tumors of these tissue, and of ischemia/real hypoxia in postlesion zones of this tissue [140,142] (reviewed in Refs. [143,144]) on one side and the usual practice to produce ex vivo the cells in cultures exposed to a high O_2 concentration (21% O_2, atmospheric one) (reviewed in Ref. [143]) should be taken into consideration. In that respect the production of these cells in an appropriate oxidative environment and/or their preconditioning either by an appropriately low O_2 concentration or by "hypoxia" mimicking factors appears as an unavoidable approach to improve these procedures in order to come closer to the clinical applications.

13.2.3.1 Neural Stem/Progenitor Cells

The beneficial ex vivo effects of physiologically relevant O_2 concentrations on proliferative capacity and viability of neural stem/progenitor cells, which were evidenced regardless of their source (adult tissue, fetal tissue, ES, and iPS), were discussed in Chapter 4, on the basis of rich literature data [145–161]. Furthermore, neural stem/progenitor cells are not only very resistant to the real hypoxia in situ, but they can be "reinforced" in this condition, as shown for those residing in the subventricular zone of one day old piglets that were preconditioned by hypoxia (8% O_2; 92% N_2 atmosphere during 3 h) before being exposed, 24 h later, to the hypoxia-ischemia (realized by a complex procedure providing pronounced hypoxemia combined with hypotension) [162]. This preconditioning resulted in a net increase in number of neural stem/progenitor colony-forming cells derived from a subventricular zone exhibiting a multipotent differentiation potential, which was in line with an increased neural progenitor proliferation in vivo and enhanced neurogenesis [162]. Exposure of mouse ES-derived neural stem/progenitor cells to 1% O_2 for 12 h followed by a reoxygenation (21% O_2 for 24 h) resulted in increased cell survival in vitro (enhanced antiapoptotic protein bcl-2 expression), promoted neurofilament and synaptophysin expression, as well as induced HIF1α expression and Epo secretion. Furthermore, this "hypoxia" preconditioning resulted in enhanced cell survival after transplantation into the ischemic rat brain (the rats were subjected to an immumosuppressive regime in order to prevent the rejection of murine cells) [163].

13.2.3.2 Mesenchymal Stromal Cells

To test the hypothesis that "hypoxia-preconditioned" MStroC may enhance the seeding efficiency and survival of neural stem/progenitor cells, Oh et al. [164] employed an interesting strategy: preconditioning of human-adipose tissue-derived MStroC exposed in 1% O_2 atmosphere during 24 h in order to protect, by these MStroC, the neuronal stem and progenitor cells. This approach was efficient both ex vivo and in vivo: (1) in cocultures designed to mimic ischemic injuries the apoptosis of neural stem/progenitor cells was greatly reduced in the presence of 1% O_2 preconditioned MStroC and (2) after in vivo coinjection of 1% O_2 preconditioned MStroC and neural stem/progenitor cells in the zone of ischemic spinal cord injury (rat model), the survival of neural stem/progenitor cells (traced by DsRed) 2 and 4 weeks after transplantation was greatly improved compared with injection of neural stem/progenitor cells only or in combination with MStroC without 1% O_2 preconditioning [164]. A similar approach was assayed by Kim et al. [165]: Human bone marrow-derived MStroC were cultured in 1% O_2 before being reoxygenated. One-day 1% O_2 culture had a negative impact on bulk MStroC viability and proliferation (for explanation see Chapter 4), but an accelerated proliferation with maintained differentiation potential after reoxygenation showed that the mesenchymal stem cells inside the MStroC population were at least maintained if not enhanced during 1% O_2 culture. One percent O_2 preconditioning also enhanced expression of prosurvival genes and migration capacity. Most importantly, the preconditioned MStroC promoted the survival of ischemic rat cortical neurons in coculture [165].

Another study evaluated the regenerative properties of MStroC in the context of neural regeneration and tested the hypothesis that "hypoxia preconditioning" could not only enhance the survival of bone marrow-derived MStroC, but also reinforce regenerative properties of these cells [166]. Bone marrow MStroC transgenic rats expressing the enhanced green fluorescent protein (eGFP) (enabling the tracking) were exposed for 24–48 h to 0.5% O_2 atmosphere, which up-regulated HIF-1α and trophic/growth factors, including brain-derived neurotrophic factor, glial cell-derived neurotrophic factor, vascular endothelial growth factor and its receptor FIK-1, erythropoietin and its receptor EpoR, and stromal derived factor-1 (SDF-1) and its CXC chemokine receptor 4 (CXCR4). Exposition to 5% O_2 also reduced inflammatory genes in bone marrow StroC and transplantation after stroke promoted anti-inflammation effect (rat transient ischemia model of middle cerebral artery occlusion). Importantly, a fraction of MStroC conditioned with 0.5% O_2 after transplantation underwent the endothelial cell differentiation and contributed to enhanced angiogenesis. Furthermore, the injected MStroC preconditioned with 0.5% O_2 even differentiated into neuronal cells. Fourteen days after transplantation, the rats treated with 5% O_2-preconditioned MStroC showed much better motor function improvement (rotarod test) compared to both control groups [166].

The same group recently administered to mice after ischemic stroke (mouse focal cerebral ischemia model) [167] and hemorrhagic stroke (collagenase IV model) [168] hypoxia preconditioned (0.1–0.3% O_2, 24h, followed by 1h "reoxygenation") rat bone marrow MStroC cells using the new noninvasive brain-targeted technique based on a hyaluronidase pretreatment aimed to disrupt the barrier function of the nasopharyngeal mucosa and facilitate cell entry to the brain. In the case of ischemic stroke, significant neuroprotection of reduced infarct volume was evidenced after transplantation of hypoxia preconditioned MStroC. Furthermore, the mice treated with these cells 3 days after transplantation performed better on a test reflecting the sensorimotor function comparing to the control mice [167]. The similar result was obtained after hemorrhagic stroke: intranasally delivered hypoxia-preconditioned MStroC migrated to peri-injury regions and secreted the factors enabling the increase of neurogenesis [168]. All these results suggest a significant importance of MStroC exposition to either physiologically relevant or really low O_2 concentrations before transplantation but are not yet translated to human preclinical and clinical studies levels.

13.2.4 Other Tissues and Systems

The experiments with preconditioning of MStroC by physiologically low O_2 concentrations or by real hypoxia, or mimicking hypoxia by inducing cellular hypoxic response, yielded promising results literally in all the tested organs and tissues. The nonexhaustive list of examples includes liver regeneration [169,170], acute ischemic kidney injury [171,172], bleomicin-induced pulmonary fibrosis [173], cell therapy of diabetes [174], and even survival of "random skin flap" (using a rat model for reconstructive surgery) [175].

13.3 Conclusions

On the basis of the literature data review presented in this chapter, several conclusions can be drawn:

1. Culturing of cells in physiologically relevant O_2 concentrations (in most cases identified by the misleading term "hypoxia") maintains the proliferative capacity of stem and progenitor cells as well as their differentiation capacities. In the case of some lineage-committed progenitors, these conditions enhance differentiation (typical example, endothelial progenitors). Due to the comparison traditionally made with to 20–21% O_2 cultures (i.e., hyperoxia with respect to physiological conditions, which lead to the enhanced commitment and differentiation in all stem/progenitor cell systems leading to the impairment of stem cells (exhaustion), this phenomenon is usually interpreted as "enhancement of stem cell capacities by hypoxia." The physiologically relevant O_2 concentrations ex vivo enable a better preservation of the secretory capacities of some cell populations present in the heterogeneous stem/committed progenitors/postprogenitor cell populations (frequently considered simply as "stem cells"). A typical example is MStroC population, containing the cells capable of extensive secretion

of cytokines and growth factors, which may not necessarily be the property of MStroC. This secretion can be important for the regeneration of the tissue since it can stimulate in situ of either tissue-specific resident stem and progenitor cells or vascular cells, to proliferate and differentiate. Thus, when the harmful effect of elevated atmospheric O_2 concentration is prevented, the functional capacities of all these cell populations are better preserved compared to the "standard" 20–21% cultures, which further results not only in better survival after injection in normal, hypoxic, and ischemic tissue, but also in a better regenerative effect and restoration of functional capacities.

2. Culturing of cells in really hypoxic or even anoxic conditions enables a selection of cells endowed by anaerobic capacities and, depending on the culture condition, their "building," leading to an enhanced survival in hypoxic and ischemic tissue after injection. This "conditioning" selects for most primitive stem cell populations and, due to this fact, provides not only a much better survival of cells in hypoxic/ischemic tissue, but is expected to give the long-term regenerative effects. It should be stressed that this not yet completely explored approach is still at the level of basic experimental work with exciting perspectives, especially in view of the data suggesting capacities of primitive stem cells to perform the anaerobic mitochondrial respiration.

3. Conditioning of stem/progenitor cells by stimulating the "hypoxic" response in atmospheric O_2 cultures ("hypoxic response mimicking cultures") represents a concept that is translated in clinical trials in hematology (ex vivo expansion of hematopoietic stem end progenitor cells) and tested in basic research and preclinical levels in other fields. This approach can either provide "protection" of hyperoxygenation (point 1) or even mimics anoxia and ischemia (point 2). It can be realized by stimulating the "hypoxic" signalization, inhibiting the aerobic metabolic processes, stimulating the anaerobic ones, acting on the "coupling" between aerobic and anaerobic processes, acting on the ROS, and so on. These manipulations can enhance the proliferative potential, modulate differentiation capacity, regulate secretion of growth factors and cytokines, and so on, which can be used as a tool in engineering the stem/progenitor cell populations corresponding to the needs of cell therapy. This approach is at the moment in the initial phase of research. At the present, a lot of mechanisms remain unexplored. We believe that the perspectives of this approach are clearly evident.

As it can be understood, some aspects of anaerobic stem cell features are translated to the clinical trial level only in hematopoietic stem and progenitor cell transplantation, while in other fields they are object of basic research and preclinical development. Concerning the results of this research, we believe that they confirm the notion of "anaerobic stem cell entity" and the "evolutionary paradigm." Integration of these results and their translation to the clinical level will provide significant advancement in cell therapy and regenerative medicine in the near future.

References

[1] Winslow RM. Oxygen: the poison is in the dose. Transfusion 2013;53:424–37.

[2] Chow DC, Wenning LA, Miller WM, Papoutsakis ET. Modeling pO_2 distributions in the bone marrow hematopoietic compartment. II. Modified Kroghian models. Biophys J 2001;81:685–96.

[3] Boveris A, Costa LE, Cadenas E, Poderoso JJ. Regulation of mitochondrial respiration by adenosine diphosphate, oxygen, and nitric oxide. Methods Enzymol 1999;301:188–98.

[4] Ivanovic Z. Hypoxia or in situ normoxia: the stem cell paradigm. J Cell Physiol 2009;219:271–5.
[5] Cipolleschi MG, Dello Sbarba P, Olivotto M. The role of hypoxia in the maintenance of hematopoietic stem cells. Blood 1993;82:2031–7.
[6] Cipolleschi MG, Rovida E, Ivanovic Z, Praloran V, Olivotto M, Dello Sbarba P. The expansion of murine bone marrow cells preincubated in hypoxia as an in vitro indicator of their marrow-repopulating ability. Leukemia 2000;14:735–9.
[7] Ivanovic Z, Bartolozzi B, Bernabei PA, Cipolleschi MG, Rovida E, Milenkovic P, et al. Incubation of murine bone marrow cells in hypoxia ensures the maintenance of marrow-repopulating ability together with the expansion of committed progenitors. Br J Haematol 2000;108:424–9.
[8] Ivanovic Z, Dello Sbarba P, Trimoreau F, Faucher JL, Praloran V. Primitive human HPCs are better maintained and expanded in vitro at 1 percent oxygen than at 20 percent. Transfusion 2000;40:1482–8.
[9] Ivanovic Z, Belloc F, Faucher JL, Cipolleschi MG, Praloran V, Dello Sbarba P. Hypoxia maintains and interleukin-3 reduces the pre-colony-forming cell potential of dividing $CD34^+$ murine bone marrow cells. Exp Hematol 2002;30:67–73.
[10] Ivanovic Z, Hermitte F, Brunet de la Grange P, Dazey B, Belloc F, Lacombe F, et al. Simultaneous maintenance of human cord blood SCID-repopulating cells and expansion of committed progenitors at low O_2 concentration (3%). Stem Cells 2004;22:716–24.
[11] Danet GH, Pan Y, Luongo JL, Bonnet DA, Simon MC. Expansion of human SCID-repopulating cells under hypoxic conditions. J Clin Invest 2003;112):126–35.
[12] Williams SF, Lee WJ, Bender JG, Zimmerman T, Swinney P, Blake M, et al. Selection and expansion of peripheral blood $CD34^+$ cells in autologous stem cell transplantation for breast cancer. Blood 1996;87:1687–91.
[13] Douay L. Experimental culture conditions are critical for ex vivo expansion of hematopoietic cells. J Hematother Stem Cell Res 2001;10:341–6.
[14] Ivanovic Z, Vlaski M. Ex vivo expansion of stem and progenitor cells using thrombopoietin. In: Hayat MA, editor. Stem cells and cancer stem cells, vol. 8. Springer Science+Business Media Dordrecht; 2012. p. 345–53.
[15] Piacibello W, Sanavio F, Garetto L, Severino A, Bergandi D, Ferrario J, et al. Extensive amplification and self-renewal of human primitive hematopoietic stem cells from cord blood. Blood 1997;89:2644–53.
[16] Luens KM, Travis MA, Chen BP, Hill BL, Scollay R, Murray LJ. Thrombopoietin, kit ligand, and flk2/flt3 ligand together induce increased numbers of primitive hematopoietic progenitors from human $CD34^+Thy\text{-}1^+Lin^-$ cells with preserved ability to engraft SCID-hu bone. Blood 1998;91:1206–15.
[17] Verfaillie CM. Can human hematopoietic stem cells be cultured ex vivo? Stem Cells 1994;12:466–76.
[18] Kaushansky K. Thrombopoietin: more than a lineage-specific megakaryocyte growth factor. Stem Cells 1997;15(Suppl. 1):97–102.
[19] Kaushansky K, Lin N, Grossmann A, Humes J, Sprugel KH, Broudy VC. Thrombopoietin expands erythroid, granulocyte-macrophage, and megakaryocytic progenitor cells in normal and myelosuppressed mice. Exp Hematol 1996;24:265–9.
[20] Huang H, Cantor AB. Common features of megakaryocytes and hematopoietic stem cells: what's the connection? J Cell Biochem 2009;107:857–64.
[21] Norol F, Drouet M, Mathieu J, Debili N, Jouault H, Grenier N, et al. Ex vivo expanded mobilized peripheral blood $CD34^+$ cells accelerate haematological recovery in a baboon model of autologous transplantation. Br J Haematol 2000;109:162–72.

[22] Reiffers J, Cailliot C, Dazey B, Attal M, Caraux J, Boiron JM. Abrogation of post-myeloablative chemotherapy neutropenia by ex-vivo expanded autologous CD34-positive cells. Lancet 1999;354:1092–3.

[23] Ivanovic Z, Duchez P, Dazey B, Hermitte F, Lamrissi-Garcia I, Mazurier F, et al. A clinical-scale expansion of mobilized CD 34$^+$ hematopoietic stem and progenitor cells by use of a new serum-free medium. Transfusion 2006;46:126–31.

[24] Boiron JM, Dazey B, Cailliot C, Launay B, Attal M, Mazurier F, et al. Large-scale expansion and transplantation of CD34$^+$ hematopoietic cells: in vitro and in vivo confirmation of neutropenia abrogation related to the expansion process without impairment of the long-term engraftment capacity. Transfusion 2006;46:1934–42.

[25] McNiece I, Jones R, Bearman SI, Cagnoni P, Nieto Y, Franklin W, et al. Ex vivo expanded peripheral blood progenitor cells provide rapid neutrophil recovery after high-dose chemotherapy in patients with breast cancer. Blood 2000;96:3001–7.

[26] Prince HM, Simmons PJ, Whitty G, Wall DP, Barber L, Toner GC, et al. Improved haematopoietic recovery following transplantation with ex vivo-expanded mobilized blood cells. Br J Haematol 2004;126:536–45.

[27] Delaney C, Heimfeld S, Brashem-Stein C, Voorhies H, Manger RL, Bernstein ID. Notch-mediated expansion of human cord blood progenitor cells capable of rapid myeloid reconstitution. Nat Med 2010;16:232–6.

[28] Horwitz ME, Chao NJ, Rizzieri DA, Long GD, Sullivan KM, Gasparetto C, et al. Umbilical cord blood expansion with nicotinamide provides long-term multilineage engraftment. J Clin Invest 2014;124:3121–8.

[29] Fan X, Gay FP, Lim FW, Ang JM, Chu PP, Bari S, et al. Low-dose insulin-like growth factor binding proteins 1 and 2 and angiopoietin-like protein 3 coordinately stimulate ex vivo expansion of human umbilical cord blood hematopoietic stem cells as assayed in NOD/SCID gamma null mice. Stem Cell Res Ther 2014;5:71.

[30] Yoshida K, Kirito K, Yongzhen H, Ozawa K, Kaushansky K, Komatsu N. Thrombopoietin (TPO) regulates HIF-1alpha levels through generation of mitochondrial reactive oxygen species. Int J Hematol 2008;88:43–51.

[31] Kirito K, Hu Y, Komatsu N. HIF-1 prevents the overproduction of mitochondrial ROS after cytokine stimulation through induction of PDK-1. Cell Cycle 2009;8:2844–9.

[32] Ivanovic Z, Boiron JM. Ex vivo expansion of hematopoietic stem cells: concept and clinical benefit. Transfus Clin Biol 2009;16:489–500.

[33] Kubo T, Nakahata T. Different responses of human marrow and circulating erythroid progenitors to stem cell factor, interleukin-3 and granulocyte/macrophage colony-stimulating factor. Int J Hematol 1993;58:153–62.

[34] Du Z, Wang Z, Zhang W, Cai H, Tan WS. Stem cell factor is essential for preserving NOD/SCID reconstitution capacity of ex vivo expanded cord blood CD34$^+$ cells. Cell Prolif 2015;48:293–300.

[35] Pedersen M, Löfstedt T, Sun J, Holmquist-Mengelbier L, Påhlman S, Rönnstrand L. Stem cell factor induces HIF-1alpha at normoxia in hematopoietic cells. Biochem Biophys Res Commun 2008;377:98–103.

[36] Benanti JA, Cheung SK, Brady MC, Toczyski DP. A proteomic screen reveals SCFGrr1 targets that regulate the glycolytic-gluconeogenic switch. Nat Cell Biol 2007;9:1184–91.

[37] Kovacevic-Filipovic M, Petakov M, Hermitte F, Debeissat C, Krstic A, Jovcic G, et al. Interleukin-6 (IL-6) and low O_2 concentration (1%) synergize to improve the maintenance of hematopoietic stem cells (pre-CFC). J Cell Physiol 2007;212:68–75.

[38] Koller MR, Bender JG, Miller WM, Papoutsakis ET. Reduced oxygen tension increases hematopoiesis in long-term culture of human stem and progenitor cells from cord blood and bone marrow. Exp Hematol 1992;20:264–70.

[39] Zhambalova AP, Darevskaya AN, Kabaeva NV, Romanov YA, Buravkova LB. Specific interaction of cultured human mesenchymal and hemopoietic stem cells under conditions of reduced oxygen content. Bull Exp Biol Med 2009;147:525–30.

[40] Hammoud M, Vlaski M, Duchez P, Chevaleyre J, Lafarge X, Boiron JM, et al. Combination of low O_2 concentration and mesenchymal stromal cells during culture of cord blood $CD34^+$ cells improves the maintenance and proliferative capacity of hematopoietic stem cells. J Cell Physiol 2012;227:2750–8.

[41] Krstic A, Vlaski M, Hammoud M, Chevaleyre J, Duchez P, Jovcic G, et al. Low O_2 concentrations enhance the positive effect of IL-17 on the maintenance of erythroid progenitors during co-culture of $CD34^+$ and mesenchymal stem cells. Eur Cytokine Netw 2009;20:10–6.

[42] Duchez P, Rodriguez L, Chevaleyre J, Lapostolle V, Vlaski M, Brunet de la Grange P, et al. Interleukin-6 enhances the activity of in vivo long-term reconstituting hematopoietic stem cells in "hypoxic-like" expansion cultures ex vivo. Transfusion 2015; 55, in press.

[43] Borsi E, Perrone G, Terragna C, Martello M, Zamagni E, Tacchetti P, et al. HIF-1α inhibition blocks the cross talk between multiple myeloma plasma cells and tumor microenvironment. Exp Cell Res 2014;328:444–55.

[44] Sanguinetti A, Santini D, Bonafè M, Taffurelli M, Avenia N. Interleukin-6 and pro inflammatory status in the breast tumor microenvironment. World J Surg Oncol 2015;13:129.

[45] Gustafsson MV, Zheng X, Pereira T, Gradin K, Jin S, Lundkvist J, et al. Hypoxia requires notch signaling to maintain the undifferentiated cell state. Dev Cell 2005;9:617–28.

[46] Cejudo-Martin P, Johnson RS. A new notch in the HIF belt: how hypoxia impacts differentiation. Dev Cell 2005;9:575–6.

[47] Wang W, Yu S, Zimmerman G, Wang Y, Myers J, Yu VW, et al. Notch receptor-ligand engagement maintains hematopoietic stem cell quiescence and niche retention. Stem Cells 2015;33:2280–93.

[48] Pistollato F, Rampazzo E, Persano L, Abbadi S, Frasson C, Denaro L, et al. Interaction of hypoxia-inducible factor-1α and Notch signaling regulates medulloblastoma precursor proliferation and fate. Stem Cells 2010;28:1918–29.

[49] Mukherjee T, Kim WS, Mandal L, Banerjee U. Interaction between Notch and Hif-alpha in development and survival of Drosophila blood cells. Science 2011;332:1210–3.

[50] Fietz T, Berdel WE, Rieder H, et al. Culturing human umbilical cord blood: a comparison of mononuclear versus CD34 selected cells. Bone Marrow Transpl 1999;23:1109–15.

[51] Ge J, Cai H, Tan WS. Chromosomal stability during ex vivo 503 expansion of UCB CD34(1) cells. Cell Prolif 2011;44:550–7.

[52] Liu W, Wen Y, Bi P, Lai X, Liu XS, Liu X, et al. Hypoxia promotes satellite cell self-renewal and enhances the efficiency of myoblast transplantation. Development 2012;139:2857–65.

[53] Fan J, Cai H, Tan WSJ. Role of the plasma membrane ROS-generating NADPH oxidase in $CD34^+$ progenitor cells preservation by hypoxia. Biotechnol 2007;130:455–62.

[54] Jang YY, Sharkis SJ. A low level of reactive oxygen species selects for primitive hematopoietic stem cells that may reside in the low-oxygenic niche. Blood 2007;110:3056–63.

[55] Ludin A, Gur-Cohen S, Golan K, Kaufmann KB, Itkin T, Medaglia C, et al. Reactive oxygen species regulate hematopoietic stem cell self-renewal, migration and development, as well as their bone marrow microenvironment. Antioxid Redox Signal 2014;21:1605–19.

[56] Fan J, Cai H, Yang S, Yan L, Tan W. Comparison between the effects of normoxia and hypoxia on antioxidant enzymes and glutathione redox state in ex vivo culture of $CD34^+$ cells. Comp Biochem Physiol B Biochem Mol Biol 2008;151:153–8.

[57] Fan JL, Cai HB, Tan WS. Effect of regulating intracellular ROS with antioxidants on the ex vivo expansion of cord blood CD34+ cells. Xi Bao Yu Fen Zi Mian Yi Xue Za Zhi 2008;24:767–70.

[58] Duchez P, Chevaleyre J, Vlaski M, Dazey B, Bijou F, Lafarge X, et al. Thrombopoietin to replace megakaryocyte-derived growth factor: impact on stem and progenitor cells during ex vivo expansion of CD34+ cells mobilized in peripheral blood. Transfusion 2011;51:313–8.

[59] Ivanovic Z, Duchez P, Chevaleyre J, Vlaski M, Lafarge X, Dazey B, et al. Clinical-scale cultures of cord blood CD34+ cells to amplify committed progenitors and maintain stem cell activity. Cell Transpl 2011;20:1453–63.

[60] Milpied N, Dazey B, Ivanovic Z, Duchez P, Vigouroux S, Tabrizi R, et al. Rapid and sustained engraftment of a single allogeneic ex-vivo expanded cord blood unit (CBU) after reduced intensity conditioning (RIC) in adults. Preliminary results of a prospective trial. Abstr A. T Am Soc Hematol Blood 2011;118:226.

[61] Peled T, Landau E, Mandel J, Glukhman E, Goudsmid NR, Nagler A, et al. Linear polyamine copper chelator tetraethylenepentamine augments long-term ex vivo expansion of cord blood-derived CD34+ cells and increases their engraftment potential in NOD/SCID mice. Exp Hematol 2004;32:547–55.

[62] Peled T, Glukhman E, Hasson N, Adi S, Assor H, Yudin D, et al. Chelatable cellular copper modulates differentiation and self-renewal of cord blood-derived hematopoietic progenitor cells. Exp Hematol 2005;33:1092–100.

[63] Prus E, Fibach E. The effect of the copper chelator tetraethylenepentamine on reactive oxygen species generation by human hematopoietic progenitor cells. Stem Cells Dev 2007;16:1053–6.

[64] Huang X, Pierce LJ, Cobine PA, Winge DR, Spangrude GJ. Copper modulates the differentiation of mouse hematopoietic progenitor cells in culture. Cell Transpl 2009;18:887–97.

[65] Peled T, Mandel J, Goudsmid RN, Landor C, Hasson N, Harati D, et al. Pre-clinical development of cord blood-derived progenitor cell graft expanded ex vivo with cytokines and the polyamine copper chelator tetraethylenepentamine. Cytotherapy 2004;6:344–55.

[66] de Lima M, McMannis J, Gee A, Komanduri K, Couriel D, Andersson BS, et al. Transplantation of ex vivo expanded cord blood cells using the copper chelator tetraethylenepentamine: a phase I/II clinical trial. Bone Marrow Transpl 2008;41:771–8.

[67] Kamat JP, Devasagayam TP. Nicotinamide (vitamin B3) as an effective antioxidant against oxidative damage in rat brain mitochondria. Redox Rep 1999;4:179–84.

[68] Kang HT, Lee HI, Hwang ES. NAM extends replicative lifespan of human cells. Aging Cell 2006;5:423–36.

[69] Kwak JY, Ham HJ, Kim CM, Hwang ES. Nicotinamide exerts antioxidative effects on senescent cells. Mol Cells 2015;38:229–35.

[70] Peled T, Shoham H, Aschengrau D, Yackoubov D, Frei G, Rosenheimer GN, et al. Nicotinamide, a SIRT1 inhibitor, inhibits differentiation and facilitates expansion of hematopoietic progenitor cells with enhanced bone marrow homing and engraftment. Exp Hematol 2012;40:342–55.

[71] Kögler G, Nürnberger W, Fischer J, Niehues T, Somville T, Göbel U, et al. Simultaneous cord blood transplantation of ex vivo expanded together with non-expanded cells for high risk leukemia. Bone Marrow Transpl 1999;24:397–403.

[72] Fernández MN, Regidor C, Cabrera R, García-Marco J, Briz M, Forés R, et al. Cord blood transplants: early recovery of neutrophils from co-transplanted sibling haploidentical progenitor cells and lack of engraftment of cultured cord blood cells, as ascertained by analysis of DNA polymorphisms. Bone Marrow Transpl 2001;28:355–63.

[73] Ivanovic Z. Interleukin-3 and ex vivo maintenance of hematopoietic stem cells: facts and controversies. Eur Cytokine Netw 2004;15:6–13.
[74] Shpall EJ, Quinones R, Giller R, Zeng C, Baron AE, Jones RB, et al. Transplantation of ex vivo expanded cord blood. Biol Blood Marrow Transpl 2002;8:368–76.
[75] de Lima M, McNiece I, Robinson SN, Munsell M, Eapen M, Horowitz M, et al. Cord-blood engraftment with ex vivo mesenchymal-cell coculture. N Engl J Med 2012;367:2305–15.
[76] Antonchuk J, Sauvageau G, Humphries RK. HOXB4-induced expansion of adult hematopoietic stem cells ex vivo. Cell 2002;109:39–45.
[77] Zhang XB, Beard BC, Beebe K, Storer B, Humphries RK, Kiem HP. Differential effects of HOXB4 on nonhuman primate short- and long-term repopulating cells. PLoS Med 2006;3(5):e173.
[78] Watts KL, Nelson V, Wood BL, Trobridge GD, Beard BC, Humphries RK, et al. Hematopoietic stem cell expansion facilitates multilineage engraftment in a nonhuman primate cord blood transplantation model. Exp Hematol 2012;40:187–96.
[79] Watts KL, Delaney C, Humphries RK, Bernstein ID, Kiem HP. Combination of HOXB4 and Delta-1 ligand improves expansion of cord blood cells. Blood 2010;116:5859–66.
[80] Watts KL, Delaney C, Nelson V, Trobridge GD, Beard BC, Humphries RK, et al. CD34$^+$ expansion with Delta-1 and HOXB4 promotes rapid engraftment and transfusion independence in a *Macaca nemestrina* cord blood transplant model. Mol Ther 2013;21:1270–8.
[81] Jakob W, Sagasser S, Dellaporta S, Holland P, Kuhn K, Schierwater B. The Trox-2 Hox/ParaHox gene of Trichoplax (Placozoa) marks an epithelial boundary. Dev Genes Evol 2004;214:170–5.
[82] Mendivil Ramos O, Barker D, Ferrier DE. Ghost loci imply Hox and ParaHox existence in the last common ancestor of animals. Curr Biol 2012;22:1951–6.
[83] Kirito K, Kaushansky K. Thrombopoietin stimulates vascular endothelial cell growth factor (VEGF) production in hematopoietic stem cells. Cell Cycle 2005;4:1729–31.
[84] Kobari L, Pflumio F, Giarratana M, Li X, Titeux M, Izac B, et al. In vitro and in vivo evidence for the long-term multilineage (myeloid, B, NK, and T) reconstitution capacity of ex vivo expanded human CD34$^+$ cord blood cells. Exp Hematol 2000;28:1470–80.
[85] Duchez P, Dazey B, Douay L, Vezon G, Ivanovic Z. An efficient large-scale thawing procedure for cord blood cells destined for selection and ex vivo expansion of CD34$^+$ cells. J Hematother Stem Cell Res 2003;12:587–9.
[86] Duchez P, Chevaleyre J, Vlaski M, Dazey B, Milpied N, Boiron JM, et al. Definitive setup of clinical scale procedure for ex vivo expansion of cord blood hematopoietic cells for transplantation. Cell Transpl 2012;21:2517–21.
[87] Forristal CE, Winkler IG, Nowlan B, Barbier V, Walkinshaw G, Levesque JP. Pharmacologic stabilization of HIF-1α increases hematopoietic stem cell quiescence in vivo and accelerates blood recovery after severe irradiation. Blood 2013;121:759–69.
[88] Takubo K, Nagamatsu G, Kobayashi CI, Nakamura-Ishizu A, Kobayashi H, Ikeda E, et al. Regulation of glycolysis by Pdk functions as a metabolic checkpoint for cell cycle quiescence in hematopoietic stem cells. Cell Stem Cell 2013;12:49–61.
[89] Forristal CE, Levesque JP. Targeting the hypoxia-sensing pathway in clinical hematology. Stem Cells Transl Med 2014;3:135–40.
[90] Liu X, Zheng H, Yu WM, Cooper TM, Bunting KD, Qu CK. Maintenance of mouse hematopoietic stem cells ex vivo by reprogramming cellular metabolism. Blood 2015;125:1562–5.
[91] Afzal MR, Samanta A, Shah ZI, Jeevanantham V, Abdel-Latif A, Zuba-Surma EK, et al. Adult bone marrow cell therapy for ischemic heart disease: evidence and Insights from Randomized Controlled trials. Circ Res July 9, 2015;117(6):558–75. PII:CIRCRESAHA.114.304792.

[92] Sanina C, Hare JM. Mesenchymal stem cells as a biological drug for heart disease: where are we with cardiac cell-based therapy? Circ Res 2015;117:229–33.
[93] Vanderkooi JM, Erecińska M, Silver IA. Oxygen in mammalian tissue: methods of measurement and affinities of various reactions. Am J Physiol 1991;260(6 Pt 1):C1131–50.
[94] Mantel CR, O'Leary HA, Chitteti BR, Huang X, Cooper S, Hangoc G, et al. Enhancing hematopoietic stem cell transplantation efficacy by mitigating oxygen shock. Cell 2015;161:1553–65.
[95] Hassan N, Tchao J, Tobita K. Concise review: skeletal muscle stem cells and cardiac lineage: potential for heart repair. Stem Cells Transl Med 2014;3:183–93.
[96] Duckers HJ, Houtgraaf J, Hehrlein C, Schofer J, Waltenberger J, Gershlick A, et al. Final results of a phase IIa, randomised, open-label trial to evaluate the percutaneous intramyocardial transplantation of autologous skeletal myoblasts in congestive heart failure patients: the SEISMIC trial. EuroIntervention 2011;6:805–12.
[97] Niagara MI, Haider HK, Jiang S, et al. Pharmacologically preconditioned skeletal myoblasts are resistant to oxidative stress and promote angiogenesis via release of paracrine factors in the infarcted heart. Circ Res 2007;100:545.
[98] Azarnoush K, Maurel A, Sebbah L, Carrion C, Bissery A, Mandet C, et al. Enhancement of the functional benefits of skeletal myoblast transplantation by means of coadministration of hypoxia-inducible factor 1alpha. J Thorac Cardiovasc Surg 2005;130:173–9.
[99] Usas A, Mačiulaitis J, Mačiulaitis R, Jakubonienė N, Milašius A, Huard J. Skeletal muscle-derived stem cells: implications for cell-mediated therapies. Med Kaunas 2011;47:469–79.
[100] Oshima H, Payne TR, Urish KL, Sakai T, Ling Y, Gharaibeh B, et al. Differential myocardial infarct repair with muscle stem cells compared to myoblasts. Mol Ther 2005;12:1130–41.
[101] Wilson HM, Welikson RE, Luo J, Kean TJ, Cao B, Dennis JE, et al. Can cytoprotective cobalt protoporphyrin protect skeletal muscle and muscle-derived stem cells from ischemic injury? Clin Orthop Relat Res June 13, 2015;473(9):2908–19. [Epub ahead of print].
[102] Scheerer N, Dehne N, Stockmann C, Swoboda S, Baba HA, Neugebauer A, et al. Myeloid hypoxia-inducible factor-1α is essential for skeletal muscle regeneration in mice. J Immunol 2013;191:407–14.
[103] Linke A, Müller P, Nurzynska D, Casarsa C, Torella D, Nascimbene A, et al. Stem cells in the dog heart are self-renewing, clonogenic, and multipotent and regenerate infarcted myocardium, improving cardiac function. Proc Natl Acad Sci USA 2005;102:8966–71.
[104] Abraham MR, Gerstenblith G. Preconditioning stem cells for cardiovascular disease: an important step forward. Circ Res 2007;100:447–9.
[105] Smith RR, Barile L, Messina E, Marbán E. Stem cells in the heart: what's the buzz all about?–Part 1: preclinical considerations. Heart Rhythm 2008;5:749–57.
[106] Sousonis V, Nanas J, Terrovitis J. Cardiosphere-derived progenitor cells for myocardial repair following myocardial infarction. Curr Pharm Des 2014;20:2003–11.
[107] Hayashi E, Hosoda T. Myocyte renewal and therapeutic myocardial regeneration using various progenitor cells. Heart Fail Rev 2014;19:789–97.
[108] Tang YL, Zhu W, Cheng M, Chen L, Zhang J, Sun T, et al. Hypoxic preconditioning enhances the benefit of cardiac progenitor cell therapy for treatment of myocardial infarction by inducing CXCR4 expression. Circ Res 2009;22(104):1209–16.
[109] Yan F, Yao Y, Chen L, Li Y, Sheng Z, Ma G. Hypoxic preconditioning improves survival of cardiac progenitor cells: role of stromal cell derived factor-1α-CXCR4 axis. PLoS One 2012;7(7):e37948.

[110] Hu S, Yan G, Xu H, He W, Liu Z, Ma G. Hypoxic preconditioning increases survival of cardiac progenitor cells via the pim-1 kinase-mediated anti-apoptotic effect. Circ J 2014;78:724–31.
[111] Muraski JA, Rota M, Misao Y, Fransioli J, Cottage C, Gude N, et al. Pim-1 regulates cardiomyocyte survival downstream of Akt. Nat Med 2007;13:1467–75.
[112] Li TS, Cheng K, Malliaras K, Matsushita N, Sun B, Marbán L, et al. Expansion of human cardiac stem cells in physiological oxygen improves cell production efficiency and potency for myocardial repair. Cardiovasc Res 2011;89:157–65.
[113] Jung SY, Choi SH, Yoo SY, Baek SH, Kwon SM. Modulation of human cardiac progenitors via hypoxia-ERK circuit improves their functional bioactivities. Biomol Ther Seoul 2013;21:196–203.
[114] Lee WC, Choi CH, Cha SH, Oh HL, Kim- YK. Role of ERK in hydrogen peroxide-induced cell death of human glioma cells. Neurochem Res 2005;30:263–70.
[115] Tan SC, Gomes RS, Yeoh KK, Perbellini F, Malandraki-Miller S, Ambrose L, et al. Preconditioning of cardiosphere-derived cells with hypoxia or prolyl-4-hydroxylase inhibitors increases stemness and decreases reliance on oxidative metabolism. Cell Transpl 2016;25, in press.
[116] Hu X, Yu SP, Fraser JL, Lu Z, Ogle ME, Wang JA, et al. Transplantation of hypoxia-preconditioned mesenchymal stem cells improves infarcted heart function via enhanced survival of implanted cells and angiogenesis. J Thorac Cardiovasc Surg 2008;135:799–808.
[117] Santos Nascimento D, Mosqueira D, Sousa LM, Teixeira M, Filipe M, Resende TP, et al. Human umbilical cord tissue-derived mesenchymal stromal cells attenuate remodeling after myocardial infarction by proangiogenic, antiapoptotic, and endogenous cell-activation mechanisms. Stem Cell Res Ther 2014;5:5.
[118] Kim HW, Haider HK, Jiang S, Ashraf M. Ischemic preconditioning augments survival of stem cells via miR-210 expression by targeting caspase-8-associated protein 2. J Biol Chem 2009;27(284):33161–8.
[119] Jaussaud J, Biais M, Calderon J, Chevaleyre J, Duchez P, Ivanovic Z, et al. Hypoxia-preconditioned mesenchymal stromal cells improve cardiac function in a swine model of chronic myocardial ischaemia. Eur J Cardiothorac Surg 2013;43:1050–7.
[120] Leroux L, Descamps B, Tojais NF, Séguy B, Oses P, Moreau C, et al. Hypoxia preconditioned mesenchymal stem cells improve vascular and skeletal muscle fiber regeneration after ischemia through a Wnt4-dependent pathway. Mol Ther 2010;18:1545–52.
[121] Chen P, Wu R, Zhu W, Jiang Z, Xu Y, Chen H, et al. Hypoxia preconditioned mesenchymal stem cells prevent cardiac fibroblast activation and collagen production via leptin. PLoS One 2014;9(8):e103587.
[122] Moriyama H, Moriyama M, Isshi H, Ishihara S, Okura H, Ichinose A, et al. Role of notch signaling in the maintenance of human mesenchymal stem cells under hypoxic conditions. Stem Cells Dev 2014;23:2211–24.
[123] Xie X, Sun A, Zhu W, Huang Z, Hu X, Jia J, et al. Transplantation of mesenchymal stem cells preconditioned with hydrogen sulfide enhances repair of myocardial infarction in rats. Tohoku J Exp Med 2012;226:29–36.
[124] Wedmann R, Bertlein S, Macinkovic I, Böltz S, Miljkovic JL, Muñoz LE, et al. Working with "H_2S": facts and apparent artifacts. Nitric Oxide 2014;41:85–96.
[125] Zhang Z, Liang D, Gao X, Zhao C, Qin X, Xu Y, et al. Selective inhibition of inositol hexakisphosphate kinases (IP6Ks) enhances mesenchymal stem cell engraftment and improves therapeutic efficacy for myocardial infarction. Basic Res Cardiol 2014;109:417.

[126] Liu XB, Wang JA, Ji XY, Yu SP, Wei L. Preconditioning of bone marrow mesenchymal stem cells by prolyl hydroxylase inhibition enhances cell survival and angiogenesis in vitro and after transplantation into the ischemic heart of rats. Stem Cell Res Ther 2014;5:111.
[127] Zhu H, Sun A, Zou Y, Ge J. Inducible metabolic adaptation promotes mesenchymal stem cell therapy for ischemia: a hypoxia-induced and glycogen-based energy prestorage strategy. Arterioscler Thromb Vasc Biol 2014;34:870–6.
[128] Muraglia A, Cancedda R, Quarto R. Clonal mesenchymal progenitors from human bone marrow differentiate in vitro according to a hierarchical model. J Cell Sci 2000;113:1161–6.
[129] Paiushina OV, Domaratskaia EI. Heterogeneity and possible structure of mesenchymal stromal cell population. Tsitologiia 2015;57:31–8.
[130] Chacko SM, Ahmed S, Selvendiran K, Kuppusamy ML, Khan M, Kuppusamy P. Hypoxic preconditioning induces the expression of prosurvival and proangiogenic markers in mesenchymal stem cells. Am J Physiol Cell Physiol 2010;299:C1562–70.
[131] Anokhina EB, Buravkova LB, Galchuk SV. Resistance of rat bone marrow mesenchymal stromal precursor cells to anoxia in vitro. Bull Exp Biol Med 2009;148:148–51.
[132] Buravkova LB, Rylova YV, Andreeva ER, Kulikov AV, Pogodina MV, Zhivotovsky B, et al. Low ATP level is sufficient to maintain the uncommitted state of multipotent mesenchymal stem cells. Biochim Biophys Acta 2013;1830:4418–25.
[133] Jian KT, Shi Y, Zhang Y, Mao YM, Liu JS, Xue FL. Time course effect of hypoxia on bone marrow-derived endothelial progenitor cells and their effects on left ventricular function after transplanted into acute myocardial ischemia rat. Eur Rev Med Pharmacol Sci 2015;19:1043–54.
[134] Lindvall O, Björklund A. Cell replacement therapy: helping the brain to repair itself. NeuroRx 2004;1:379–81.
[135] Lindvall O. Developing dopaminergic cell therapy for Parkinson's disease–give up or move forward? Mov Disord 2013;28:268–73.
[136] Lindvall O, Kokaia Z, Martinez-Serrano A. Stem cell therapy for human neurodegenerative disorders-how to make it work. Nat Med 2004;10(Suppl):S42–50.
[137] Hoeber J, Trolle C, Konig N, Du Z, Gallo A, Hermans E, et al. Human embryonic stem cell-derived progenitors assist functional sensory axon regeneration after dorsal root avulsion injury. Sci Rep 2015;5:10666.
[138] Raspa A, Pugliese R, Maleki M, Gelain F. Recent therapeutic approaches for spinal cord injury. Biotechnol Bioeng 2015;112, in press.
[139] Janowski M, Wagner DC1, Boltze J. Stem cell-based tissue replacement after stroke: factual necessity or notorious fiction? Stroke June 23, 2015;46(8):2354–63. PII:STROKEAHA.114.007803. [Epub ahead of print].
[140] Liu Y, Weick JP, Liu H, Krencik R, Zhang X, Ma L, et al. Medial ganglionic eminence-like cells derived from human embryonic stem cells correct learning and memory deficits. Nat Biotechnol 2013;31:440–7.
[141] Acharya MM, Martirosian V, Christie LA, Riparip L, Strnadel J, Parihar VK, et al. Defining the optimal window for cranial transplantation of human induced pluripotent stem cell-derived cells to ameliorate radiation-induced cognitive impairment. Stem Cells Transl Med 2015;4:74–83.
[142] Dings J, Meixensberger J, Jager A, Roosen K. Clinical experience with 118 brain tissue oxygen partial pressure catheter probes. Neurosurgery 1998;43:1082–95.
[143] Panchision DM. The role of oxygen in regulating neural stem cells in development and disease. J Cell Physiol 2009;220:562–8.

[144] Crossin KL. Oxygen levels and the regulation of cell adhesion in the nervous system: a control point for morphogenesis in development, disease and evolution? Cell Adh Migr 2012 Jan-Feb;6(1):49–58.

[145] Studer L, Csete M, Lee SH, Kabbani N, Walikonis J, Wold B, et al. Enhanced proliferation, survival, and dopaminergic differentiation of CNS precursors in lowered oxygen. J Neurosci 2000;20:7377–83.

[146] Morrison SJ, Csete M, Groves AK, Melega W, Wold B, Anderson DJ. Culture in reduced levels of oxygen promotes clonogenic sympathoadrenal differentiation by isolated neural crest stem cells. J Neurosci 2000;20:7370–6.

[147] Zhang CP, Zhu LL, Zhao T, Zhao H, Huang X, Ma X, et al. Characteristics of neural stem cells expanded in lowered oxygen and the potential role of hypoxia-inducible factor-1Alpha. Neuro-Signals 2006;15:259–65.

[148] Storch A, Paul G, Csete M, Boehm BO, Carvey PM, Kupsch A, et al. Long-term proliferation and dopaminergic differentiation of human mesencephalic neural precursor cells. Exp Neurol 2001;170:317–25.

[149] Chen HL, Pistollato F, Hoeppner DJ, Ni HT, McKay RD, Panchision DM. Oxygen tension regulates survival and fate of mouse central nervous system precursors at multiple levels. Stem Cells 2007;25:2291–301.

[150] De Filippis L, Lamorte G, Snyder EY, Malgaroli A, Vescovi AL. A novel, immortal, and multipotent human neural stem cell line generating functional neurons and oligodendrocytes. Stem Cells 2007;25:2312–21.

[151] Santilli G, Lamorte G, Carlessi L, Ferrari D, Rota Nodari L, Binda E, et al. Mild hypoxia enhances proliferation and multipotency of human neural stem cells. PLoS One 2010;5(1):e8575.

[152] Pistollato F, Chen HL, Schwartz PH, Basso G, Panchision DM. Oxygen tension controls the expansion of human CNS precursors and the generation of astrocytes and oligodendrocytes. Mol Cell Neurosci 2007;35:424–35.

[153] Francis KR, Wei L. Human embryonic stem cell neural differentiation and enhanced cell survival promoted by hypoxic preconditioning. Cell Death Dis 2010;1:e22.

[154] Jensen P, Gramsbergen JB, Zimmer J, Widmer HR, Meyer M. Enhanced proliferation and dopaminergic differentiation of ventral mesencephalic precursor cells by synergistic effect of FGF2 and reduced oxygen tension. Exp Cell Res 2011;317(12):1649–62.

[155] Krabbe C, Bak ST, Jensen P, von Linstow C, Martinez Serrano A, Hansen C, et al. Influence of oxygen tension on dopaminergic differentiation of human fetal stem cells of midbrain and forebrain origin. PLoS One 2014;9(5):e96465.

[156] Krabbe C, Courtois E, Jensen P, Jorgensen JR, Zimmer J, Martinez-Serrano A, et al. Enhanced dopaminergic differentiation of human neural stem cells by synergistic effect of Bcl-xL and reduced oxygen tension. J Neurochem 2009;110:1908–20.

[157] Liu S, Tian Z, Yin F, Zhao Q, Fan M. Generation of dopaminergic neurons from human fetal mesencephalic progenitors after co-culture with striatal-conditioned media and exposure to lowered oxygen. Brain Res Bull 2009;80:62–8.

[158] Stacpoole SR, Bilican B, Webber DJ, Luzhynskaya A, He XL, Compston A, et al. Efficient derivation of NPCs, spinal motor neurons and midbrain dopaminergic neurons from hESCs at 3% oxygen. Nat Protoc 2011;6:1229–40.

[159] Stacpoole SR, Webber DJ, Bilican B, Compston A, Chandran S, Franklin RJ. Neural precursor cells cultured at physiologically relevant oxygen tensions have a survival advantage following transplantation. Stem Cells Transl Med 2013;2:464–72.

[160] Milosevic J, Schwarz SC, Krohn K, Poppe M, Storch A, Schwarz J. Low atmospheric oxygen avoids maturation, senescence and cell death of murine mesencephalic neural precursors. J Neurochem 2005;92:718–29.
[161] Bilican B, Livesey MR, Haghi G, Qiu J, Burr K, Siller R, et al. Physiological normoxia and absence of EGF is required for the long-term propagation of anterior neural precursors from human pluripotent cells. PLoS One 2014;9(1):e85932.
[162] Ara J, De Montpellier S. Hypoxic-preconditioning enhances the regenerative capacity of neural stem/progenitors in subventricular zone of newborn piglet brain. Stem Cell Res 2013;11:669–86.
[163] Theus MH, Wei L, Cui L, Francis K, Hu X, Keogh C, et al. In vitro hypoxic preconditioning of embryonic stem cells as a strategy of promoting cell survival and functional benefits after transplantation into the ischemic rat brain. Exp Neurol 2008;210:656–70.
[164] Oh JS, Ha Y, An SS, Khan M, Pennant WA, Kim HJ, et al. Hypoxia-preconditioned adipose tissue-derived mesenchymal stem cell increase the survival and gene expression of engineered neural stem cells in a spinal cord injury model. Neurosci Lett 2010;472:215–9.
[165] Kim YS, Noh MY, Cho KA, Kim H, Kwon MS, Kim KS, et al. Hypoxia/Reoxygenation-Preconditioned human bone marrow-derived mesenchymal stromal cells Rescue ischemic rat cortical neurons by enhancing trophic factor release. Mol Neurobiol 2015;52:792–803.
[166] Wei L, Fraser JL, Lu ZY, Hu X, Yu SP. Transplantation of hypoxia preconditioned bone marrow mesenchymal stem cells enhances angiogenesis and neurogenesis after cerebral ischemia in rats. Neurobiol Dis 2012;46:635–45.
[167] Wei N, Yu SP, Gu X, Taylor TM, Song D, Liu XF, et al. Delayed intranasal delivery of hypoxic-preconditioned bone marrow mesenchymal stem cells enhanced cell homing and therapeutic benefits after ischemic stroke in mice. Cell Transpl 2013;22:977–91.
[168] Sun J, Wei ZZ, Gu X, Zhang JY, Zhang Y, Li J, et al. Intranasal delivery of hypoxia-preconditioned bone marrow-derived mesenchymal stem cells enhanced regenerative effects after intracerebral hemorrhagic stroke in mice. Exp Neurol 2015;273, in press.
[169] Yu J, Yin S, Zhang W, Gao F, Liu Y, Chen Z, et al. Hypoxia preconditioned bone marrow mesenchymal stem cells promote liver regeneration in a rat massive hepatectomy model. Stem Cell Res Ther 2013;4:83.
[170] Garg A, Newsome PN. Bone marrow mesenchymal stem cells and liver regeneration: believe the hypoxia! Stem Cell Res Ther 2013;4(5):108.
[171] Yu X, Lu C, Liu H, Rao S, Cai J, Liu S, et al. Hypoxic preconditioning with cobalt of bone marrow mesenchymal stem cells improves cell migration and enhances therapy for treatment of ischemic acute kidney injury. PLoS One 2013;8(5):e62703.
[172] Zhang W, Liu L, Huo Y, Yang Y, Wang Y. Hypoxia-pretreated human MSCs attenuate acute kidney injury through enhanced angiogenic and antioxidative capacities. Biomed Res Int 2014;2014:462472.
[173] Lan YW, Choo KB, Chen CM, Hung TH, Chen YB, Hsieh CH, et al. Hypoxia-preconditioned mesenchymal stem cells attenuate bleomycin-induced pulmonary fibrosis. Stem Cell Res Ther 2015;6:97.
[174] Mottaghi S, Larijani B, Sharifi AM. Apelin 13: a novel approach to enhance efficacy of hypoxic preconditioned mesenchymal stem cells for cell therapy of diabetes. Med Hypotheses 2012;79:717–8.
[175] Wang JC, Xia L, Song XB, Wang CE, Wei FC. Transplantation of hypoxia preconditioned bone marrow mesenchymal stem cells improves survival of ultra-long random skin flap. Chin Med J Engl 2011;124:2507–11.

Cancer Stem Cell Case and Evolutionary Paradigm

14

Chapter Outline

14.1 Concept of Cancer Stem Cell

One and a half century since Robert Remak defined a concept of neoplasm arising from the various specific tissues of the body by progressive cell division (cited in Ref. [1]) and since the Friedrich's (assistant of Rudolf Virchow) statement that the origin of cancer can be traced to the presence of embryonic tissue residues [2] (cited in Ref. [3]), and also one century after farseeing Steven Paget's hypothesis [4] of "seed and soil," John Dick's group isolated a cell entity that corresponds by functional criteria to a leukemic stem cell [5]. One decade later, the cancer stem cells (CSCs) were evidenced on the basis of their self-renewal capacity and multipotent potential and enriched on the basis of surface markers in brain cancer cell populations [6–8], as well as in breast cancer [9]. Later, the CSCs were evidenced in melanoma [10], gastric cancer [11], hepatic cancer (reviewed in Ref. [12]), osteosarcoma [13], and others. These discoveries were preceded for two to three decades by the works of Pierce and colleagues [14,15], who suggested that breast cancer would appear to be caused and maintained by stem cells (cancer "neoplastic" stem cells).

The CSCs, thus, possess characteristics associated with normal stem cells: the abilities of self-renewal and differentiation into multiple cell types resulting in a capacity to give rise to all cell types found in a particular cancer sample (organization into a

Anaerobiosis and Stemness. http://dx.doi.org/10.1016/B978-0-12-800540-8.00014-4

specific hierarchy) as well as in a resistance to apoptosis and drugs [16], enabling the CSCs to perpetuate the same type of cancer. In contrast to the vast majority of cancer cells, CSCs are tumorigenic. Indeed, only a small fraction of the tumor cells (0.2–1%) are endowed with "tumor regenerating" capabilities [17].

So, CSC exhibit the typical features of primitiveness as high G_0-phase cell proportion, self-renewal capacity, differentiation potential, capacity of migration, drug resistance, and so on. Due to these characteristics, they can be resistant to anticancer therapy and persist when the main tumor mass is significantly reduced and be the cause of tumor relapse; CSCs seem to be the cells responsible for metastasis; that is, a "grain" of cancer disseminated by tissue migration ("invasiveness") as well as by blood and lymph circulation to give rise to new tumors in tissues and organs other than the one in which the cancer originated [18].

Depending on the context of their detection, the terms "cancer initiating cells," "cancer propagating cells," "cancer maintaining cells," and "cancer repopulating cells" were and still are frequently used in the operational context to describe the CSC conceptual entity. Similar to normal stem cell functional assays, the CSC ability to generate and propagate a malignant cell population should retrospectively be detected using an in vivo or ex vivo approach. In the current experimental practice these are (1) in vivo assays based on transplantation of cells to immunodeficient mice in order to evidence the CSC capable of "regenerating" a malignant population that resembles the "parental" cancer, and (2) ex vivo sphere formations and colony-forming assays in cultures [8,11,19]. In view of a large genomic and phenotypic heterogeneity of malignant populations, phenotypic and transcriptional marker profiles may not show the high level of specificity for CSC. Given that the normal stem cells with extensive self-renewal potential are intrinsically heterogeneous with respect to control of differentiation and self-renewal in different developmental stages [20,21], the heterogeneity among CSC in a single tumor type cannot be considered as abnormal or as a consequence of a DNA mutation. As normal stem cells produce a large number of transit-amplifying cells (committed progenitors and precursors), the neoplastic stem cells (modern term including premalignant stem cell population and CSC) generate a similar hierarchy-organized structure. This hierarchic organization is absent when all or most of the tumor cells have CSC properties.

14.2 Metabolic Aspect of Cancer and Cancer Stem Cells versus Normal Tissue Stem Cells

14.2.1 *Warburg Effect*

Nine decades ago, Otto Warburg and colleagues published their study "The metabolism of tumors in the body" [22], in which, among other features, they investigated respiration of the tumor in the body; in other words, how glucose consumption of the tumor is divided between "respiration" and "fermentation" (in this case the term "respiration" was used for oxidative phosphorylation and "fermentation" for glycolysis). They found that the tumor yields a lot of lactate and consumes lot of glucose: 66%

of the glucose available is used in "fermentation" and only 33% in "respiration." The predominant glycolysis found in cancer cells even in aerobic conditions ("aerobic glycolysis") over oxidative phosphorylation becomes generally accepted as one of the main cancer-specific features. Furthermore, Warburg formulated so-called "metabolic hypothesis of cancer," which continues in the following statement: "Cancer, above all other diseases, has countless secondary causes. But, even for cancer, there is only one prime cause. Summarized in a few words, the prime cause of cancer is the replacement of the respiration of oxygen in normal body cells by a fermentation of sugar" [23]. In view of the fact that after 3 decades, a "genocentric view" based therapeutic anticancer approach did not resolve the problem of the cancer therapy and in the light of the CSC concept, Menendez et al. reactualized the "metabolic cancer hypothesis," associating the Warburg effect with the stemness and tumor-forming capacity [24].

Warburg statements "All normal body cells meet their energy needs by respiration of oxygen, whereas cancer cells meet their energy needs in great part by fermentation" and "All normal body cells are thus obligate aerobes, whereas all cancer cells are partial anaerobes" [23] became some kind of dogma in oncology. In a correspondence in *Science* in 1956, Weinhouse listed the valuable arguments against the specificity of aerobic glycolysis for cancer cells [25], and Warburg even argued against using the anaerobic glycolysis as a specific test for cancer cells [26]. In spite of this and the fact that it is established today that all normal cells are not obligate aerobes, since several normal tissues, including retina, kidney medulla, cartilage, bone marrow, skin, fibroblasts, intestinal mucosa, placenta, and proliferating thymocytes exhibit high rates of aerobic glycolysis [27] as well as the highly proliferative normal cells in general [28], many authors still consider "glycolysis-addiction," explicitly or implicitly, as exclusive cancer cell property. For example, it is possible to find such statements in literature from 2012 as "Hypoxia is not tolerated by normal metabolizing human cell" [29]. Recent findings, however, clearly demonstrated that normal somatic stem cells, as well as embryonic stem cells (ESCs) and induced pluripotent stem cells (iPSCs), exhibit these metabolic features still considered by some to be cancer-specific (see Chapters 6 and 9). Initially, glycolytic metabolic profile and HIF-1α expression were considered to be related to the "hypoxic" bone marrow niche [30–32]. However, while this metabolic profile enables the hematopoietic stem cells (HSCs) to reside in extremely low O_2 areas, it is not corollary to this niche but represents rather an intrinsic property of HSC and is maintained when HSC is out of the "hypoxic niche" (Chapter 8) [33–35]. Indeed, HSC from bone marrow (environment with an O_2 gradient varying from nearly anoxic to sufficiently oxygenated to allow a full aerobiosis) or mobilized in peripheral blood (aerobic environment) exhibit a glycolytic metabolic profile even if cultured in the presence of nonphysiologically high O_2 concentrations (ex vivo cultures in atmospheric air O_2 concentrations, several-hours long measurement of metabolic parameters) [31,32,35]. Interestingly, the intensive anaerobic glycolysis in aerobic conditions was evidenced a long time ago for some hematopoietic stem/progenitor nonneoplastic cell lines [36]. It is important to stress that this metabolic set-up, long considered as an exclusive cancer cell property, is emerging as an important determinant of stemness [37]: in normal HSC, it was shown that the inhibition of OXPHOS and stimulation of anaerobic glycolysis enhances the stemness regardless of how this action is operated, [31,38,39] and, vice versa, commitment and differentiation are paralleled by intensification of oxidative phosphorylation, resulting in loss of the self-renewal capacity [31,40,41]. This principle seems to be applicable to adult stem cell populations exhibiting other differentiation potentials (Chapter 13). In line with this principle, during the "reprogramming" of mature cells toward iPSCs the part of oxidative phosphorylation decreases as the glycolysis increases and vice versa, commitment and differentiation of iPSC is associated with the shift toward oxidative phosphorylation [42,43].

In view of this nonexhaustive list of arguments, it is clear that the Warburg's statement "From the standpoint of the physics and chemistry of life this difference (author refers to "aerobic glycolysis" as dominant metabolic feature of cancer cells) between normal and cancer cells is so great that one can scarcely picture a greater difference" [23], made before the knowledge on normal stem cell metabolism was available, does not have the same value today, when the peculiarities of stem cell metabolism (Chapter 6) are known, as it used to have in 1966. On the basis of the data available today, it can be concluded that the "addiction to glycolysis" (Warburg effect) is rather a hallmark of stemness both in the case of normal and of CSCs. The same message emerges from considerations of Pacini and Borziani [44] who, upon elaboration of a pertinent line of arguments stated, "The specific metabolic phenotype known here as Warburg effect is not considered as a metabolic signature that is acquired during the oncogenesis process but represents an aberrant expression of a metabolic layout that is typical of the undifferentiated state…Therefore, we state that as contributory cause of the neoplastic development that is related to undifferentiated cells that there is a gradual and irreversible establishment of an undifferentiated state, with a gradual or complete loss of oxidative phosphorylation, rather than respiration in itself, which, as we have noted, is often present in neoplasm." In other words, the Warburg effect represents a "metabolic sign" reflecting the stem cell origin of a cancer cell.

14.2.2 Anaerobic Mitochondrial Respiration: Fumarate Respiration

The phenomenon of "anaerobic mitochondrial respiration"—ATP synthesis involving a part of respiratory chain (complex I and II) and using an electron acceptor other than oxygen—is a widespread metabolic feature in many anaerobic free-living organisms and parasites (for review see the comprehensive work of Müller et al. [45,46]). For this consideration, we will focus on the fumarate used as electron acceptor, in a reaction concerning fumarate reductase activity, which is the reverse reaction of succinate-ubiquinone reductase. The fumarate reduction is coupled to an anaerobically functioning electron transport chain in which electrons are transferred from NADH to fumarate via complex I, rhodoquinone (RQ), and a membrane-associated fumarate reductase (reviewed in Ref. [45]). In this reaction, oxidative phosphorylation occurs via proton pumping at complex I alone and without O_2 as the terminal electron acceptor; the resulting proton gradient is harnessed by the mitochondrial ATP synthase. This way, the endproduct of reaction is succinate. In some organisms (i.e. *Fasciola*), one additional ATP may be gained by decarboxylation of succinate to propionate (reviewed in Ref. [45]). In context of the mammalian and specifically human biology, this feature is formally evidenced in cancer cells [47–49], confirming earlier cues provided by findings of succinate, fumarate, and malate accumulation in cancer tissues [50], and the fact that pyrvinium pamoate (used as an antihelmintic drug), an inhibitor of fumarate reductase, is revealed to exhibit anticancer activity [48,49,51]. This anaerobic respiration (known also as "fumarate respiration") can take place in anoxia when glucose is not available (thus in ischemic conditions) since it can be "fueled" by malate and α-ketoglutarate. Due to this, it may represent an adaptation mechanism enabling the cells capable of performing it, to survive in ischemic niches when the glycolysis cannot be exploited. Indeed, the human mitochondrial succinate-ubiquinone reductase (SQR, complex II) with flavoprotein subunit type II (type I and type II exist) has lower optimal pH than complex II with type I. Flavoprotein subunit type II is induced by ischemia as demonstrated using the human colon cancer cell line DDL-1 in culture with glucose-free medium (with or without serum) exposed to 0.1% O_2 [52].

The increased glycolysis is the principal explanation for how cancer cells generate energy in the absence of oxygen. However, in actual human tumor microenvironments, hypoxia is often associated with hypoglycemia because of the poor blood supply (i.e., ischemia). The fact that some cancer lines under hypoxia or poor nutrition conditions predominantly express type II flavoprotein mRNA [47,52] and that tumor microenvironment in general mimics hypoglycemic/hypoxic conditions, in addition to abovementioned data concerning the pyrvinium-pamoate effect on cancer cells and demonstration that they effectively respire this way, additionally strengthened the association of the fumarate respiration phenomenon with the cancer.

However, both flavoprotein isoforms are expressed in normal tissues (liver, heart, skeletal muscle, brain, and kidney) [47] and predominant expression of isoform II (similar to that in cancer lines) was found in fetal tissues [52]. Of course these data were provided on the basis of whole tissue homogenate analysis and were not done separately on differentiated and undifferentiated cells from the same tissue. The anaerobic mitochondrial respiration may be promoted by malate and fumarate (in combination with α-keto glutarate) in normal cells of freshly isolated kidney proximal tubules and its endproduct is succinate [53,54]. As already stated in Chapter 9, accumulation of succinate in the context of heart tissue ischemia results from reversal of succinate dehydrogenase, which is driven by fumarate overflow from purine nucleotide breakdown and partial reversal of the malate/aspartate shuttle [55]. We supposed that somatic stem cells may use the fumarate respiration, if necessary, and tested this hypothesis on mesenchymal stromal cells (MStroC; containing as subpopulation mesenchymal stem cells (MSC); see Chapter 1) in conditions of anoxia and glucose deprivation. The results (our unpublished data) confirm that this way the MSC for their completely anaerobic and/or ischemic proliferation may provide a nonnegligible amount of mitochondrial ATP production, which is related to fumarate respiration (as showed by pyrvinium-pamoate inhibition of fumarate reductase).

All these data suggest that the fumarate respiration is not specific for cancer cells. We believe that the capacity for this type of anaerobic respiration, an ancestral and primitive feature, is especially conserved in stem cell populations (see Chapter 9).

Currently, there are no available data concerning expression of type II flavoprotein for the populations enriched in stem cells nor the functional data concerning stem cells with differentiation potential other than mesenchymal.

14.3 Multidrug Resistance Phenomenon

The phenomenon of multidrug resistance (MDR), traditionally associated with cancer and related to similar bacterial and parasite features [56], is extensively elaborated in literature. This point was discussed in Chapter 12, in view of its evolutionary aspects and its relation to anaerobiosis and stemness. On the basis of multiple arguments and a series of evidence, we concluded that it is associated with the phenomenon of stemness, with undifferentiated cell phenotype both of normal stem cells and CSC as heritage in the context of the early evolutionary origin of stemness.

14.4 Tissue Migration ("Invasiveness"), Circulation, and "Seeding": A General Stem Cell Property

It is generally known that the malignant cells acquire the ability to metastasize, a process that relies upon cell motility and regulation of expression of membrane proteins allowing to the CSC to "abandon" the "parental" tumor tissue as well as to "anchor" in some other tissue. One of the factors on which this process of "living" and "seeding" depends is related to regulation of CXCR4, as shown for many cancers [57–61]. Typically, this feature is described in the "cancer-centric" manner as the following example: "Malignant cells acquire the ability to leave the original site and spread to other locations to establish new colonies. Metastasis is a highly coordinated phenomenon: malignant cells have to down-regulate expression of adhesion molecules and detach from tumor primary site" (literal citation from Ref. [29]).

For the purpose of further discussion, we have to stress that cell motility and capacity for migration after releasing from the "paternal" tissue and implantation in distant, the same, or another tissue is a phenomenon largely present in physiology and nonmalignant physiopathology. In the context of normal stem cell biology, this feature is sine qua non condition for regeneration phenomenon.

As mentioned in Chapter 10, the studies using the first approach enabling a "retrospective detection" of a stem/progenitor cell population—colony-forming unit–spleen (CFU-S) [62]—established that hematopoietic cells are capable of migrating from intact body parts to irradiated ones in order to regenerate hematopoiesis in destructed or heavily damaged bone marrow [63–65]. Hematopoietic stem/progenitor cells, after sublethal cytotoxic insults, are "mobilized" in peripheral blood and repopulate distant areas of hematopoietic tissue [66–68]. These were experimental confirmations of a phenomenon deduced decades before in parabiosis experiments [69,70]. Stem/progenitor cells exhibiting hematopoietic, endothelial, and mesenchymal differentiation capacity are present in the steady state neonatal and adult human peripheral blood [71–76].

Stem and progenitor cells can be "mobilized" from bone marrow to blood using certain cytokines and CXCR4 antagonists [77–79] and after their intravenous injection they can "seed" in bone marrow and other tissues and participate in their repopulation and regeneration. Stem and progenitor cells can migrate through tissue in the steady state and in pathophysiological conditions [80–82]. These are only some out of a huge amount of data showing that the adult stem cells are dynamic, motile, and mobile entities, capable of migrating through the tissues, of entering the circulation, and of "anchoring" in tissues to repopulate them. These properties are an integral part of the stem cell functional nature.

In most studied stem cell models through evolution, cell migration turns out to be a feature that is required as a part of the definition of a stem cell entity. Among numerous examples can be mentioned one of migration of neoblasts in planarians. As detailed in Chapter 10, the planarian stem cells, neoblasts, are able to migrate from intact parts through the intact or irradiated parts of the body to reach the part that should be regenerated (cut parts). After grafting unirradiated tissue into irradiated host the dividing viable neoblasts migrate from grafted unirradiated tissue into the tissue of the irradiated host. Thus, in any case, the regeneration capacity depends on the viable neoblasts and their migration faculties [83–85]. The migration as a crucial feature of a stem cell is evidenced in basal metazoans (Chapter 10). In Cnidaria, the example of colonial hydroids Hydractinia is very obvious: the "I-cells" of these organisms are totipotent and ensure the cell homeostasis of the whole colony of polyps. These cells are able to migrate from one polyp to another and one "I-cell" is able to repopulate not only one polyp, but the whole colony [86,87]. Furthermore, although considered as a specifically metazoan phenomenon (and usually related to cancer invasiveness), the invasive cell growth and migration are among the most conserved evolutionary features since they result from mechanisms of cytoskeletal rearrangements, membrane trafficking, and signaling processes, which may be traced until the last eukaryotic common ancestor (LECA) [88].

In view of these arguments, we believe that, in the context of stem cell biology, "the invasive cell growth and migration" can be considered one of the primary features of stemness and, although unequivocally they do take part in the phenomenon of cancer invasiveness, are not specific for cancer.

14.5 Evolutionary Roots of Cancer: Link with Stemness

Merlo et al. [89] noticed an analogy between the evolution of life and evolution of cancer as well as between ecology and conditions of the internal environment of the organism in which a cancer appears and evolves. The authors consider the neoplasm as "microcosms of evolution": "Within a neoplasm, a mosaic of mutant cells compete for space and resources, evade predation by the immune system and can even cooperate to disperse and colonize new organs" [89]. The interest for an evolutionary model of cancer enhanced with the notions that initial transformation may occur at a single-cell level—the CSC—the heterogeneity is the ultimate expression of growth, random variation, and differentiation natural selection [90]. It has been proposed that the tumor adaptation and therapeutic failure occur through Darwinian selection [91] and that all malignant cancers, whether inherited or sporadic, are fundamentally governed by "Darwinian" dynamics [92].

14.5.1 Cancer Hypotheses Evoking a Recapitulative Reverse Evolution Principle "Back to the Roots"

Concerning the evolutionary roots of cancer, three recent hypotheses are of particular interest for our subject:

1. Hypothesis of "prokaryotic homolog toolbox" engagement of malignancy. The establishment of unicellular behavior is supported by overexpression of prokaryotic homolog proteins related to defense against environment and virus homologs related to proliferation and survival—the more aggressive the cancer is, the more reminiscent of unicellular behavior employing the prokaryote-like survival strategies its metabolic machinery is [29].
2. Cancer is an atavistic manifestation that occurs when genetic or epigenetic malfunction unlocks an ancient "toolkit" of preexisting adaptations, reestablishing the dominance of an earlier layer of genes that controlled loose-knit colonies of only partially differentiated cells; the tumors, are, in fact, reminiscent of the premetazoan colony structures composed of single-celled eukaryotes (termed "metazoan 1.0" by the authors) [93].
3. Cancer cell represents a phenotypic reversion to the earliest stage of eukaryotic evolution, especially a facultative anaerobic metabolic profile and heterotrophy. This phenotype matches the phenotype of LECA that resulted from the endosymbiosis between α-proteobacterium (which later became mitochondrion) and an archeobacterium. This hypothesis considers the evolution of cancer as an inverse recapitulation of the evolution of the eukaryotic cell from the fully differentiated cells to LECA [94].

14.5.1.1 "Prokaryotic Homolog Toolbox" Hypothesis

This hypothesis is based on the observation that malignant cells display some features of single-celled organism behavior (the authors specify "prokaryotic," but we consider that this can also be relevant for single-celled eukaryotes; i.e., their first common ancestors who inherited these genes directly from prokaryotes). Among these features are "survival under adverse environments," principally enabled by "metabolic alterations" as chemo-resistance mediated by efflux pumps and aerobic glycolysis Warburg effect. The other feature conferred to the cancer cell, an "independent behavior," is loss of cell adhesion and gain of mobility and metastases [29]. We completely agree with the observation that the cells responsible for cancer (CSCs) exhibit these features, which are reminiscent of the ancestral single-celled organisms. What we do not agree with is that these features represent cancer-specific "metabolic alterations" since both the MDR phenotype (see Chapter 12, entirely dedicated to this issue) and Warburg effect (see Section 14.2.1) are nonmalignant features involved in physiology and physiopathology from prokaryotes and protist to the mammals. Both of these features are related to the anaerobiosis/facultative anaerobiosis and, in our opinion, belong to the general features of stemness from its prototype features in protist to the mammalian stem/progenitor cell systems (see Chapter 10). Although these features, as well as those ensured by homeostasis-holding proteins (p53 and RB1) and RAS, Akt, Myc, and EGFR were inherited from bacteria and viruses, respectively (as detailed in Ref. [29]), the functional "reversion" is going back to the first eukaryotic common ancestor (FECA) and not to the prokaryotes since the cancer cells are still eukaryote cells. These points, as well as the whole line of evidence presented by the authors [29] strengthens, indeed, the relation between the cancer and primitive ancestral single-cell function. The considerations presented in this book picture stemness as an ancestral functional level conserved in metazoan and higher animals including mammals and humans as a condition required for the physiological tissue homeostasis and regeneration. This ancestral functional level is required also for cancer but is not specific to cancer. Also, we cannot agree with the statement of the authors that "Detachment of a cell from normal multicellular structure (normal tissue) causes normal cells to die" and that this "detachment" is specific for metastatic cells escaping from tumor mass [29]. Even if this statement is correct for some mature cells of solid tissues and it makes part of the metastasizing process, this is an evolutionary conserved feature of normal stemness documented in the physiology of basic metazoan and traced back to the most primitive ancestral single-celled eukaryotes (see Section 14.4).

14.5.1.2 "Atavistic" Cancer Hypothesis

Davies and Lineweaver [93] developed an extremely interesting hypothesis based on the postulate that "the principal mechanism causing cancer is the accumulation of mutations, which destroy the genetic regulation that evolved during the evolution of metazoan multicellularity, thereby reactivating an ancient genetic toolkit of pre-programmed behaviors." This behavior resumes the evolutionary step in which the single-celled eukaryotes were organized in colonies consisting of networks of adhering cells forming self-organized assemblages, exchanging information chemically. These proto-metazoans, which have preceded metazoan, appeared ~600 million years ago (see Chapter 8) and were small loosely-knit ecosystems and, according the authors, a prototype of tumor-like neoplasms [93]. The genetic apparatus of proto-metazoan was that of single-celled metazoan living in colonies; not very different from LECA but probably some genes enabling a more complex life cycle (an evolutionary prototype of differentiation: see Chapter 10) were gained. The real metazoans exhibited a richer repertoire of biological processes and above all were capable of coordinating a larger number of differentiated and nondifferentiated cell types (see Chapter 10, concerning Thrichoplax, Sponges, and Cnidaria). In fact, the regulatory network in metazoan had to manage the use of proto-metazoan genes with their minimal adaptations or to suppress them, if necessary. The authors consider that a lot of these proto-metazoan genes are still present in metazoan and even mammalian cells and that they represent a "toolkit" for "survival and propagation of non-differentiated or weakly-differentiated cells—'tumors'—and when things go wrong (often in senescence of the organism) with the nuanced overlay that characterizes metazoan, the system may revert to the ancient, more robust way of building multicellular assemblages—proto-metazoa. The result is cancer" [93].

We believe that this kind of event occurs in cancer; not as a specific cause of cancer, but as a phenomenon related to a general context of stemness, associated with cancer. As we proposed 6 years ago, the normal stem cell self-renewal is reminiscent of the first single-cell eukaryotes simple cell division ([95]; elaborated in Chapters 10 and 11) while prototype for differentiation is given in the complexification of the life cycle during the diversification of single-celled eukaryotes after LECA (Chapter 10). We postulated that the passage from self-renewal to commitment and consequent differentiation is reminiscent of evolution from FECA (simple mitotic divisions and reproduction *à l'identique*; i.e., self-renewal) to LECA and maybe to divergence of eukaryotes. The forthcoming acquisitions of molecular extensions during the colonial single-celled eukaryotes evolution and early evolution of metazoans represent "capital" of differentiation "accumulated" along with evolution and are still available in stem cells of higher organisms. In the context of normal mammalian stem cell full (symmetric division) self-renewal, we consider the activation ("disinhibition") of the "minimal essential genome" on which the self-renewing divisions should rely; that is, the expression of those genes that are strictly necessary for simple anaerobic/nanoaerobic (low-energy) cell cycle regulation and survival. We believe that this "minimal essential eukaryote genome" was not much changed from LECA to human and that it can be expressed only if "lockers," "inhibitors," and "breaks" inhibit more recent genes (acquired during the eukaryote diversification, colonial organization, and metazoan evolution). Thus, our "minimal essential genome" refers to the same collection of genes, which is termed "prokaryotic" one in a previous hypothesis [29] and which is an essential basis for the LECA and colonial metazoan "upgrades." This highly evolutionary conserved collection of gene served for vital, essential functions in ancestral single-celled eukaryotes [96] (probably survival and proliferation); that's why deregulations of these genes result in proliferative disorders that resemble in appearance the new single-celled eukaryote species, in cancer lines—the reason why they are termed "oncogenes."

In our "integrative theory of stemness," we propose that this primitive functional mode should be established to enable the full stem cell self-renewal (division without commitment and differentiation), but how far the regulation modality will go back in evolutionary recapitulation depends on which kind of phenomenon will be reproduced: asymmetric self-renewal/differentiation or colonial-like organization. In fact, the "colonial like organization" is not at all specific for CSCs able to metastasize. The numerous examples of colony-forming ability, inherent to normal stem/progenitor cell populations out of malignant context (i.e., in the context of physiology and nonmalignant pathophysiology), perfectly mimic the phenomenon of ancestral colonial organization of single-celled eukaryotes. The first retrospective experimental model for detection of HSCs—CFU-S [63]—perfectly resumes the single-celled eukaryote colonial model: a single stem cell from circulation seeds and implants into the spleen tissue, proliferates and forms a macroscopically visible colony that is, de facto, a tumor. This occurs both upon injection of syngeneic cells after lethal cytotoxic treatment and during the endogenous regeneration of hematopoiesis on/sublethal cytotoxic insult ("endogenous CFU-S") [66]. After several days these colonies, composed mainly of morphologically recognizable precursor cells of three blood lineages and a much lower proportion of committed progenitors and stem cells, along with differentiation of their cell content, diminish and finally completely vanish. A typical colonial organization was initiated by a single cell, even passing by a tumor stadium, but not giving a cancer. The colonies of undifferentiated cells on the stromal layer, known as "cobblestone areas" [97–99], are typical examples of mimicking the single-celled eukaryote colonial organization without yielding a cancer, as well as being the CFU-F (Colony Forming Unit—Fibroblastoid) [100–102]. The colony-formation, mimicking the evolutionary step of colonial single-celled eukaryote organization is, in fact, one of the steps in tissue regeneration. The well-known paracrine regulation of stem/progenitor cells represents literally the same phenomenon as "chemical communication" considered to be the only way allowing communication between proto-metazoan colony members. It is conceivable that during colony formation, the stem cells behave as primitive single-cell colony members and that their molecular set-up mimics the primitive one. Thus, the functional passage to "pro-metazoan essential genome" ("ancient toolkit" as termed by Davies and Lineweaver [93]) or to "LEKA essential genome" or even to "minimal essential genome" (FECA genome) (our terms; see Chapter 10) is not sufficient to cause cancer but is crucial for acquiring the stemness, an essential feature for full manifestation of a "cancer entity" (Figure 14.1). Thus, the point is not that this "ancient toolkit" can be simply "turned on" (as it happens in normal stem cells) by a genetic alteration, but this alteration should provoke its deregulation.

14.5.1.3 "Return to the LECA" Hypothesis

The Davilla and Zamorano hypothesis published under the title "Mitochondria and the Evolutionary Roots of Cancer" [94] also proposes that the cancer represents a phenotypic

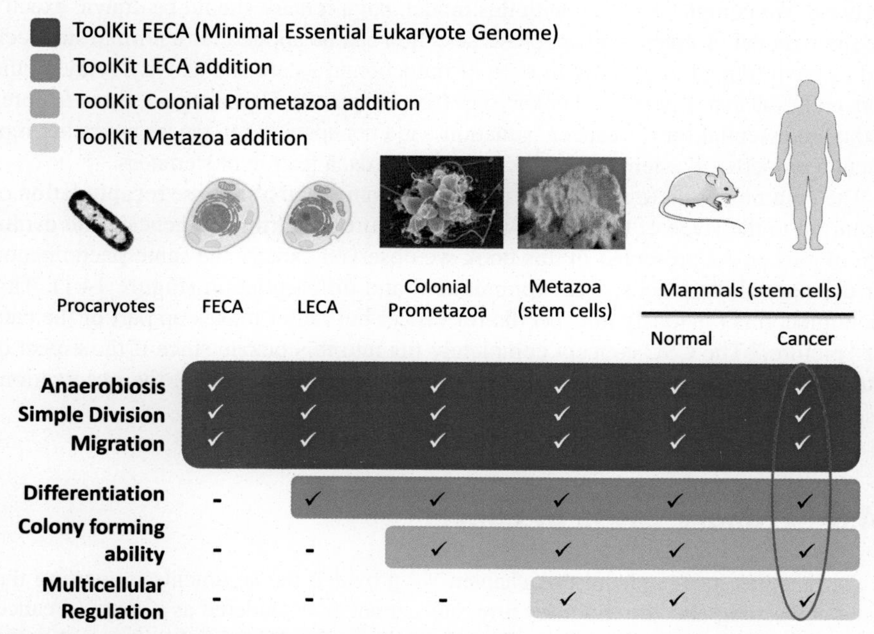

Figure 14.1 Schematic representation of the crucial gene toolkits related to the main features associated with stemness, with respect to the evolutionary stage of eukaryotes. The manifestation of primitiveness of normal and cancer stem cells is related to the activation of ancestral gene toolkits back to the FECA.

reversion to the earliest stage of eukaryotic evolution—until LECA. Briefly, the authors, by tracing the evolution of energetic metabolism and mitochondria, consider as common features of cancer cell and LECA: (1) the "return" to the glycolytic metabolic profile (i.e., Warburg effect); (2) hypoxia, the optimal environment for cancer cells and LECA; (3) apoptosis, a selection mechanism in tumors in view of an evolutionary paradigm; and (4) migration capacity, invasiveness, and metastasizing [94]. As already detailed in this book, all of these points are also valid for stem cells in general; that is, they are not specific for cancer. The issue of the Warburg effect is elaborated in Section 14.2.1, the issue of "hypoxia" in all chapters of this book, and the issue concerning the migration capacity and metastases in Section 14.4. Concerning the apoptosis issue, the authors [94] evoked the evolutionary hypothesis of apoptosis [103]. It proposes that apoptosis is nothing but an evolutionary vestige of conflict within the eukaryotic cell between the endosymbiont and host, rooting from the capacity of α-proteobacteria symbiont to kill the host cells in stress situations (such as nutrient starvation) and metabolize its remains [103]. Again, the authors related the "positive" apoptosis selection in cancer ("in tumors") to the glycolytic metabolic profile and hypoxia as "positive indicators" of tumor fitness. This is in line with the "atavistic model" [93], which stipulates that, in the case of cancer, the mechanism of selection in cancer is based on the fitness of individual cells rather than of a mechanism to synchronize multicellular growth; consequently, apoptosis in tumors should select cancer cells with an advantage to proliferate in the tumor microenvironment, along with an increase in glycolysis rate, while the hypoxia apoptosis rate should decrease. We completely agree with this model, but a remark should be drawn: exactly the same model of fitness and apoptosis selection can be applied for a normal stem cell and its environment, especially in view of mitochondria state and activity, relaying the stem cells' ancestral single-cell eukaryotes (see Chapter 9). Hence, again, these features (although essential for the cancer "situation") are not specific for cancer but rather represent a property of a stem cell entity; in other words, a feature of stemness.

The common point to all these hypotheses is some kind of reverse recapitulation of evolution to the stage of the first eukaryotes, occurring during the genesis and evolution of cancer. As presented in this book we observed exactly the same phenomenon for the normal stem cells, their commitment and differentiation (Figure 14.1). This phenomenon is obviously not specific for cancer but rather makes up part of the cancer "picture." The CSC concept completely fits into this puzzle since if the aspect of "stemness" is taken as an integral part of the cancer phenomenon, the observations discussed above become completely and logically plausible.

14.6 Primary Cause of Cancer

In the previous paragraphs of this chapter, we provided the arguments suggesting that metabolic setting and "evolutionary reversal" cannot be considered as a cause of cancer but as general hallmarks of the stemness that is an integral part of the cancer phenomenon. In that respect, two possibilities exist: either that stem cells themselves could be targets of transforming mutations, or dedifferentiation of transformed, terminally differentiated cells results in the emergence of CSCs and thereby, disease manifestation [104].

In the latter, the mature cells should acquire the features of stemness. Thus, we believe that the requiring trigger should hit genetic material either in a strictly genetic or epigenetic way. In addition to the "minimal essential genome" or "ancient toolkit" (i.e., "oncogenes") alterations *sui generis* there are epigenetic factors capable of provoking the triggering event. More than 600 genes (tumor suppressor genes, oncogenes, and cancer-associated viral genes) have been reported to be regulated by epigenetic mechanisms [105]. The epigenetic changes concern chromatin remodeling complexes and DNA methylation while genetic alterations account for microinstability (mitochondrial DNA (mtDNA) mutations, and nuclear gene mutations) and chromosomal instability and aneuploidy [106]. As concluded by Bapat [106]: "Essentially, all the epigenetic and genetic events described above are possible in both—the forward transition (transformation of stem cells), as well as dedifferentiation hypothesis (transformation of differentiated cells) that lead to the emergence of a CSC."

In conclusion, a series of mutations in CSC genome and/or cell cycle and progression driven by epigenetic deregulations induces amplification of genetically altered populations that give rise to a heterogeneous primary tumor. Then, progression of a tumor is driven by "internal" Darwinian selection. This "internal Darwinism," although heavily criticized [93], cannot be completely rejected as a concept since it is a fact that the cancer evolution represents adaptation allowing an advantage in growth and perpetuation of the cell line in question. The similar alterations in single-celled eukaryotes ensuring the competitive advantages, if they were free-living in nature or as parasites, might give new lines (i.e., species). Of course, in a mammalian organism, this outcome is not possible because the "ecosystem" is limited and, with the growth of cancer over certain measure, the "host" cannot support this new "species" behaving as a parasite and dies. This represents the end of both the "host" organism and of the cancer line. Although mimicking primitive single-celled eukaryote, the mammalian CSC is not a protist and is not able to survive and perpetuate as a free-living organism in the current geological era natural environment. However, cancer can be propagated for decades through cell cultures at 37 °C and CO_2 concentrations significantly above the atmospheric air ones, witnessing that the CSC acquired a highly autonomous behavior.

References

[1] Lagunoff D. Omnis cellula e cellula : Virchow and Remak on cell theory and neoplasia. ASIP Bull 2004;7(2). pages not numbered.

[2] Cohnheim J. Ueber entzundung und eiterung. Pathol Anat Physiol Klin Med 1867;40:1–79.

[3] Houghton J, Morozov A, Smirnova I, Wang TC. Stem cells and cancer. Semin Cancer Biol 2007;17:191–203.

[4] Paget S. The distribution of secondary growths in cancer of the breast. Cancer Metastasis Rev 1989;8:98–101.

[5] Lapidot T, Sirard C, Vormoor J, Murdoch B, Hoang T, Caceres-Cortes J, et al. A cell initiating human acute myeloid leukaemia after transplantation into SCID mice. Nature 1994;367:645–8.

[6] Hemmati HD, Nakano I, Lazareff JA, Masterman-Smith M, Geschwind DH, Bronner-Fraser M, et al. Cancerous stem cells can arise from pediatric brain tumors. Proc Natl Acad Sci USA 2003;100:15178–83.
[7] Nakano I, Kornblum HI. Brain tumor stem cells. Pediatr Res 2006;59:R54–8.
[8] Singh SK, Hawkins C, Clarke ID, Squire JA, Bayani J, Hide T, et al. Identification of human brain tumour initiating cells. Nature 2004;432:396–401.
[9] Al-Hajj M, Wicha MS, Benito-Hernandez A, Morrison SJ, Clarke MF. Prospective identification of tumorigenic breast cancer cells. Proc Natl Acad Sci USA 2003;100:3983–8.
[10] Schatton T, Murphy GF, Frank NY, Yamaura K, Waaga-Gasser AM, Gasser M, et al. Identification of cells initiating human melanomas. Nature 2008;451:345–9.
[11] Han ME, Jeon TY, Hwang SH, Lee YS, Kim HJ, Shim HE, et al. Cancer spheres from gastric cancer patients provide an ideal model system for cancer stem cell research. Cell Mol Life Sci 2011;68:3589–605.
[12] Chiba T, Iwama A, Yokosuka O. Cancer stem cells in hepatocellular carcinoma: therapeutic implications based on stem cell biology. Hepatol Res 2015;45, in press.
[13] Yu L, Liu S, Zhang C, Zhang B, Simões BM, Eyre R, et al. Enrichment of human osteosarcoma stem cells based on hTERT transcriptional activity. Oncotarget 2013;4:2326–38.
[14] Pierce GB. Neoplasms, differentiation, mutations. Am J Pathol 1974;77:103–18.
[15] Pierce GB. Neoplastic stem cells. Adv Pathobiol 1977;6:141–52.
[16] Romano G. The role of adult stem cells in carcinogenesis. Drug News Perspect 2005;18:555–9.
[17] Wicha MS, Liu S, Dontu G. Cancer stem cells: an old idea—a paradigm shift. Cancer Res 2006;66:1883–90.
[18] Sampieri K, Fodde R. Cancer stem cells and metastasis. Semin Cancer Biol 2012;22:187–93.
[19] Ponti D, Costa A, Zaffaroni N, Pratesi G, Petrangolini G, Coradini D, et al. Isolation and in vitro propagation of tumorigenic breast cancer cells with stem/progenitor cell properties. Cancer Res 2005;65:5506–11.
[20] Copley MR, Beer PA, Eaves CJ. Hematopoietic stem cell heterogeneity takes center stage. Cell Stem Cell 2012;10:690–7.
[21] Bixby S, Kruger GM, Mosher JT, Joseph NM, Morrison SJ. Cell-intrinsic differences between stem cells from different regions of the peripheral nervous system regulate the generation of neural diversity. Neuron 2002;35:643–56.
[22] Warburg O, Wind F, Negelein E. The metabolism of tumors in the body. J Gen Physiol 1927;8:519–30.
[23] Warburg O. The prime cause and prevention of cancer - part 1 with two prefaces on prevention. In: Revised lecture at the meeting of the Nobel-Laureates on June 30, 1966 at Lindau, Lake Constance, Germany. 1966. Available at: http://healingtools.tripod.com/primecause1.html/.
[24] Menendez JA, Joven J, Cufí S, Corominas-Faja B, Oliveras-Ferraros C, Cuyàs E, et al. The Warburg effect version 2.0: metabolic reprogramming of cancer stem cells. Cell Cycle 2013;12:1166–79.
[25] Weinhouse S. On respiratory impairment in cancer cells. Science 1956;124:267–9.
[26] Warburg O. On respiratory impairment in cancer cells. Science 1956;124:269–70.
[27] Hockenbery DM, Tom M, Abikoff C, Margineantu D. The Warburg effect and beyond: metabolic dependencies for cancer cells. In: Johnson DE, editor. Cell death signaling in cancer biology and treatment. New York: Springer Science+Business Media; 2013. p. 35–51.

[28] Lunt SY, Vander Heiden MG. Aerobic glycolysis: meeting the metabolic requirements of cell proliferation. Annu Rev Cell Dev Biol 2011;27:441–64.
[29] Fernandes J, Guedes PG, Lage CL, Rodrigues JC, Lage Cde A. Tumor malignancy is engaged to prokaryotic homolog toolbox. Med Hypotheses 2012;78:435–41.
[30] Simsek T, Kocabas F, Zheng JK, DeBerardinis RJ, Mahmoud AI, Olson EN, et al. The distinct metabolic profile of hematopoietic stem cells reflects their location in a hypoxic niche. Cell Stem Cell 2010;7:380–90.
[31] Takubo K, Nagamatsu G, Kobayashi CI, Nakamura-Ishizu A, Kobayashi H, Ikeda E, et al. Regulation of glycolysis by Pdk functions as a metabolic checkpoint for cell cycle quiescence in hematopoietic stem cells. Cell Stem Cell 2013;12:49–61.
[32] Kim CG, Lee JJ, Jung DY, Jeon J, Heo HS, Kang HC, et al. Profiling of differentially expressed genes in human stem cells by cDNA microarray. Mol Cells 2006;21:343–55.
[33] Nombela-Arrieta C, Pivarnik G, Winkel B, Canty KJ, Harley B, Mahoney JE. Quantitative imaging of haematopoietic stem and progenitor cell localization and hypoxic status in the bone marrow microenvironment. Nat Cell Biol 2013;15:533–43.
[34] Piccoli C, D'Aprile A, Ripoli M, Scrima R, Boffoli D, Tabilio A, et al. The hypoxia-inducible factor is stabilized in circulating hematopoietic stem cells under normoxic conditions. FEBS Lett 2007;581:3111–9.
[35] Kocabas F, Xie L, Xie J, Yu Z, DeBerardinis RJ, Kimura W, et al. Hypoxic metabolism in human hematopoietic stem cells. Cell Biosci July 17, 2015;5:39. http://dx.doi.org/10.1186/s13578-015-0020-3. eCollection 2015.
[36] Whetton AD, Bazill GW, Dexter TM. Haemopoietic cell growth factor mediates cell survival via its action on glucose transport. EMBO J 1984;3:409–13.
[37] Kohli L, Passegué E. Surviving change: the metabolic journey of hematopoietic stem cells. Trends Cell Biol 2014;24:479–87.
[38] Forristal CE, Winkler IG, Nowlan B, Barbier V, Walkinshaw G, Levesque JP. Pharmacologic stabilization of HIF-1α increases hematopoietic stem cell quiescence in vivo and accelerates blood recovery after severe irradiation. Blood 2013;121:759–69.
[39] Liu X, Zheng H, Yu WM, Cooper TM, Bunting KD, Qu CK. Maintenance of mouse hematopoietic stem cells ex vivo by reprogramming cellular metabolism. Blood 2015;125:1562–5.
[40] Maryanovich M, Zaltsman Y, Ruggiero A, Goldman A, Shachnai L, Zaidman SL, et al. An MTCH2 pathway repressing mitochondria metabolism regulates haematopoietic stem cell fate. Nat Commun 2015;6:7901.
[41] Daud H, Browne S, Al-Majmaie R, Murphy W, Al-Rubeai M. Metabolic profiling of hematopoietic stem and progenitor cells during proliferation and differentiation into red blood cells. N Biotechnol 2015;33, in press.
[42] Varum S, Rodrigues AS, Moura MB, Momcilovic O, Easley 4th CA, Ramalho-Santos J, et al. Energy metabolism in human pluripotent stem cells and their differentiated counterparts. PLoS One 2011;6(6):e20914.
[43] Panopoulos AD, Yanes O, Ruiz S, Kida YS, Diep D, Tautenhahn R, et al. The metabolome of induced pluripotent stem cells reveals metabolic changes occurring in somatic cell reprogramming. Cell Res 2012;22:168–77.
[44] Pacini N, Borziani F. Cancer stem cell theory and the Warburg effect, two sides of the same coin? Int J Mol Sci 2014;15:8893–930.
[45] Müller M, Mentel M, van Hellemond JJ, Henze K, Woehle C, Gould SB, et al. Biochemistry and evolution of anaerobic energy metabolism in eukaryotes. Microbiol Mol Biol Rev 2012;76:444–95.

[46] Inaoka DK, Shiba T, Sato D, Balogun EO, Sasaki T, Nagahama M, et al. Structural insights into the molecular design of flutolanil derivatives targeted for fumarate respiration of parasite mitochondria. Int J Mol Sci 2015;16:15287–308.
[47] Tomitsuka E, Kita K, Esumi H. Regulation of succinate–ubiquinone reductase and fumarate reductase activities in human complex II by phosphorylation of its flavoprotein subunit. Proc Jpn Acad Ser B Phys Biol Sci 2009;85:258–65.
[48] Tomitsuka E, Kita K, Esumi H. The NADH-fumarate reductase system, a novel mitochondrial energy metabolism, is a new target for anticancer therapy in tumor microenvironments. Ann N Y Acad Sci 2011;201:44–9.
[49] Tomitsuka E, Kita K, Esumi H. An anticancer agent, pyrvinium pamoate inhibits the NADH-fumarate reductase system–a unique mitochondrial energy metabolism in tumour microenvironments. J Biochem 2012;152:171–83.
[50] Hirayama A, Kami K, Sugimoto M, Sugawara M, Toki N, Onozuka H, et al. Quantitative metabolome profiling of colon and stomach cancer microenvironment by capillary electrophoresis time-of-flight mass spectrometry. Cancer Res 2009;69:4918–25.
[51] Esumi H, Lu J, Kurashima Y, Hanaoka T. Antitumor activity of pyrvinium pamoate, 6-(di methylamino)-2-[2-(2,5-dimethyl-1-phenyl-1H-pyrrol-3-yl)ethenyl]- 1-methyl-quinolinium pamoate salt, showing preferential cytotoxicity during glucose starvation. Cancer Sci 2004;95:685–90.
[52] Sakai C, Tomitsuka E, Miyagishi M, Harada S, Kita K. Type II Fp of human mitochondrial respiratory complex II and its role in adaptation to hypoxia and nutrition-deprived conditions. Mitochondrion 2013;13:602–9.
[53] Weinberg JM, Venkatachalam MA, Roeser NF, Saikumar P, Dong Z, Senter RA, et al. Anaerobic and aerobic pathways for salvage of proximal tubules from hypoxia-induced mitochondrial injury. Am J Physiol Ren Physiol 2000;279:F927–43.
[54] Weinberg JM, Venkatachalam MA, Roeser NF, Nissim I. Mitochondrial dysfunction during hypoxia/reoxygenation and its correction by anaerobic metabolism of citric acid cycle intermediates. Proc Natl Acad Sci USA 2000;97:2826–31.
[55] Chouchani ET, Pell VR, Gaude E, Aksentijević D, Sundier SY, Robb EL, et al. Ischaemic accumulation of succinate controls reperfusion injury through mitochondrial ROS. Nature 2014;515:431–5.
[56] Avner BS, Fialho AM, Chakrabarty AM. Overcoming drug resistance in multi-drug resistant cancers and microorganisms: a conceptual framework. Bioengineered 2012;3:262–70.
[57] Darash-Yahana M, Pikarsky E, Abramovitch R, Zeira E, Pal B, Karplus R, et al. Role of high expression levels of CXCR4 in tumor growth, vascularization, and metastasis. FASEB J 2004;18:1240–2.
[58] Kato M, Kitayama J, Kazama S, Nagawa H. Expression pattern of CXC chemokine receptor-4 is correlated with lymph node metastasis in human invasive ductal carcinoma. Breast Cancer Res 2003;5:R144–50.
[59] Arya M, Patel HR, McGurk C, Tatoud R, Klocker H, Masters J, et al. The importance of the CXCL12-CXCR4 chemokine ligand-receptor interaction in prostate cancer metastasis. J Exp Ther Oncol 2004;4:291–303.
[60] Berghuis D, Schilham MW, Santos SJ, Savola S, Knowles HJ, Dirksen U, et al. The CXCR4-CXCL12 axis in Ewing sarcoma: promotion of tumor growth rather than metastatic disease. Clin Sarcoma Res 2012;2:24.
[61] Mazur G, Butrym A, Kryczek I, Dlubek D, Jaskula E, Lange A, et al. Decreased expression of CXCR4 chemokine receptor in bone marrow after chemotherapy in patients with non-Hodgkin lymphomas is a good prognostic factor. PLoS One 2014;9(5): e98194.

[62] Till JE, McCulloch EA. A direct measurement of the radiation sensitivity of normal mouse bone marrow cells. Radiat Res 1961;14:213–22.
[63] Boggs DR, Chervenick PA. Migration of transplanted hematopoietic stem cells to the spleen of irradiated mice. Transplantation 1971;112:191–2.
[64] Petrov RV, Khaitov RM. Migration of stem cells from screened bone marrow following non-uniform irradiation. Radiobiologiia 1972;12:69–76.
[65] Croizat H, Frindel E, Tubiana M. The effect of partial body irradiation on haemopoietic stem cell migration. Cell Tissue Kinet 1980;13:319–25.
[66] Ueno Y. Kinetics of endogenous CFU-s in mice receiving divided-dose irradiation. J Radiat Res 1975;16:10–8.
[67] Meredith RF, Okunewick JP, Kovacs CJ, Evans MJ, Dicke KA. Effects of piperazinedione (NSC 135785) plus irradiation on endogenous CFU-S, intestinal crypts, and as a possible ablative protocol for marrow transplantation as evaluated in Rauscher leukemic mice. Radiat Res 1982;89:559–74.
[68] Yonezawa M, Horie K, Kondo H, Kubo K. Increase in endogenous spleen colonies without recovery of blood cell counts in radioadaptive survival response in C57BL/6 mice. Radiat Res 2004;161:161–7.
[69] Woenckhaus E. Beitrag zur Allgemeinwirkung der Roentgenstrahlen. Arch Exp Path Pharm 1930;150:82.
[70] Brecher G, Cronkite EP. Post-radiation parabiosis and survival in rats. Proc Soc Exp Biol Med 1951;77:292–4.
[71] Udomsakdi C, Lansdorp PM, Hogge DE, Reid DS, Eaves AC, Eaves CJ. Characterization of primitive hematopoietic cells in normal human peripheral blood. Blood 1992;80:2513–21.
[72] Hirayama F, Yamaguchi M, Yano M, Yasui K, Horie Y, Matsumoto K, et al. Spontaneous and rapid reexpression of functional CXCR4 by human steady-state peripheral blood $CD34^+$ cells. Int J Hematol 2003;78:48–55.
[73] Brunet de la Grange P, Vlaski M, Duchez P, Chevaleyre J, Lapostolle V, Boiron JM, et al. Long-term repopulating hematopoietic stem cells and "side population" in human steady state peripheral blood. Stem Cell Res 2013;11:625–33.
[74] Ivanovic Z, Duchez P, Chevaleyre J, Vlaski M, Lafarge X, Dazey B, et al. Clinical-scale cultures of cord blood CD34(+) cells to amplify committed progenitors and maintain stem cell activity. Cell Transpl 2011;20:1453–63.
[75] Weil BR, Maceneaney OJ, Stauffer BL, Desouza CA. Habitual short sleep duration and circulating endothelial progenitor cells. J Cardiovasc Dis Res 2011;2:110–4.
[76] Ha CW, Park YB, Chung JY, Park YG. Cartilage repair using composites of human umbilical cord blood-derived mesenchymal stem cells and hyaluronic acid hydrogel in a minipig model. Stem Cells Transl Med August 3, 2015;4(9):1044–51. pii: sctm.2014-0264. [Epub ahead of print].
[77] Ivanovic Z, Kovacevic-Filipovic M, Jeanne M, Ardilouze L, Bertot A, Szyporta M, et al. $CD34^+$ cells obtained from "good mobilizers" are more activated and exhibit lower ex vivo expansion efficiency than their counterparts from "poor mobilizers". Transfusion 2010;50:120–7.
[78] Patel B, Pearson H, Zacharoulis S. Mobilisation of haematopoietic stem cells in paediatric patients, prior to autologous transplantation following administration of plerixafor and G-CSF. Pediatr Blood Cancer 2015;62:1477–80.
[79] Hoggatt J, Speth JM, Pelus LM. Concise review: sowing the seeds of a fruitful harvest: hematopoietic stem cell mobilization. Stem Cells 2013;31:2599–606.
[80] Joo S, Yeon Kim J, Lee E, Hong N, Sun W, Nam Y. Effects of ECM protein micropatterns on the migration and differentiation of adult neural stem cells. Sci Rep 2015;5:13043.

[81] Wollank Y, Ramer R, Ivanov I, Salamon A, Peters K, Hinz B. Inhibition of FAAH confers increased stem cell migration via PPARα. J Lipid Res 2015;56, in press.
[82] Lee SH, Jin KS, Bang OY, Kim BJ, Park SJ, Lee NH, et al. Differential migration of mesenchymal stem cells to ischemic regions after middle cerebral artery occlusion in rats. PLoS One 2015;10(8):e0134920.
[83] Dubois F. Contribution a l'etude de la migration des cellules de regeneration chez les planaires dulcicoles. Bull Biol 1949;83:213–83.
[84] Baguñà J, Saló E, Auladell MC. Regeneration and pattern formation in planarians. III. Evidence that neoblasts are totipotent stem-cells and the source of blastema cells. Development 1989;107:77–86.
[85] Reddien PW, Sánchez Alvarado A. Fundamentals of planarian regeneration. Annu Rev Cell Dev Biol 2004;20:725–57.
[86] Müller WA, Teo R, Frank U. Totipotent migratory stem cells in a hydroid. Dev Biol 2004;275:215–24.
[87] Künzel T, Heiermann R, Frank U, Müller W, Tilmann W, Bause M, et al. Migration and differentiation potential of stem cells in the cnidarian Hydractinia analysed in eGFP-transgenic animals and chimeras. Dev Biol 2010;348:120–9.
[88] Vaškovičová K, Žárský V, Rösel D, Nikolič M, Buccione R, Cvrčková F, et al. Invasive cells in animals and plants: searching for LECA machineries in later eukaryotic life. Biol Direct 2013;8:8.
[89] Merlo LM, Pepper JW, Reid BJ, Maley CC. Cancer as an evolutionary and ecological process. Nat Rev Cancer 2006;6:924–35.
[90] Purushotham AD, Sullivan R. Darwin, medicine and cancer. Ann Oncol 2010;21:199–203.
[91] Gerlinger M, Rowan AJ, Horswell S, Larkin J, Endesfelder D, Gronroos E, et al. Intratumor heterogeneity and branched evolution revealed by multiregion sequencing. N Engl J Med 2012;366:883–92.
[92] Gillies RJ, Verduzco D, Gatenby RA. Evolutionary dynamics of carcinogenesis and why targeted therapy does not work. Nat Rev Cancer 2012;12:487–93.
[93] Davies PC, Lineweaver CH. Cancer tumors as Metazoa 1.0: tapping genes of ancient ancestors. Phys Biol 2011;8:015001.
[94] Davila AF, Zamorano P. Mitochondria and the evolutionary roots of cancer. Phys Biol 2013;10:026008.
[95] Ivanovic Z. Hypoxia or in situ normoxia: the stem cell paradigm. J Cell Physiol 2009;219:271–5.
[96] Weinberg RA. Oncogenes and the molecular biology of cancer. J Cell Biol 1983;97:1661–2.
[97] Fruehauf S, Breems DA, Knaän-Shanzer S, Brouwer KB, Haas R, Löwenberg B, et al. Frequency analysis of multidrug resistance-1 gene transfer into human primitive hematopoietic progenitor cells using the cobblestone area-forming cell assay and detection of vector-mediated P-glycoprotein expression by rhodamine-123. Hum Gene Ther 1996;7:1219–31.
[98] Schrezenmeier H, Jenal M, Herrmann F, Heimpel H, Raghavachar A. Quantitative analysis of cobblestone area-forming cells in bone marrow of patients with aplastic anemia by limiting dilution assay. Blood 1996;88:4474–80.
[99] van Os RP, Dethmers-Ausema B, de Haan G. In vitro assays for cobblestone area-forming cells, LTC-IC, and CFU-C. Methods Mol Biol 2008;430:143–57.
[100] Friedenstein AJ, Chailakhjan RK, Lalykina KS. The development of fibroblast colonies in monolayer cultures of guinea-pig bone marrow and spleen cells. Cell Tissue Kinet 1970;3:393–403.

[101] Luria EA, Panasyuk AF, Friedenstein AY. Fibroblast colony formation from monolayer cultures of blood cells. Transfusion 1971;11:345–9.
[102] Mabuchi Y, Houlihan DD, Akazawa C, Okano H, Matsuzaki Y. Prospective isolation of murine and human bone marrow mesenchymal stem cells based on surface markers. Stem Cells Int 2013;2013:507301.
[103] Frade JM, Michaelidis TM. Origin of eukaryotic programmed cell death: a consequence of aerobic metabolism? Bioessays 1997;19:827–32.
[104] Reya T, Morrison SJ, Clarke MF, Weissman IL. Stem cells, cancer, and cancer stem cells. Nature 2001;414:105–11.
[105] Wilson CB, Makar KW, Shnyreva M, Fitzpatrick DR. DNA methylation and the expanding epigenetics of T cell lineage commitment. Semin Immunol 2005;17:105–19.
[106] Bapat SA. Evolution of cancer stem cells. Semin Cancer Biol 2007;17:204–13.

Index

Note: Page numbers followed by "f" indicate figures and "t" indicate tables.

I

K

L

Printed in the United States
By Bookmasters